Student Solutions Manual

for

Finite Mathematics

Sixth Edition

Howard L. Rolf
Baylor University

THOMSON

BROOKS/COLE

Australia • Canada • Mexico • Singapore • Spain • United Kingdom • United States

Printed in Canada
1   2   3   4   5   6   7   08   07   06   05   04

Printer: Webcom Limited

ISBN: 0-534-49162-6

For more information about our products, contact us at:
Thomson Learning Academic Resource Center
1-800-423-0563

For permission to use material from this text or product, submit a request online at
http://www.thomsonrights.com.
Any additional questions about permissions can be submitted by email to **thomsonrights@thomson.com.**

Thomson Brooks/Cole
10 Davis Drive
Belmont, CA 94002-3098
USA

Asia
Thomson Learning
5 Shenton Way #01-01
UIC Building
Singapore 068808

Australia/New Zealand
Thomson Learning
102 Dodds Street
Southbank, Victoria 3006
Australia

Canada
Nelson
1120 Birchmount Road
Toronto, Ontario M1K 5G4
Canada

Europe/Middle East/South Africa
Thomson Learning
High Holborn House
50/51 Bedford Row
London WC1R 4LR
United Kingdom

Latin America
Thomson Learning
Seneca, 53
Colonia Polanco
11560 Mexico D.F.
Mexico

Spain/Portugal
Paraninfo
Calle/Magallanes, 25
28015 Madrid, Spain

Contents

Answers to Odd-Numbered Exercises

Preface

This Student's Solutions Manual accompanies **Finite Mathematics, Sixth Edition** by Howard L. Rolf All references to chapters, sections, and exercises refer to the textbook. This manual contains worked out solutions to odd-numbered exercises in the textbook.

Do not use this manual as a substitute for working the homework problems yourself. In one sense, you should use this manual as a last resort in solving a problem.

You may find it useful to compare your solution with the solution in this manual. Be aware that your solution may be different, yet correct. In fact, you may have a clever solution that is better. Quite often the steps in a solution may occur in a different order and still be correct. Therefore, do not assume your solution is incorrect if it differs from the one in this manual.

Chapter 1
Functions and Lines

Section 1.1 Functions

1. $y = 15x + 20$ is the rule. The numbers in the domain represent the number of hours worked.
The numbers in the range represent the number of dollars of fee.

3.   **(a)** $f(5) = \$23.75$ **(b)** $f(3) = \$14.25$

5.   **(a)** $f(1) = 4(1) - 3 = 4 - 3 = 1$ **(b)** $f(-2) = 4(-2) - 3 = -8 - 3 = -11$
       (c) $f(1/2) = 4(1/2) - 3 = 2 - 3 = -1$ **(d)** $f(a) = 4a - 3$

7.   **(a)** $f(5) = \dfrac{5+1}{5-1} = \dfrac{6}{4} = \dfrac{3}{2}$ **(b)** $f(-6) = \dfrac{-6+1}{-6-1} = \dfrac{5}{7}$

       (c) $f(0) = \dfrac{0+1}{0-1} = \dfrac{1}{-1} = -1$ **(d)** $f(2c) = \dfrac{2c+1}{2c-1}$

9.   **(a)** $p(1995) = 1.32(1995) - 2589.5 = 2633.4 - 2589.5 = 43.9$, an estimated 43.9 thousand people.
            $p(2010) = 1.32(2010) - 2589.5 = 2653.2 - 2589.5 = 63.7$, an estimated 63.7 thousand people.
       (b)   When will $p(t) = 75.0$?
            $75.0 = 1.32t - 2589.5$
            $2664.5 = 1.32t$
            $t = \dfrac{2664.5}{1.32} = 2018.6$
            The number is estimated to reach 75,000 in 2018.

11.   **(a)** $f(60) = 9(60) = 540$ calories
       (b)   Solve $9x = 750$ $x = 83.3$ minutes

13.   Let $x =$ number of hamburgers.
      $y = 2.40x + 25$

Section 1.1 Functions

15.   Let x = regular price.
      y = x - 0.20x or y = 0.80x

17.   Let x = number of loads.
      y = 0.60x + 12

19.   Let x = number of students.
      y = 3500x + 5,000,000

21.   Let x = list price.
      y = 0.88x

23.   **(a)** $S(2.5) = 11.25(2.5) + 300 = 328.125$ rounded to $328.13.
      (b)   h = 5 so S(h) = 11.25(5) + 300 = 356.25   Her weekly salary was $356.25.
      (c)   S(h) = 395.63 so
            395.63 = 11.25h + 300
            11.25h = 395.63 - 300 = 95.63
            h = 95.63/11.25 = 8.5004 which we round to 8.5.
            She worked 8.5 hours overtime.

25.   **(a)** $A = \pi r^2$ is a function
      (b)   Domain: positive numbers. Range: positive numbers

27.   **(a)**   p = price per pound times w is a function.
      (b)   Domain: positive numbers. Range: positive numbers.

29.   **(a)** $y = x^2$ is a function
      (b)   Domain: all real numbers.   Range: All nonnegative numbers.

31.   y is not a function of x. There can be more than one person with a given
      family name. x is function of y.

33.   Not a function because two classes can have the same number of boys, but the
      with combined weights different.

35.   Not a function because two families with the same number of children can
      have a different number of boys.

37.   The domain is the set of numbers in the interval [-2, 4}. The range is the set
      of numbers in the interval [-1, 3].

39.   The domain is the set of numbers in the intervals [0, 4] or [7, 12] and the range
      is the set of numbers in the interval [-4, 8].

2

43. Let x = a person's age, y = pulse rate
    (a) $p = 0.40(220 - x)$ **(b)** $p = 0.70(220 - x)$

45.   **(a)** $d = 1.1(30) + 0.055(30)^2 = 33 + 49.5 = 82.5$ About 83 feet are required.
      (b) $d = 1.1(60) + 0.055(60)^2 = 66 + 198.0 = 264$ About 264 feet are required to stop.

47.   **(a)** $f(225) = 12.47$ **49.**   **(a)** $f(4.5) = 57.6625$
      (b) $f(416) = 23.93$ **(b)** $f(3.3) = 32.4205$
      (c) $f(367) = 20.99$ **(c)** $f(8.2) = 258.776$

51.   **(a)** 68.1%   **(b)** 40.5%
      (c) 75.0%   **(d)** 81.9%

53.   **(a)** 272.4 million   **(b)** 154.3 million
      (c) 70.0 million   **(d)** 19.6 million
      (e) 3.0 million   **(f)** 329.1 million
      (g) 424.4 million
      (h) The population is estimated to reach 400 million in 2042.7, in the year 2042.
           The population is estimated to reach 500 million in 2071.5, during the year 2071.

55. $y = 6, 3, -6, -18, -45$ **57.** $y = 39.28, 46.48, 104.24$

59. $y = 16, 19, 22, 25, 28$

Using Your TI-83

1. $y = 3, 27, 51$ **3.** $y = 9, 9, 14, 30$

5. $y = 0, -0.25, -1.6,$ and 8.9231

Using EXCEL

1. =A4+B4   3. =C4+C5

5. =B2*B3   7. =(B1+B2)/2

9. =2.1*A5-1.8

Section 1.2

1. $f(x) = 3x + 8$
    $f(0) = 8, f(1) = 11$

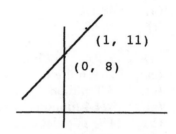

3. $f(x) = x + 7$
    $f(1) = 8, f(-1) = 6$

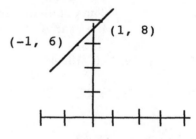

5. $f(x) = -3x - 1$
    $f(0) = -1, f(-1) = 2$

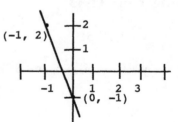

7. Slope = 7, y-intercept = 22

9. Slope = -2/5, y-intercept = 6

11.
$5y = -2x + 3$
$y = -\dfrac{2}{5}x + \dfrac{3}{5}$
Slope = -2/5, y-intercept = 3/5

13.
$3y = x + 6$
$y = \dfrac{1}{3}x + 2$
Slope = 1/3, y-intercept = 2

15. $m = \dfrac{4-2}{3-1} = \dfrac{2}{2} = 1$

17. $m = \dfrac{-5-(-1)}{-1-(-4)} = \dfrac{4}{3}$

19. Negative

21. Positive

23. $y = -2$

25. $y = 0$

27.
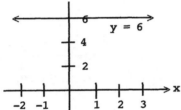

29.
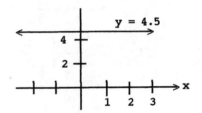

31. $m = \dfrac{5-2}{3-3} = \dfrac{3}{0}$ m is undefined so the graph is a vertical line $x = 3$

33. Vertical line $x = 10$

35.
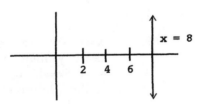

37.
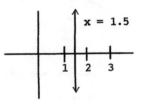

39. $y = 4x + 3$

41. $y = -x + 6$

43. $y = \dfrac{1}{2}x$

45. $y = -4x + b$
$1 = -4(2) + b$
$b = 9$
$y = -4x + 9$

47.
$$y = \frac{1}{2}x + b$$
$$4 = \frac{1}{2}(5) + b$$
$$b = \frac{3}{2}$$
$$y = \frac{1}{2}x + \frac{3}{2}$$

49.
$$y - 5 = 7(x - 1)$$
$$y = 7x - 7 + 5$$
$$y = 7x - 2$$

51.
$$y - 6 = \frac{1}{5}(x - 9)$$
$$y = \frac{1}{5}x - \frac{9}{5} + 6$$
$$y = \frac{1}{5}x + \frac{21}{5}$$

53.
$$m = \frac{1 - 0}{2 + 1} = \frac{1}{3}$$
$$y - 0 = \frac{1}{3}(x + 1)$$
$$y = \frac{1}{3}x + \frac{1}{3}$$

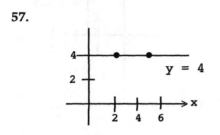

55.
$$m = \frac{2 - 0}{1 - 0} = \frac{2}{1} = 2$$
$$y - 0 = 2(x - 0)$$
$$y = 2x$$

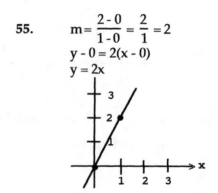

57.

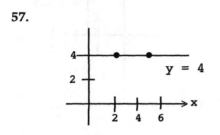

y = 4

59.
x = 0: -3y = 15 so y = -5
is the y-intercept
y = 0: 5x = 15 so x = 3
is the x-intercept

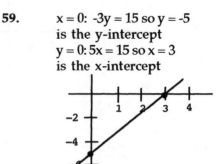

6

Section 1.2 Graphs and Lines

61. When $x = 0$, $-5y = 25$ so $y = -5$ is the
    y-intercept
When $y = 0$, $2x = 25$, so $x = 12.5$ is the
    x-intercept

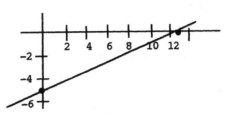

63. Line through (8, 2) and (3, -3) has slope $m_1 = \dfrac{-3-2}{3-8} = \dfrac{-5}{-5} = 1$.

Line through (6, -1) and (16, 9) has slope $m_2 = \dfrac{9+1}{16-6} = \dfrac{10}{10} = 1$.

The lines are parallel.

65. Line through (5, 4) and (1, -2) has slope $m_1 = \dfrac{-2-4}{1-5} = \dfrac{-6}{-4} = \dfrac{3}{2}$

Line through (1, 2) and (6, 8) has slope $m_2 = \dfrac{8-2}{6-1} = \dfrac{6}{5}$

The lines are not parallel.

67. $m_1 = 6 = m_2$ (parallel)

69. The first line may be written $y = \dfrac{1}{2}x - \dfrac{3}{2}$ so $m_1 = \dfrac{1}{2}$

The second line may be written $y = -2x + 1$ so $m_2 = -2$

The lines are not parallel.

71. The product of the slopes is $-2 \times 0.5 = -1$ so the lines are perpendicular.

73. The product of slopes is $3 \times 5 = 15 \neq -1$ so the lines are not perpendicular.

75. $m = 3$

$y - 5 = 3(x + 1)$

$y = 3x + 3 + 5$

$y = 3x + 8$

77. $m = -\dfrac{5}{7}$

$y = -\dfrac{5}{7}x + 8$

79. For Exercise 76 $y = -\dfrac{3}{2}x + 9$ is written $3x + 2y = 18$

For Exercise 77 $y = -\dfrac{5}{7}x + 8$ is written $5x + 7y = 56$

For Exercise 78 $y = \dfrac{5}{2}x - 13$ is written as $5x - 2y = 26$

7

81. $y - 5 = \dfrac{2}{3}(x - 2)$

When $x = 0$, $y = \dfrac{2}{3}(0 - 2) + 5 = -\dfrac{4}{3} + 5 = \dfrac{11}{3}$, so the y-intercept is $\dfrac{11}{3}$

83. The slope of the given line is $m = -\dfrac{5}{3}$ so the perpendicular line has slope $\dfrac{3}{5}$

$y - 3 = \dfrac{3}{5}(x - 2)$

$5y - 15 = 3x - 6$

$3x - 5y = -9$

85. Let x = no. of weeks from the start, y = weight. Then $m = -3$, and $(14, 196)$ is a point on the line.

(a) A point and a slope.

(b) $y - 196 = -3(x - 14)$

$y = -3x + 42 + 196$

$y = -3x + 238$

(c) $y = -3(0) + 238 = 238$ pounds

87. Let x = number of items and y = the cost. Then $(500, 1340)$ and $(800, 1760)$ are points on the line, so

$m = \dfrac{1760 - 1340}{800 - 500} = \dfrac{420}{300} = 1.4$

$y - 1340 = 1.4(x - 500)$

$y = 1.4x - 700 + 1340$

$y = 1.4x + 640$

89. (a) $\dfrac{y - 3}{-1 - 2} = 3$

$y - 3 = -9$

$y = -6$

(b) $\dfrac{3 - 2}{x - 1} = -4$

$-4x + 4 = 1$

$-4x = -3$

$x = \dfrac{3}{4}$

(c) $\dfrac{y - 0}{5 + 2} = \dfrac{3}{4}$

$y = 7\left(\dfrac{3}{4}\right)$

$y = \dfrac{21}{4}$

(d) $\dfrac{4 + 3}{x + 1} = -\dfrac{1}{2}$

$-\dfrac{1}{2}x - \dfrac{1}{2} = 7$

$-\dfrac{1}{2}x = \dfrac{15}{2}$

$x = -15$

Section 1.2 Graphs and Lines

91. **(a)** For 1991 x = 0 so we have two points on the line, (0, 13892) and
(9, 21423).

$$m = \frac{21423 - 13892}{9} = 836.78$$

b = 13,892

y = 836.78x + 13,892

(b) For 2005 x = 14.

y = 836.78(14) + 13,892 = 25,606.92

The estimated average cost for 2005 is $25,607.

93. Let x = the year and y = tuition per semester hour.

m = 50 and (0, 375) is a point on the line.

y − 375 = 50(x)

y = 50x + 375

95. **(a)** A point (0, 5.00) and slope 0.078.

(b) x = KWH used, y = amount of bill

y = 0.078x + 5

97. **(a)** Increases 4   **(b)** Decreases 3

(c) Increases 2/3   **(d)** Decreases 1/2

(e) $y = -\frac{2}{3}x + \frac{4}{3}$ so it decreases 2/3 **(f)** No change

99. Let x = number of miles and y = cost. Then the points (125, 35.75) and
(265, 51.15) are on the line, so

$$m = \frac{51.15 - 35.75}{265 - 125} = \frac{15.40}{140} = 0.11$$

y - 35.75 = 0.11(x - 125)

y = 0.11x - 13.75 + 35.75

y = 0.11x + 22

101. Let x = number of years since 1997 and y = number of cars.

(a) The slope is -15,000 and for year 0, y = 536,000 which is the
y-intercept.

y = −15,000x + 536,000 x years after 1997.

(b) The slope is 4500 and for year 0, y = 536,000 which is the y-intercept.

y = 4500x + 536,000 x years after 1997.

103. Let x = taxable income. The slope of the line m = 0.27 and (27950, 3892.5) is a
point on the line.

y - 3892.5 = 0.27(x - 27950)

y = 0.27(x - 27950) + 3892.5 or y = 0.27x - 3654

105. Let x = taxable income. The slope of the line m = 0.15 and (12000, 1200) is a point on the line.

$$y - 1200 = 0.15(x - 12000)$$
$$y = 0.15(x - 12000) + 1200 \text{ or } y = 0.15x - 600$$

This equation holds for $12{,}000 \leq x \leq 46{,}700$.

107. **(a)** Let x = number of years with x = 0 for 1980.
Let y = birth rate
We are given two points (0, 13.7) and (22, 10.03). The slope of the line through these points is -0.167 and the y-intercept is 13.7 so the linear function is $y = -0.167x + 13.7$ for x years after 1980.

(b) For 1985 x = 5 so the birth rate for 1985 is estimated to be
$$y = -0.167(5) + 13.7 = 12.9$$
The linear function gives a high estimate for 1985.

(c) The birth rate will reach zero when y = 0 so
$$0 = -0.167x + 13.7$$
$$x = 13.7/0.167 = 82.04$$
This function estimates that Japan's birth rate will drop to zero in the year 1980 + 82 = 2062. This conclusion is based on the assumption that birth rates will drop in a linear manner at the same rate they dropped in 1980-2002. It is unrealistic to expect that no babies will be born in an entire year so the linear function is not a valid long term estimate.

109. **(a)** Let x be the admission price and y be the estimated attendance. The given information provides two points on a line, (5, 185) and (6, 140). The slope of the line through these points is -45 and the equation of the line is $y - 185 = -45(x - 5)$ which reduces to $y = -45x + 410$.

(b) When admission is $7, x = 7 and attendance = $-45(7) + 410 = 95$.

(c) When attendance is 250
$$250 = -45x + 410$$
$$x = 3.555$$
For an estimated attendance of 250, the manager would likely round the admission of 3.555 to $3.55.

(d) For an attendance of zero
$$0 = -45x + 410$$
$$x = 9.111$$
An admission of $9.11, or more, would result in no attendance.

(e) If admission were free, x = 0 and the estimated attendance would be
$$y = -45(0) + 410 = 410$$

Section 1.3 Mathematical Models

111. Let x = depth in feet and y = water pressure in pounds per square inch.
We have two points on the line, (18, 8) and (90, 40).
$$m = \frac{40-8}{90-18} = \frac{32}{72} = 0.4444$$
y - 8 = 0.4444(x - 18)
y = 0.4444x
At 561 feet y = 0.4444(561) = 249.3 so the pressure is approximately 249 pounds per square inch.

115. About 4:30 am.

117. All have y-intercepts of 4, but they are not parallel.

119. All go through the origin. **121.** m = -10/3

123. m = 1.240

125. m = 12/5 **127.** m = 0.351
y - 4 = (12/5)(x - 5) y - 13.42 = 0.351(x - 21.65)
y = (12/5)x - 8 y = 0.351x + 5.821

Using Excel

1. m = 0.25 **3.** m = 2.73

5. y = -0.67 + 4.33 **7.** y = -0.56x + 4.24

1. y = 1.4x + 0.8 **3.** y = 1.25x + 1.35

Section 1.3

1. **(a)** C(180) = 43(180) + 2300 = $10,040
(b) Solve 43x + 2300 = 11,889
43x = 9589   x = 223 bikes
(c) Unit cost is $43, fixed cost is $2,300

3. **(a)** Fixed cost is $400, unit cost is $3
(b) For 600 units, C(600) = 3(600) + 400 = $2,200
For 1,000 units, C(1000) = 3(1000) + 400 = $3,400

Section 1.3 Mathematical Models

5.   **(a)**   R(x) = 32x   **(b)**   R(78) = 32(78) = $2,496
      (c)   Solve 32x = 672 x = 21 pairs

7.   **(a)**   R(x) = 3.39x
      (b)   R(834) = 3.39(834) = $2,827.26

9.   **(a)**   C(x) = 57x + 780   **(b)**   R(x) = 79x
      (c)   79x = 57x + 780   22x = 780
             x = 35.45, so the break-even number is 36 coats.

11.   C(x) = 4x + 500   C(800) = 4(800) + 500 = $3,700

13.   **(a)**   Let x = number of T-shirts and C = the cost.
             Then the points (600, 1400) and (700, 1600)
             lie on the line, so
             $$m = \frac{1600 - 1400}{700 - 600} = \frac{200}{100} = 2$$
             $$y - 1600 = 2(x - 700)$$
             $$y = 2x - 1400 + 1600$$
             $$C(x) = 2x + 200$$
      (b) $200   **(c)** $2

15.   **(a)**   C(x) = 649x + 1500
      (b)   R(x) = 899x
      (c)   C(37) = 649(37) + 1500 = $25,513
      (d)   R(37) = 899(37) = $33,263
      (e)   899x = 649x + 1500
             250x = 1500   x = 6 computers

17.   **(a)**   Let x = number of years and BV = the book value. Then the points
             (0, 425) and (8, 25) are on the line, so
             $$m = \frac{425 - 25}{0 - 8} = \frac{400}{-8} = -50$$
             $$BV - 425 = -50x$$
             $$BV = -50x + 425$$
      (b)   Annual depreciation is $50
      (c)   BV(3) = -50(3) + 425 = -150 + 425 = $275

19.   **(a)**   Let x = number of years and BV = the book value. Then the points (0,
             9750) and (6, 300) lie on the line, so
             $$m = \frac{9750 - 300}{0 - 6} = \frac{9450}{-6} = -1575$$

$$BV - 9750 = -1575x$$
$$BV = -1575x + 9750$$

(b) $1,575

(c) $BV(2) = -1575(2) + 9750 = \$6600$
        $BV(5) = -1575(5) + 9750 = \$1875$

(d)   The auto might have been abused or been in a wreck.

21.   Revenue must be greater than costs, so
        $0.45x > 0.23x + 475$
        $0.22x > 475$
        $x > 2159.09$ at break-even, so at least 2,160 cookies must be sold to make a profit.

23.   The cost function is $y = 12x + 845$ and the revenue function is $y = 21x$ where x is the number of ties sold in a week.
      Profit $= 21x - (12x + 845) = 9x - 845$
      Profit occurs when
        $9x - 845 > 0$
        $9x > 845$
        $x > 93.889$
      The shop must sell at least 94 ties per week to make a profit.

25.   Cost function: $y = 0.32x + (300 + 130 + 1800 + 90) = 0.32x + 2320$ where x is the number of bagels sold per week.
      Revenue function: $y = 0.95x$
      Profit $= 0.95x - (0.32x + 2320) = 0.63x - 2320$
      Profit occurs when
        $0.63x - 2320 > 0$
        $0.63x > 2320$
        $x > 3682.5$
      He must sell at least 3683 bagels per week to make a profit

27.   Let x = number of miles traveled. The weekly costs for Company A, $C(x) = 0.14x + 105$, must be less than the weekly costs for Company B, $C(x) = 0.10x + 161$.
        $0.14x + 105 < 0.10x + 161$
        $0.04x < 56$
        $x < 1400$
      Company A is the better deal when the weekly mileage is less than 1400 miles.

29.   Let x = number of copies sold. A profit occurs when
        $0.85x + 0.20x > 0.40x + 1400$
        $0.65x > 1400$
        $x > 2153.8$, so at least 2,154 copies must be sold to make a profit.

Section 1.3 Mathematical Models

31. Let x = number of books. The unit cost gives m = 12.65. The point (2700, 36295) lies on the line so

y - 36295 = 12.65(x - 2700)

y= 12.65x + 2140

33. Let x = number of tickets.

(a) 6x = 650 + 45 + 2.20x

3.8x = 695

x = 182.89, so 183 tickets must be sold to break even.

(b) 6x = 650 + 45 + 2.20x + 700

3.8x = 1395

x = 367.105, so 368 tickets must be sold to clear $700.

(c) 7.5x = 2.20x + 1395

5.3x = 1395

x = 263.208, so 264 tickets must be sold to clear $700.

35. Let x = number of members

(a) R(x) = 35x   **(b)** R(1238) = 35(1238) = $43,330

(c) Solve 35x = 595

x = 17

37. **(a)** Let x = number of years and BV = book value, so the points (3, 14175) and (7, 8475) lie on the line, so

$$m = \frac{8475 - 14175}{7 - 3} = \frac{-5700}{4} = -1425$$

BV - 14175 = -1425(x - 3)

BV = -1425x + 18,450

(b) $1,425   **(c)** BV(0) = $18,450

39. Let x = number of memberships.

(a) Solve a(260) = 3120

a = 12   so R(x) = 12x

(b) Since the break-even membership is 260 and the break-even revenue is 3120, the point (260, 3120) is on the cost line. The revenue for 200 memberships is $2400 , $330 less than the cost, so the point

(200, 2730) is on the cost line. From these two points m = $\dfrac{3120 - 2730}{260 - 200}$

$= \dfrac{390}{60} = 6.5$

y - 2730 = 6.5(x - 200)

y = 6.5x - 1300 + 2730

C(x) = 6.5x + 1430

Section 1.3 Mathematical Models

41.　**(a)**　Solve a(1465) = 32962.50

　　　　　　a = 22.5, so R(x) = 22.5x

　　(b)　The points (1465, 26405.50) and (940, 17638) lie on the line, so

$$m = \frac{26405.50 \ - \ 17638}{1465 \ - \ 940}$$

$$= \frac{8767.5}{525} = 16.7$$

　　　　　　y - 17638 = 16.7(x - 940)

　　　　　　y = 16.7x - 15698 + 17638

　　　　　　C(x) = 16.7x + 1940

　　(c)　Solve 22.5x = 16.7x + 1940

　　　　　　5.8x = 1940

　　　　　　x = 334.48 so use 335 for break-even quantity.

43.　Let x = number of minutes called and y = total monthly cost.

　　The cost of the first option is y = 0.06x + 7.95

　　The cost of the second option is y = 0.09x

　　(a)　The two options cost the same when

　　　　　　0.09x = 0.06x + 7.95

　　　　　　0.03x = 7.95

　　　　　　x = 265

　　　　The two options cost the same for 265 minutes per month.

　　(b)　We are to solve

　　　　　　0.06x + 7.95 < 0.09x

　　　　　　7.95 < 0.03x

$$x > \frac{7.95}{0.03} = 265$$

　　　　The first plan is less costly when calls total more than 265 minutes
　　　　a month.

45.　Each plan can be written with a linear equation where x = number of checks
　　per month written and y = total monthly charge.

　　Plan 1 y = 15

　　Plan 2 y = 5 + 0.08x

　　Plan 3 y = 0.16x

　　The graph of Plan 2 lies below the graphs of both Plan 1 and Plan 3 between
　　x = 62.5 and x = 125 so Plan 2 is better when the number of checks written in a
　　month is more than 62 and less than 125.

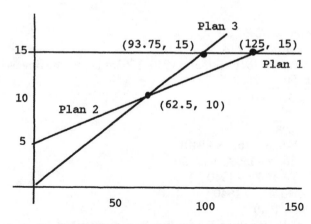

47. Let x be the number of knives produced daily. The cost function for the current process is $C(x) = 3.85x + 1400$. The cost function for the proposed process is $CP(x) = 2.70x + 1725$.

Graph both functions.

The lines intersect when

$$3.85x + 1400 = 2.70x + 1725$$
$$1.15x = 325$$
$$x = 282.6$$

The new process is more economical when the plant produces more than 282 knives per day.

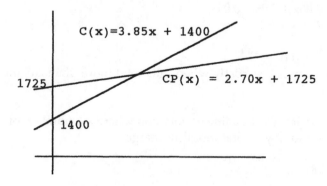

49. **(b)** Shortage   **(c)** Surplus

      (d) When the demand is 20, the price is -$40 so people must be paid to purchase. It is unrealistic to expect a demand of 20.

51. Because costs cannot exceed $100,000,

$$16700 + 140x \le 100,000$$
$$140x \le 83300$$
$$x = 595 \text{ is the maximum production.}$$

16

Section 1.3 Mathematical Models

53. Let x = monthly sales. Then the three plans are
Plan 1: $y = 2500$
Plan 2: $y = 0.17x + 2000$
Plan 3: $y = 0.25x + 1700$

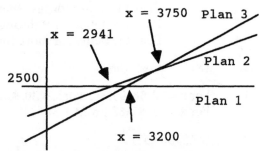

Plan 1 and Plan 2 intersect when $2500 = 2000 + 0.17x$, when x = $2,941.
Plan 1 and Plan 3 intersect when $2500 = 1700 + 0.25x$, when x = $3,200.
Plan 2 and Plan 3 intersect when $2000 + 0.17x = 1700 + 0.25x$, when x = $3,750.
Plan 1 is better when sales are less than $2941.
Plan 2 is better when sales are between $2941 and $3750.
Plan 3 is better when sales are greater than $3750.

55. **(a)** Let x = number of tickets sold. The cost at the Convention Center is
$C(x) = 20x + 600$. The cost at the Ferrell Center is $C(x) = 17x + 1300$.
Break even at the Convention Center occurs when $35x = 20x + 2000 + 700 + 600 = 20x + 3300$. $x = 220$.

(b) Break even at the Ferrell Center occurs when $35x = 17x + 2000 + 700 + 1300 = 17x + 4000$. $x = 222.22$. They must sell 223 to break even.

(c) At the Convention Center the profit is
$P(x) = 35x - 20x - 3300 = 15x - 3300$
At the Ferrell Center the profit is
$P(x) = 35x - 17x - 4000 = 18x - 4000$
These lines intersect at $x = 233.3$. The Ferrell Center is more profitable when more than 233 tickets are sold, otherwise the Convention Center is more profitable.

57. **(a)** For 10 trips, Plan 1 = $2500, Plan 2 = $1600, and Plan 3 = $16500. Plan 2 is better at a cost of $1600.
For 15 trips, Plan 1 = $2500, Plan 2 = $2250, and Plan 3 = $2475. Plan 2 is better at a cost of $2250.

(b) Let x = number of trips and y = total cost.
The linear equations for the three plans are:
Plan 1: $y = 2500$
Plan 2: $y = 130x + 300$
Plan 3: $y = 165x$
Plan 1 and Plan 2 intersect at $x = 16.9$.
Plan 1 and Plan 3 intersect at $x = 15.15$.
Plan 2 and Plan 3 intersect at $x = 8.57$.

Plan 3 is better for less than 9 trips.
Plan 2 is better for 9 through 16 trips.
Plan 1 is better for more than 16 trips.

59. **(a)** Cost = 5600   **(b)** Cost = 7925

61. For x = 10, $840 loss. For x = 30, $240 profit. For x = 45, $1050 profit.
For x = 62, $1968 profit.

63. (27.2, 587.52)   **65.** **(a)** (64, 5088)
                                                    (b) x = 121.14
                                                    (c) x = 47.23

TI-83

1. (4, 7)   **3.** (4.27, 0.91)

5. (2, 6)

Using EXCEL

1. -35, 215, 590, 1015, and 1490   **3.** -1396; 16709; 52,919; 77,059;
127,753

5. (25, 1220)

Review Exercises, Chapter 1

1. **(a)** $f(5) = \dfrac{7 \times 5 - 3}{2} = 16$ **(b)** $f(1) = 2$

   (c) $f(4) = 12.5$ **(d)** $f(b) = \dfrac{7b - 3}{2}$

3. $f(2) + g(3) = \dfrac{2 + 2}{2 - 1} + 5(3) + 3 = \dfrac{4}{1} + 15 + 3 = 22$

5. **(a)** $f(3.5) = 1.20(3.5) = \$4.20$ **(b)** Solve $1.20x = 3.30$
                                                    $x = 2.75$ pounds

7. **(a)** $f(x) = 29.95x$ **(b)** $f(x) = 1.25x + 40$

9. **(a)**

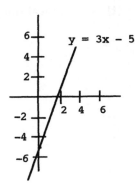

(b)

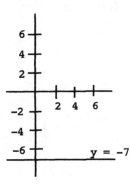

(c)

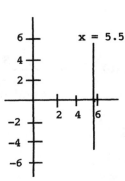

(d)

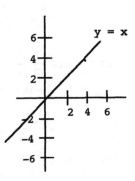

11. **(a)** Slope is -2, y-intercept is 3

(b) Slope is 2/3, y-intercept is -4

(c) $y = \dfrac{5}{4}x + \dfrac{3}{2}$ slope is $\dfrac{5}{4}$, y-intercept is $\dfrac{3}{2}$

(d) $y = -\dfrac{6}{7}x - \dfrac{5}{7}$ slope is $-\dfrac{6}{7}$, y-intercept is $-\dfrac{5}{7}$

13. $y = -\dfrac{6}{5}x + 3$

(a) $m = -\dfrac{6}{5}$ **(b)** y-intercept = 3

(c) When y = 0, x = 5/2, so x-intercept = 5/2

15. **(a)** $y = -\dfrac{3}{4}x + 5$ **(b)** $y = 8x - 3$

(c) $y + 1 = -2(x - 5)$ **(d)** Horizontal line, y = 6
$y = -2x + 10 - 1$
$y = -2x + 9$

(e) $m = \dfrac{4-3}{-1-5} = \dfrac{1}{-6}$

(f)   Vertical line, x = -2

$$y - 3 = -\frac{1}{6}(x - 5)$$

$$y = -\frac{1}{6}x + \frac{5}{6} + 3$$

$$y = -\frac{1}{6}x + \frac{23}{6}$$

(g) $y = \dfrac{4}{3}x - \dfrac{22}{3}$, so $m = \dfrac{4}{3}$

$$y - 7 = \frac{4}{3}(x - 2)$$

$$y = \frac{4}{3}x + \frac{13}{3} \text{ or } 4x - 3y = -13$$

17.   (a) $m = \dfrac{2-2}{-3-6} = 0$

$$y - 2 = 0(x - 6)$$
$$y = 2$$

(b) $m = \dfrac{-2-5}{-4+4}$

slope undefined, x = -4

(c) $m = \dfrac{10-0}{5-5}$ slope undefined, x = 5

(d) $m = \dfrac{6-6}{7+7} = 0$, y = 6

19.   The line through (5, 19) and (-2, 7) has slope $m_1 = \dfrac{19-7}{5+2} = \dfrac{12}{7}$

The line through (11, 3) and (-1, -5) has slope $m_2 = \dfrac{3+5}{11+1} = \dfrac{8}{12} = \dfrac{2}{3}$ The lines are not parallel.

21.   The given line has slope m = -2. The line through (8, 6) and (-3, 14) has slope
$$m = \frac{14-6}{-3-8} = \frac{8}{-11} = -\frac{8}{11}$$
The lines are not parallel.

23.   The given line has slope $m = \dfrac{3}{2}$. The line through (9, 10) and (5, 6) has slope
$$m = \frac{10-6}{9-5} = 1 \text{ The lines are not parallel.}$$

25.   Let x = number of items produced.
C(x) = 36x + 12,800

27.   (a) $C(580) = 3.60(580) + 2850 = \$4{,}938$

       (b)   Solve $3.60x + 2850 = 5208$

$$3.60x = 2358$$
$$x = 655 \text{ bags}$$

29.   (a) $R(x) = 11x$ (b) $C(x) = 6.5x + 675$

       (c)   Solve $11x = 6.5x + 675$

$$4.5x = 675$$
$$x = 150$$

31. $R(x) = 19.5x$

The points $(1840, 25260)$ and $(2315, 31102.5)$ lie on the cost line, so, m =

$$\frac{31102.5 - 25260}{2315 - 1840} = \frac{5842.5}{475} = 12.3$$

$y - 25260 = 12.3(x - 1840)$

$C(x) = 12.3x + 2628$

To find the break-even quantity, solve

$19.5x = 12.3x + 2628$

$7.2x = 2628$

$x = 365 \text{ watches}$

33.   (a)   The points $(0, 17500)$ and $(8, 900)$ lie on the line so

$$m = \frac{900 - 17500}{8 - 0} = \frac{-16600}{8} = -2075$$

BV $= -2075x + 17500$

       (b) $\$2{,}075$

       (c) $BV(5) = -2075(5) + 17500 = \$7{,}125$

35.   The points $(0, 1540)$ and $(5, 60)$ lie on the line, so

$$m = \frac{1540 - 60}{0 - 5} = \frac{1480}{-5} = -296$$

BV $= -296x + 1540$

37.   The slope of the given line is $4/5$, so $\dfrac{k - 9}{-3 - 2} = \dfrac{4}{5}$ $k - 9 = -4$ $k = 5$

39.   The points $(0, 22000)$ and $(5, 3000)$ lie on the line, so

$$m = \frac{3000 - 22000}{5 - 0} = \frac{-19000}{5} = -3800 \qquad BV = -3800x + 22{,}000$$

41.   Let x = number of hamburgers and y = cost. Then $m = 0.67$, and the point $(1150, 1250.5)$ lies on the line, so

$y - 1250.5 = 0.67(x - 1150)$ $y = 0.67x - 770.5 + 1250.5$

$C(x) = 0.67x + 480$

43. Let x = number of items sold. The second option is better when
$$0.75x + 2000 > 17000$$
$$0.75x > 15000 \quad x > 20,000$$
The second plan is better when sales exceed 20,000 items.

45. $\dfrac{k-4}{-2-9} = -2$ $k - 4 = 22$ $k = 26$

47. Let x = number of tapes and C(x) = total cost. Since the unit cost is $6.82, m = 6.82, and the point (1730, 12813.60) lies on the line,
$$y - 12813.60 = 6.82(x - 1730) \qquad C(x) = 6.82x + 1015$$
The fixed cost is $1015.

49. Let x = sales, then the plans can be represented by:

Plan 1: $y = 2500$
Plan 2: $y = 0.15x + 2000$
Plan 3: $y = 0.30x + 1700$

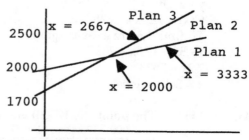

Plan 1 and Plan 2 intersect at $2500 = 2000 + 0.15x$, $x = 3,333$.
Plan 1 and Plan 3 intersect at $2500 = 1700 + 0.30x$, $x = 2,667$.
Plan 2 and Plan 3 intersect at $2000 + 0.15x = 1700 + 0.30x$, $x = 2000$.
Plan 1 is better when sales are less than $2,667.
Plan 2 is never better.
Plan 3 is better when sales are larger than $2,667.

51. **(a)** $y = 120x + 8400$
      (b) $10,000 = 120x + 8400$ $x = 13.33$
           The library holdings will reach 10,000 books during the 14th month.

53. $y = 500x - 1700$

Chapter 2
Linear Systems

Section 2.1

1. $4x - y = 5$
      $x + 2y = 8$

$x = 8 - 2y$
$4(8 - 2y) - y = 5$
$32 - 8y - y = 5$
$-9y = -27$
$y = 3$
$x = 8 - 2(3) = 2$ $(2, 3)$

3. $5x - y = -15$
      $x + y = -3$

$x = -3 - y$
$5(-3 - y) - y = -15$
$-15 - 5y - y = -15$
$-6y = 0$
$y = 0$
$x = -3 - 0 = -3$ $(-3, 0)$

5. $y = 5x$
      $6x - 2y = 12$

$6x - 2(5x) = 12$
$-4x = 12$
$x = -3$
$y = 5(-3) = -15$ $(-3, -15)$

7. $7x - y = 32$
      $2x + 3y = 19$

$y = 7x - 32$
$2x + 3(7x - 32) = 19$
$2x + 21x - 96 = 19$
$23x = 115$
$x = 5$
$y = 7(5) - 32 = 3$ $(5, 3)$

9. $5x + 2y = 14$
      $x - 3y = 30$

$x = 3y + 30$
$5(3y + 30) + 2y = 14$
$15y + 150 + 2y = 14$
$17y = -136$
$y = -8$
$x = 3(-8) + 30 = 6$ $(6, -8)$

11.　　$22x + y = 81$
　　　　　$8x - 3y = 16$

$y = 81 - 22x$
$8x - 3(81 - 22x) = 16$
$8x - 243 + 66x = 16$
$74x = 259$
$x = 3.5$
$y = 81 - 22(3.5) = 4$　　　(3.5, 4)

13.　　$6x - 3y = 9$
　　　　　$9x - 15y = 31$

$3y = 6x - 9$
$y = 2x - 3$
$9x - 15(2x - 3) = 31$
$9x - 30x + 45 = 31$
$-21x = -14$
$x = 2/3$
$y = 2(2/3) - 3 = 4/3 - 3 = -5/3$　　　(2/3, -5/3)

15.　　$y = 3x - 5$
　　　　　$8x - 4y - 30 = 0$

$8x - 4(3x - 5) - 30 = 0$
$8x - 12x + 20 - 30 = 0$
$-4x - 10 = 0$
$x = -2.5$
$y = -12.5$　　　　　(-2.5, -12.5)

17.　　$3x - 4y = 22$
　　　　　$2x + 5y = 7$

$6x - 8y = 44$
$\underline{-6x - 15y = -21}$
　　　$-23y = 23$
$y = -1$
$3x - 4(-1) = 22$
$3x + 4 = 22$
$x = 6$　　　　　(6, -1)

19.　　$6x - y = 18$
　　　　　$2x + y = 2$

$6x - y = 18$
$\underline{2x + y = 2}$
$8x\ \ = 20$
$x = 5/2$
$2(5/2) + y = 2$
$5 + y = 2$
$y = -3$　　　　　(5/2, -3)

21.　　$-2x + y = 7$
　　　　　$6x + 12y = 24$

$-6x + 3y = 21$
$\underline{6x + 12y = 24}$
　　　$15y = 45$
$y = 3$
$-2x + 3 = 7$
$-2x = 4$
$x = -2$　　　　　(-2, 3)

23. $2x + y = -9$
$4x + 3y = 1$

$-4x - 2y = 18$
$\underline{4x + 3y = 1}$
$\quad\quad y = 19$
$2x + 19 = -9$
$2x = -28$
$x = -14$ $\qquad\qquad$ (-14, 19)

25. $7x + 3y = -1.5$
$2x - 5y = -30.3$

$35x + 15y = -7.5$
$\underline{6x - 15y = -90.9}$
$41x \quad\quad = -98.4$
$x = -2.4$
$2(-2.4) - 5y = -30.3$
$-5y = -25.5$
$y = 5.1$ $\qquad\qquad$ (-2.4, 5.1)

27. $2x - 3y = -0.27$
$5x - 2y = 0.04$

$-4x + 6y = 0.54$
$\underline{15x - 6y = 0.12}$
$11x \quad\quad = 0.66$
$x = 0.06$
$2(0.06) - 3y = -0.27$
$-3y = -0.39$
$y = 0.13$ $\qquad\qquad$ (0.06, 0.13)

29. $6x - 9y = 8$
$10x - 15y = -20$

$30x - 45y = 40$
$\underline{-30x + 45y = 60}$
$\quad\quad 0 = 100$ $\qquad$ No solution

31. $8x + 10y = 2$
$12x + 15y = 3$

$4x + 5y = 1$
$\underline{-4x - 5y = -1}$
$\quad 0 = 0$ $\qquad$ Infinite number

33. $x - 6y = 4$
$5x - 30y = 20$

$x - 6y = 4$
$\underline{-x + 6y = -4}$
$\quad 0 = 0$ $\qquad$ Infinite number

35. $p = -3x + 15$
$p = 2x - 5$

$2x - 5 = -3x + 15$
$5x = 20$
$x = 4$
$p = -3(4) + 15 = 3$ $\qquad$ (4, 3)

37. $p = -4x + 130$
$p = x - 20$

$x - 20 = -4x + 130$
$5x = 150$
$x = 30$
$p = 30 - 20 = 10$ $\qquad$ (30, 10)

39. $p = -5x + 83$ $4x - 52 = -5x + 83$
          $p = 4x - 52$ $9x = 135$
                                    $x = 15$
                                    $p = 4(15) - 52 = 8$ $(15, 8)$

41. $p = -2.5x + 148$ $1.7x + 43 = -2.5x + 148$
          $p = 1.7x + 43$ $4.2x = 105$
                                    $x = 25$
                                    $p = 1.7(25) + 43 = 85.5$ $(25, 85.5)$

43.   Let x = number of oranges and y = number of apples.
          $50x + 8y = 151$ $50x + 8y = 151$
          $0.5x + 0.4y = 2.55$ $\underline{-50x - 40y = -255}$
                                    $-32y = -104$
                                    $y = 3.25$
                                    $50x + 8(3.25) = 151$
                                    $50x = 125$
                                    $x = 2.5$ 2.5 oranges, 3.25 apples

45.   **(a)** $C(x) = 0.85x + 160,000, R(x) = 5x$
                  The break-even point occurs when
                        $5x = 0.85x + 160,000$
                        $4.15x = 160,000$
                        $x = 38,554.2$ which we round to 38555
                  They must sell 38,555 CD's in order to break even.
          (b)   If 20,000 CD's are sold, the costs are
                  $0.85(20,000) + 160,000 = 177,000$. The revenue is $5(20,000) = 100,000$.
                  Thus, a loss of $77,000 occurs.
          (c)   If 50,000 CD's are sold, the costs are
                  $0.85(50,000) + 160,000 = 202,500$ and the revenue is $5(50,000) =$
                  $250,000$. Thus, a profit of $47,500 occurs.

47.   Let n number of nickels and d = number of dimes.
          $n + d = 165$ $d = 165 - n$
          $5n + 10d = 1435$ $5n + 10(165 - n) = 1435$
                                    $5n + 1650 - 10n = 1435$
                                    $-5n = -215$
                                    $n = 43$
                                    $d = 165 - 43 = 122$
                                    43 nickels, 122 dimes

49.   Let x = the amount of the first drink, y = the amount of the second drink.
  x + y = 600 (Ounces of drink requested)
  5.0x + 5.8y = 5.5(600) (Total grams in the drink)
        Substitute x = 600 - y into the second equation
        5(600 - y) + 5.8y = 5.5(600) = 3300
        3000 - 5y + 5.8y = 3300
        0.8y = 300
        y = 375
        x = 600 - y = 225
Mix 225 ounces of the first drink and 375 of the second drink.

51.   Let x = number produced at McGregor and y = number produced at Ennis.
The McGregor cost function is C = 8.4x + 7480 and the Ennis cost function is C = 7.8y + 5419. We have the system
        x + y = 1500
        8.4x + 7480 + 7.8y + 5419 = 24956   which reduces to
        x + y = 1500
        8.4x + 7.8y = 12057
By substitution we have
        8.4(1500 - y) + 7.8y = 12057
        -0.6y = -543
        y = 905
        x = 1500 - 905 = 595   595 at McGregor, 905 at Ennis

53.   Let x = number of boxes of oranges and y = number of boxes of grapefruit.
  x + y = 502 (Number of boxes sold)
  14x + 16y = 7570 (Income)
        Substitute y = 502 - x in the second equation.
        14x + 16(502 - x) = 7570
        14x + 8032 - 16x = 7570
        -2x = -462
        x = 231
        231 + y = 502
        y = 271
They sold 231 boxes of oranges and 271 boxes of grapefruit.

55.   Let x = cases of Golden Punch and y = cases of Light Punch.
  4x + 7y = 142   12x + 21y = 426
  6x + 3y = 108   -12x - 6y = -216
                                        15y = 210
                                          y = 14
                                          x = 11
                        11 cases of Golden, 14 cases of Light Punch

57.   Let x = number of two-seat tables and y = number of four-seat tables.
x + y = 20 (Number of tables)
2x + 4y = 66 (Number of seats)

      Eliminate x from the second equation.

      2x + 2y = 40
      <u>2x + 4y = 66</u>
       -2y = -26
      y = 13
      x + 13 = 20
      x = 7

There are 7 two-place tables and 13 four-place tables.

59.   Let x = amount in tax-free and y = amount in money market.

x + y = 50000   x = 50000 - y
0.074x + 0.088y = 4071   74(50000 - y) + 88y = 4071000
                                 14y = 371,000
                                  y = 26,500
                                  x = 23,500
           $23,500 in tax-free, $26,500 in money market

61.   Let x = amount of federal tax and y = amount of state tax.
Federal tax = x = 0.20(198000 - y)
State tax = y = 0.05(198000 - x)
These equations reduce to

  x + 0.2y = 39600   0.05x + 0.01y = 1980
  0.05x +   y = 9900   <u>-0.05x -   y = -9900</u>
                              -0.99y = -7920
                              y = 8,000
                            x = 0.2(190,000)= 38,000
                            Federal tax is $38,000, state tax is $8,000

63.   **(a)**   Let H = the amount budgeted for Habitat for Humanity and A the amount budgeted for The Family Abuse Center.
H = 0.60(5000 - A) = 3000 - 0.60A
A = 0.40(5000 - H) = 2000 - 0.40H

      Then H = 3000 - 0.60(2000 - 0.40H) = 3000 - 1200 + 0.24H
      0.76H = 1800
      H = 2368.42 which we round to 2370.
      A = 0.40(5000 - 2370) = 0.40(2630) = 1052
      which we round to 1050.

They budgeted $2370 to Habitat for Humanity, $1050 to the Family Abuse Center.

    (b)   The remainder, $5000 - $2370 - $1050 = $1580, was budgeted for their church.

67.   Let x = number of years later and y = total deposits.
For Central Bank $y = 250{,}000x + 3{,}000{,}000$.
For Citizens Bank $y = -300{,}000x + 9{,}000{,}000$.
Their deposits are equal when
$$250{,}000x + 3{,}000{,}000 = -300{,}000x + 9{,}000{,}000.$$
$$550{,}000x = 6{,}000{,}000$$
$$x = 10.91$$
Their deposits will be equal in a little less than 11 years.

69.   **(a)** $x = 4.54, y = 7.74$ $\qquad$ **(b)** $x = 0.75, y = 9.25$
$\qquad$ **(c)** $x = 1.26, y = 4.80$ $\qquad$ **(d)** $x = 3.02, y = -0.72$

71.   20.55 tricycles at a price of \$117.53. If we round to 21 tricycles, the supply price is \$118.80 and the demand price is \$115.50.

73.   **(a)**   Equilirium occurs at (27. 126)
$\qquad$ **(b)**   Demand = 35, supply = 11, shortage of 24
$\qquad$ **(c)**   Demand = 22.5, supply = 36, surplus of 13.5, or about 14

75.   (5.32, 2.26) $\qquad\qquad$ **77.**   (2.5, 6.25)

Section 2.2

1. $\quad x + 2y = 7$ $\qquad\qquad\qquad$ **3.** $\quad 2x + 5y = -1$
$\quad 3x + 5y = 19$ $\qquad\qquad\qquad\qquad\quad 6x - 4y = 16$

$$
\begin{array}{ll}
3x + 6y = 21 & \qquad 6x + 15y = -3 \\
\underline{3x + 5y = 19} & \qquad \underline{6x - 4y = 16} \\
y = 2 & \qquad\qquad\quad 19y = -19 \\
x + 2(2) = 7 & \qquad\qquad\quad y = -1 \\
x = 3 \qquad (3,\,2) & \qquad 2x + 5(-1) = -1 \\
 & \qquad\qquad x = 2 \qquad (2,\,-1)
\end{array}
$$

5. $\quad x + y - z = -1$
$\quad x - y + z = 5$
$\quad x - y - z = 1$

$$
\begin{array}{l}
x + y - z = -1 \\
\underline{x - y + z = 5} \quad \text{subtract second from first} \\
 2y - 2z = -6 \\
x - y + z = 5 \\
\underline{x - y - z = 1} \quad \text{subtract third from second} \\
 2z = 4
\end{array}
$$

$z = 2$

Substitute $z = 2$

$2y - 2(2) = -6$

$2y = -2$

$y = -1$

Substitute $y = -1$ and

$z = 2$ into the first equation.

$x + (-1) - 2 = -1$

$x = 2$ $\qquad\qquad$ $x = 2, y = -1, z = 2$

7. $\quad x + 4y - 2z = 21$

$3x - 6y - 3z = -18$

$2x + 4y + z = 37$

$3x + 12y - 6z = 63$ $\qquad$ multiply first equation by 3

$\underline{3x - 6y - 3z = -18}$ $\qquad$ subtract second

$\quad 18y - 3z = 81$

$2x + 8y - 4z = 42$ multiply first equation by 2

$\underline{2x + 4y + z = 37}$ subtract third

$\quad 4y - 5z = 5$

Use these two new equations

$18y - 3z = 81$

$4y - 5z = 5$

$36y - 6z = 162$

$\underline{36y - 45z = 45}$

$\quad 39z = 117$

$\quad z = 3$

Substitute $z = 3$

$18y - 3(3) = 81$

$18y = 90$

$\quad y = 5$

Substitute $z = 3$ and $y = 5$ into the first equation.

$x + 4(5) - 2(3) = 21$

$x = 7$ $\qquad\qquad$ (7, 5, 3)

9. $\quad 2x + 4y - 6z = -2$

$4x - 3y + z = 11$

$3x + 2y - 2z = 7$

$4x + 8y - 12z = -4$ $\quad$ multiply first equation by 2

$\underline{4x - 3y + \ z = 11}$ $\quad$ subtract second

$\quad 11y - 13z = -15$

$6x + 12y - 18z = -6$ $\quad$ multiply first equation by 3

$\underline{6x + 4y - 4z = 14}$ $\quad$ multiply equation 3 by 2

$\quad 8y - 14z = -20$ $\quad$ subtract

Now solve $11y - 13z = -15$
$$8y - 14z = -20$$
$88y - 104z = -120$
$\underline{88y - 154z = -220}$
$$50z = 100$$
$$z = 2$$
$11y - 13(2) = -15$
$11y = 11$
$y = 1$
$2x + 4(1) - 6(2) = -2$
$2x = 6$
$x = 3$ 　　　　　　$x = 3, y = 1, z = 2$

11.　　(a)　　$a_{11} = 2,\ a_{22} =,\ a_{33} = 6,\ a_{43} = 11$

　　　　(b)　　$(2, 3)$

　　　　(c)　　$a_{12} = 4,\ a_{32} = 0,\ a_{41} = 9$

13.　　coeff: $\begin{bmatrix} 5 & -2 \\ 3 & 1 \end{bmatrix}$ 　　　　aug: $\begin{bmatrix} 5 & -2 & | & 1 \\ 3 & 1 & | & 7 \end{bmatrix}$

15.　　coeff: $\begin{bmatrix} 1 & 1 & -1 \\ 3 & 4 & -2 \\ 2 & 0 & 1 \end{bmatrix}$ 　　aug: $\begin{bmatrix} 1 & 1 & -1 & | & 14 \\ 3 & 4 & -2 & | & 9 \\ 2 & 0 & 1 & | & 7 \end{bmatrix}$

17.　　coeff: $\begin{bmatrix} 1 & 5 & -2 & 1 \\ 1 & -1 & 2 & 4 \\ 6 & 3 & -11 & 1 \\ 5 & -3 & -7 & 1 \end{bmatrix}$ 　　aug: $\begin{bmatrix} 1 & 5 & -2 & 1 & | & 12 \\ 1 & -1 & 2 & 4 & | & -5 \\ 6 & 3 & -11 & 1 & | & 14 \\ 5 & -3 & -7 & 1 & | & 22 \end{bmatrix}$

19.　　$5x + 3y = -2$

　　　　$-x + 4y = 4$

21.　　$5x_1 + 2x_2 - x_3 = 3$

　　　　$-2x_1 + 7x_2 + 8x_3 = 7$

　　　　$3x_1 \quad + x_3 = 5$

23.　　$3x_1 \quad + 2x_3 + 6x_4 = 4$

　　　　$-4x_1 + 5x_2 + 7x_3 + 2x_4 = 2$

　　　　$x_1 + 3x_2 + 2x_3 + 5x_4 = 0$

　　　　$-2x_1 + 6x_2 - 5x_3 + 3x_4 = 4$

25.　　$\begin{bmatrix} 1 & 2 & -4 & | & 6 \\ 4 & 2 & 5 & | & 7 \\ 1 & -1 & 0 & | & 4 \end{bmatrix}$

27. $\begin{bmatrix} 1 & 3 & 2 & | & -4 \\ 0 & -7 & -1 & | & 3 \\ 0 & -6 & -10 & | & 9 \end{bmatrix}$

29. $\begin{bmatrix} 1 & -3 & 2 & | & -6 \\ 0 & 1 & -2 & | & 4 \\ 0 & 4 & 3 & | & 8 \end{bmatrix}$

31. $\begin{bmatrix} 2 & 3 & | & 5 \\ 1 & -2 & | & -1 \end{bmatrix} R1 \leftrightarrow R2$ $\begin{bmatrix} 1 & -2 & | & -1 \\ 2 & 3 & | & 5 \end{bmatrix} -2R1 + R2 \to R2$

$\begin{bmatrix} 1 & -2 & | & -1 \\ 0 & 7 & | & 7 \end{bmatrix} (1/7)R2 \to R2$

$\begin{bmatrix} 1 & -2 & | & -1 \\ 0 & 1 & | & 1 \end{bmatrix} 2R2 + R1 \to R1$ $\begin{bmatrix} 1 & 0 & | & 1 \\ 0 & 1 & | & 1 \end{bmatrix}$ $x = 1, y = 1$

33. $\begin{bmatrix} 1 & -3 & | & -1 \\ 4 & 5 & | & 30 \end{bmatrix} -4R1 + R2 \to R2$ $\begin{bmatrix} 1 & -3 & | & -1 \\ 0 & 17 & | & 34 \end{bmatrix} (1/17)R2 \to R2$

$\begin{bmatrix} 1 & -3 & | & -1 \\ 0 & 1 & | & 2 \end{bmatrix} 3R2 + R1 \to R1$ $\begin{bmatrix} 1 & 0 & | & 5 \\ 0 & 1 & | & 2 \end{bmatrix} x = 5, y = 2$

35. $\begin{bmatrix} 2 & 4 & | & -7 \\ 1 & -3 & | & 9 \end{bmatrix} R1 \leftrightarrow R2$ $\begin{bmatrix} 1 & -3 & | & 9 \\ 2 & 4 & | & 7 \end{bmatrix} -2R1 + R2 \to R2$

$\begin{bmatrix} 1 & -3 & | & 9 \\ 0 & 10 & | & -25 \end{bmatrix} (1/10)R2 \to R2$ $\begin{bmatrix} 1 & -3 & | & 9 \\ 0 & 1 & | & -5/2 \end{bmatrix} 3R2 + R1 \to R1$

$\begin{bmatrix} 1 & 0 & | & 3/2 \\ 0 & 1 & | & -5/2 \end{bmatrix}$ $x = 3/2, y = -5/2$

37. $\begin{bmatrix} 1 & 2 & -1 & | & 3 \\ 1 & 3 & -1 & | & 4 \\ 1 & -1 & 1 & | & 4 \end{bmatrix} \begin{matrix} \\ -R1+R2 \to R2 \\ -R1+R3 \to R3 \end{matrix}$ $\begin{bmatrix} 1 & 2 & -1 & | & 3 \\ 0 & 1 & 0 & | & 1 \\ 0 & -3 & 2 & | & 1 \end{bmatrix} \begin{matrix} -2R2+R1 \to R1 \\ \\ 3R2+R3 \to R3 \end{matrix}$

$\begin{bmatrix} 1 & 0 & -1 & | & 1 \\ 0 & 1 & 0 & | & 1 \\ 0 & 0 & 2 & | & 4 \end{bmatrix} (1/2)R3 \to R3$

$\begin{bmatrix} 1 & 0 & -1 & | & 1 \\ 0 & 1 & 0 & | & 1 \\ 0 & 0 & 1 & | & 2 \end{bmatrix} R3 + R1 \to R1$ $\begin{bmatrix} 1 & 0 & 0 & | & 3 \\ 0 & 1 & 0 & | & 1 \\ 0 & 0 & 1 & | & 2 \end{bmatrix}$ $x_1 = 3, x_2 = 1, x_3 = 2$

39. $\begin{bmatrix} 2 & 4 & 2 & | & 6 \\ 2 & 1 & 1 & | & 16 \\ 1 & 1 & 2 & | & 9 \end{bmatrix} R1 \leftrightarrow R3$ $\begin{bmatrix} 1 & 1 & 2 & | & 9 \\ 2 & 1 & 1 & | & 16 \\ 2 & 4 & 2 & | & 6 \end{bmatrix} \begin{matrix} \\ -2R1+R2 \to R2 \\ -2R1+R3 \to R3 \end{matrix}$

$\begin{bmatrix} 1 & 1 & 2 & | & 9 \\ 0 & -1 & -3 & | & -2 \\ 0 & 2 & -2 & | & -12 \end{bmatrix} \begin{matrix} R2+R1 \to R1 \\ -R2 \to R2 \\ 2R2+R3 \to R3 \end{matrix}$

$$\begin{bmatrix} 1 & 0 & -1 & | & 7 \\ 0 & 1 & 3 & | & 2 \\ 0 & 0 & -8 & | & -16 \end{bmatrix} -(1/8)R3 \rightarrow R3 \qquad \begin{bmatrix} 1 & 0 & -1 & | & 7 \\ 0 & 1 & 3 & | & 2 \\ 0 & 0 & 1 & | & 2 \end{bmatrix} \begin{matrix} R3+R1 \rightarrow R1 \\ -R3+R2 \rightarrow R2 \end{matrix}$$

$$\begin{bmatrix} 1 & 0 & 0 & | & 9 \\ 0 & 1 & 0 & | & -4 \\ 0 & 0 & 1 & | & 2 \end{bmatrix} \qquad (9, -4, 2)$$

41.
$$\begin{bmatrix} 1 & 2 & -1 & | & -1 \\ 2 & -3 & 2 & | & 15 \\ 0 & 1 & 4 & | & -7 \end{bmatrix} -2R1+R2 \rightarrow R2 \qquad \begin{bmatrix} 1 & 2 & -1 & | & -1 \\ 0 & -7 & 4 & | & 17 \\ 0 & 1 & 4 & | & -7 \end{bmatrix} R2 \leftrightarrow R3$$

$$\begin{bmatrix} 1 & 2 & -1 & | & -1 \\ 0 & 1 & 4 & | & -7 \\ 0 & -7 & 4 & | & 17 \end{bmatrix} \begin{matrix} -2R2+R1 \rightarrow R1 \\ \\ 7R2+R3 \rightarrow R3 \end{matrix}$$

$$\begin{bmatrix} 1 & 0 & -9 & | & 13 \\ 0 & 1 & 4 & | & -7 \\ 0 & 0 & 32 & | & -32 \end{bmatrix} \qquad \begin{bmatrix} 1 & 0 & -9 & | & 13 \\ 0 & 1 & 4 & | & -7 \\ 0 & 0 & 1 & | & -1 \end{bmatrix} \qquad \begin{bmatrix} 1 & 0 & 0 & | & 4 \\ 0 & 1 & 0 & | & -3 \\ 0 & 0 & 1 & | & -1 \end{bmatrix}$$
$$(4, -3, -1)$$

43.
$$\begin{bmatrix} 1 & -5 & 1 & | & 1 \\ 4 & -2 & -3 & | & 6 \\ -3 & 1 & 3 & | & -3 \end{bmatrix} \begin{bmatrix} 1 & 5 & -1 & | & 1 \\ 0 & -22 & 1 & | & 2 \\ 0 & 16 & 0 & | & 0 \end{bmatrix} \begin{bmatrix} 1 & 0 & -1 & | & 1 \\ 0 & 0 & 1 & | & 2 \\ 0 & 1 & 0 & | & 0 \end{bmatrix} \begin{bmatrix} 1 & 0 & 0 & | & 3 \\ 0 & 0 & 1 & | & 2 \\ 0 & 1 & 0 & | & 0 \end{bmatrix}$$
$$(3, 0, 2)$$

45.
$$\begin{bmatrix} 20 & 20 & 40 & | & 161 \\ 10 & 24 & -10 & | & 60 \\ -31 & 12 & -2 & | & 0 \end{bmatrix} \begin{bmatrix} 1 & 1 & 2 & | & 8 \\ 0 & 1 & -3 & | & -2 \\ 0 & 1 & 1 & | & 6 \end{bmatrix} \begin{bmatrix} 1 & 0 & 5 & | & 10 \\ 0 & 1 & -3 & | & -2 \\ 0 & 0 & 1 & | & 2 \end{bmatrix} \begin{bmatrix} 1 & 0 & 0 & | & 0 \\ 0 & 1 & 0 & | & 4 \\ 0 & 0 & 1 & | & 2 \end{bmatrix}$$
$$(0, 4, 2)$$

47.
$$\begin{bmatrix} 1 & 2 & -3 & | & -6 \\ 1 & -3 & -7 & | & 10 \\ 1 & -1 & 1 & | & 10 \end{bmatrix} \begin{bmatrix} 1 & 2 & -3 & | & -6 \\ 0 & 5 & 4 & | & -16 \\ 0 & -3 & 4 & | & 16 \end{bmatrix} \begin{bmatrix} 1 & 0 & -23/5 & | & 2/5 \\ 0 & 1 & 4/5 & | & -16/5 \\ 0 & 0 & 32/5 & | & 32/5 \end{bmatrix}$$

$$\begin{bmatrix} 1 & 0 & 0 & | & 5 \\ 0 & 1 & 0 & | & -4 \\ 0 & 0 & 1 & | & 1 \end{bmatrix} \qquad (5, -4, 1)$$

49.
$$\begin{bmatrix} 1 & 1 & 1 & 1 & | & 4 \\ 1 & 2 & -1 & -1 & | & 7 \\ 2 & -1 & -1 & -1 & | & 8 \\ 1 & -1 & 2 & -2 & | & -7 \end{bmatrix} \begin{matrix} -R1+R2 \rightarrow R2 \\ -2R1+R3 \rightarrow R3 \\ -R1+R4 \rightarrow R4 \end{matrix} \begin{bmatrix} 1 & 1 & 1 & 1 & | & 4 \\ 0 & 1 & -2 & -2 & | & 3 \\ 0 & -3 & -3 & -3 & | & 0 \\ 0 & -2 & 1 & -3 & | & -11 \end{bmatrix} \begin{matrix} 3R2+R3 \rightarrow R3 \\ 2R2+R4 \rightarrow R4 \end{matrix}$$

$$\begin{bmatrix} 1 & 0 & 3 & 3 & | & 1 \\ 0 & 1 & -2 & -2 & | & 3 \\ 0 & 0 & -9 & -9 & | & 9 \\ 0 & 0 & -3 & -7 & | & -5 \end{bmatrix} (-1/9)R3 \rightarrow R3 \begin{bmatrix} 1 & 0 & 3 & 3 & | & 1 \\ 0 & 1 & -2 & -2 & | & 3 \\ 0 & 0 & 1 & 1 & | & -1 \\ 0 & 0 & -3 & -7 & | & -5 \end{bmatrix} \begin{matrix} -3R3+R1 \rightarrow R1 \\ 2R3+R2 \rightarrow R2 \\ \\ 3R3+R4 \rightarrow R4 \end{matrix}$$

$$\begin{bmatrix} 1 & 0 & 0 & 0 & | & 4 \\ 0 & 1 & 0 & 0 & | & 1 \\ 0 & 0 & 1 & 1 & | & -1 \\ 0 & 0 & 0 & -4 & | & -8 \end{bmatrix} (-1/4)R4 \to R4 \qquad \begin{bmatrix} 1 & 0 & 0 & 0 & | & 4 \\ 0 & 1 & 0 & 0 & | & 1 \\ 0 & 0 & 1 & 1 & | & -1 \\ 0 & 0 & 0 & 1 & | & 2 \end{bmatrix} -R4 + R3 \to R3$$

$$\begin{bmatrix} 1 & 0 & 0 & 0 & | & 4 \\ 0 & 1 & 0 & 0 & | & 1 \\ 0 & 0 & 1 & 0 & | & -3 \\ 0 & 0 & 0 & 1 & | & 2 \end{bmatrix} \qquad (4, 1, -3, 2)$$

51.

$$\begin{bmatrix} 2 & 6 & 4 & -2 & | & 18 \\ 1 & 4 & -2 & -1 & | & -1 \\ 3 & -1 & -1 & 2 & | & 6 \\ -1 & -2 & -5 & 0 & | & -20 \end{bmatrix} \qquad \begin{bmatrix} 1 & 3 & 2 & -1 & | & 9 \\ 0 & 1 & -4 & 0 & | & -10 \\ 0 & -10 & -7 & 5 & | & -21 \\ 0 & 1 & -3 & -1 & | & -11 \end{bmatrix}$$

$$\begin{bmatrix} 1 & 0 & 14 & -1 & | & 39 \\ 0 & 1 & -4 & 0 & | & -10 \\ 0 & 0 & -47 & 5 & | & -121 \\ 0 & 0 & 1 & -1 & | & -1 \end{bmatrix} \begin{bmatrix} 1 & 0 & 14 & -1 & | & 39 \\ 0 & 1 & -4 & 0 & | & -10 \\ 0 & 0 & 1 & -1 & | & -1 \\ 0 & 0 & -47 & 5 & | & -121 \end{bmatrix} \begin{bmatrix} 1 & 0 & 0 & 13 & | & 53 \\ 0 & 1 & 0 & -4 & | & -14 \\ 0 & 0 & 1 & -1 & | & -1 \\ 0 & 0 & 0 & -42 & | & -168 \end{bmatrix}$$

$$\begin{bmatrix} 1 & 0 & 0 & 0 & | & 1 \\ 0 & 1 & 0 & 0 & | & 2 \\ 0 & 0 & 1 & 0 & | & 3 \\ 0 & 0 & 0 & 1 & | & 4 \end{bmatrix} \qquad (1, 2, 3, 4)$$

53.   Let x_1 = number cases of Regular, x_2 = number cases of Premium, x_3 = number cases of Classic.

$$4x_1 + 4x_2 + 5x_3 = 316 \quad \text{(Apple juice)}$$
$$5x_1 + 4x_2 + 2x_3 = 292 \quad \text{(Pineapple juice)}$$
$$x_1 + 2x_2 + 3x_3 = 142 \quad \text{(Cranberry juice)}$$

55.   Let x_1 = number of student tickets, x_2 = number of faculty tickets, x_3 = number of general public tickets.

$$3x_1 + 5x_2 + 8x_3 = 2542$$
$$x_1 = 3x_2$$
$$x_3 = 2x_1$$

57.   Let x = number shares of X, y = number shares of Y, z = number shares of Z.

$$44x + 22y + 64z = 20{,}480$$
$$42x + 28y + 62z = 20{,}720$$
$$42x + 30y + 60z = 20{,}580$$

59.   Let x_{11} = cost of a Six-pack, x_{12} = cost of a bag of chips, x_{13} = cost of a package of cookies.

$$3x_{11} + 2x_{12} + 4x_{13} = 22.00 \quad (\text{Andrew's cost})$$
$$2x_{11} + 4x_{12} + 5x_{13} = 26.30 \quad (\text{Cutter's cost})$$
$$x_{11} + 2x_{12} + 4x_{13} = 17.20 \quad (\text{Madeline's cost})$$

The augmented matrix of this system is

$$\left[\begin{array}{ccc|c} 3 & 2 & 4 & 22.00 \\ 2 & 4 & 5 & 26.30 \\ 1 & 2 & 4 & 17.20 \end{array}\right]$$

Interchange rows 1 and 3 and pivot on the (1, 1) entry.

$$\left[\begin{array}{ccc|c} 1 & 2 & 4 & 17.20 \\ 2 & 4 & 5 & 26.30 \\ 3 & 2 & 4 & 22.00 \end{array}\right] \begin{array}{l} \\ -2R1 + R2 \to R2 \\ -3R1 + R3 \to R3 \end{array}$$

$$\left[\begin{array}{ccc|c} 1 & 2 & 4 & 17.20 \\ 0 & 0 & -3 & -8.10 \\ 0 & -4 & -8 & -29.60 \end{array}\right] R2 \leftrightarrow R3$$

$$\left[\begin{array}{ccc|c} 1 & 2 & 4 & 17.20 \\ 0 & -4 & -8 & -29.60 \\ 0 & 0 & -3 & -8.10 \end{array}\right] \begin{array}{l} \\ -1/4R2 \to R2 \\ -1/3R3 \to R3 \end{array}$$

$$\left[\begin{array}{ccc|c} 1 & 2 & 4 & 17.20 \\ 0 & 1 & 2 & 7.40 \\ 0 & 0 & 1 & 2.70 \end{array}\right] \begin{array}{l} -2R2 + R1 \to R1 \\ \\ -2R3 + R2 \to R2 \end{array}$$

$$\left[\begin{array}{ccc|c} 1 & 0 & 0 & 2.40 \\ 0 & 1 & 0 & 2.00 \\ 0 & 0 & 1 & 2.70 \end{array}\right]$$

A six-pack cost \$2.40, a bag of chips cost \$2.00, and a package of cookies cost \$2.70.

61.   Let A = amount invested in stock A, B = amount invested in stock B, C = amount invested in stock C.

$$A + B + C = 40,000$$
$$0.06A + 0.07B + 0.08C = 2730$$
$$0.03A + 0.04B + 0.02C = 1080$$

$$\left[\begin{array}{ccc|c} 1 & 1 & 1 & 40000 \\ 0.06 & 0.07 & 0.08 & 2730 \\ 0.03 & 0.04 & 0.02 & 1080 \end{array}\right] \left[\begin{array}{ccc|c} 1 & 1 & 1 & 40000 \\ 0 & 1 & 2 & 33000 \\ 0 & 1 & -1 & -12000 \end{array}\right] \left[\begin{array}{ccc|c} 1 & 0 & -1 & 7000 \\ 0 & 1 & 2 & 33000 \\ 0 & 0 & -3 & -45000 \end{array}\right]$$

$$\left[\begin{array}{ccc|c} 1 & 0 & -1 & 7000 \\ 0 & 1 & 2 & 33000 \\ 0 & 0 & 1 & 15000 \end{array}\right] \left[\begin{array}{ccc|c} 1 & 0 & 0 & 22000 \\ 0 & 1 & 0 & 3000 \\ 0 & 0 & 1 & 15000 \end{array}\right]$$

\$22,000 in stock A, \$3,000 in stock B, \$15,000 in stock C

63.   Let x_1 = number high school tickets, x_2 = number college tickets, x_3 = number adult tickets.

$$4x_1 + 6x_2 + 8x_3 = 14980$$
$$4x_1 + 5x_2 + 9x_3 = 14430$$
$$2x_1 + 7x_2 + 7x_3 = 14450$$

$$\begin{bmatrix} 4 & 6 & 8 & | & 14980 \\ 4 & 5 & 9 & | & 14430 \\ 2 & 7 & 7 & | & 14450 \end{bmatrix} \begin{bmatrix} 1 & 3/2 & 2 & | & 3745 \\ 0 & -1 & 1 & | & -550 \\ 0 & 4 & 3 & | & 6960 \end{bmatrix} \begin{bmatrix} 1 & 0 & 7/2 & | & 2920 \\ 0 & 1 & -1 & | & 550 \\ 0 & 0 & 7 & | & 4760 \end{bmatrix}$$

$$\begin{bmatrix} 1 & 0 & 7/2 & | & 2920 \\ 0 & 1 & -1 & | & 550 \\ 0 & 0 & 1 & | & 680 \end{bmatrix} \begin{bmatrix} 1 & 0 & 0 & | & 540 \\ 0 & 1 & 0 & | & 1230 \\ 0 & 0 & 1 & | & 680 \end{bmatrix}$$

540 to high school students, 1230 to college students, 680 to adults

65.

$$\begin{bmatrix} 1 & 2 & 3 & | & 142 \\ 5 & 4 & 2 & | & 292 \\ 4 & 4 & 5 & | & 316 \end{bmatrix} \begin{bmatrix} 1 & 2 & 3 & | & 142 \\ 0 & -6 & -13 & | & -418 \\ 0 & -4 & -7 & | & -252 \end{bmatrix} \begin{bmatrix} 1 & 0 & -4/3 & | & 8/3 \\ 0 & 1 & 13/6 & | & 209/3 \\ 0 & 0 & 5/3 & | & 80/3 \end{bmatrix}$$

$$\begin{bmatrix} 1 & 0 & -4/3 & | & 8/3 \\ 0 & 1 & 13/6 & | & 209/3 \\ 0 & 0 & 1 & | & 16 \end{bmatrix} \begin{bmatrix} 1 & 0 & 0 & | & 24 \\ 0 & 1 & 0 & | & 35 \\ 0 & 0 & 1 & | & 16 \end{bmatrix}$$ 24 cases of Regular
35 cases of Premium
16 cases of Classic

67.

$$\begin{bmatrix} 1 & -3 & 0 & | & 0 \\ 2 & 0 & 1 & | & 0 \\ 3 & 5 & 8 & | & 2542 \end{bmatrix} \begin{bmatrix} 1 & -3 & 0 & | & 0 \\ 0 & -6 & 1 & | & 0 \\ 0 & 14 & 8 & | & 2542 \end{bmatrix} \begin{bmatrix} 1 & 0 & -1/2 & | & 0 \\ 0 & 1 & -1/6 & | & 0 \\ 0 & 0 & 31/3 & | & 2542 \end{bmatrix}$$

$$\begin{bmatrix} 1 & 0 & 0 & | & 123 \\ 0 & 1 & 0 & | & 41 \\ 0 & 0 & 1 & | & 246 \end{bmatrix}$$ 123 students, 41 faculty, 246 general public

69.

$$\begin{bmatrix} 44 & 22 & 64 & | & 20480 \\ 42 & 28 & 62 & | & 20720 \\ 42 & 30 & 60 & | & 20580 \end{bmatrix} \begin{bmatrix} 1 & 1/2 & 16/11 & | & 5120/11 \\ 0 & 7 & 10/11 & | & 12880/11 \\ 0 & 9 & -12/11 & | & 11340/11 \end{bmatrix}$$

$$\begin{bmatrix} 1 & 0 & 107/77 & | & 4200/11 \\ 0 & 1 & 10/77 & | & 1840/11 \\ 0 & 0 & -174/77 & | & -5220/11 \end{bmatrix} \begin{bmatrix} 1 & 0 & 0 & | & 90 \\ 0 & 1 & 0 & | & 140 \\ 0 & 0 & 1 & | & 210 \end{bmatrix}$$ 90 shares of X
140 shares of Y
210 shares of Z

71.   Let x_1 = number of days A operates, x_2 = number of days B operates, x_3 = number of days C operates.

$$300x_1 + 700x_2 + 400x_3 = 39,500$$
$$500x_1 + 900x_2 + 400x_3 = 52,500$$
$$200x_1 + 100x_2 + 800x_3 = 12,500$$

75. Madeline is correct. The solution is (7, -4, -4, 3)

77. (3, 10) **79.** (2.5, 1)

81. **(b)** (12.2, 5.6, -7.4) **83.** (5.149, -0.990)

TI-83 Exercises
1. (1, -2, 3) **3.** (6, -1, 5)

Using EXCEL

1. (1, -2, 3) **3.** (6, -1, 5)

5. (2, -3, 5, 7)

TI-83 Exercises
1. (1, -2, 3) **3.** (6, -1, 5)

Using EXCEL

1. (1, -2, 3) **3.** (6, -1, 5)

5. (2, -3, 5, 7)

Section 2.3

1. In reduced echelon form

3. Not in reduced echelon form because column 3 does not contain a zero in row 2.

5. In reduced echelon form.

7. Not in reduced echelon form because the leading 1 in row 3 is to the left of the leading 1 in row 2, and the 3 in row 3 should be 0.

9. Perform the following row operations to reduce the third column.
$\frac{1}{4}$ R3 → R3, -2R3 + R1 → R1, and -3R3 + R2 → R2 and obtain $\begin{bmatrix} 1 & 0 & 0 & | & 1 \\ 0 & 1 & 0 & | & -8 \\ 0 & 0 & 1 & | & 2 \end{bmatrix}$

$$\begin{bmatrix} 1 & -1 & 0 & 0 & | & -6 \\ 0 & 0 & 1 & 0 & | & 1 \\ 0 & 0 & 0 & 1 & | & 3 \\ 0 & 0 & 0 & 0 & | & -25 \end{bmatrix} \qquad \begin{bmatrix} 1 & -1 & 0 & 0 & | & 0 \\ 0 & 0 & 1 & 0 & | & 0 \\ 0 & 0 & 0 & 1 & | & 0 \\ 0 & 0 & 0 & 0 & | & 1 \end{bmatrix}$$

21. $x_1 = 3$

$x_2 = -2$

$x_3 = 5$

The solution is (3, -2, 5).

23. $x_1 \quad + 3x_3 = 4$

$x_2 + x_3 = -6$

$x_4 = 2$

An infinite number of solutions of the form (4 - 3k, -6 - k, k, 2).

25. $x_1 = 0$

$x_2 = 0$

$0 = 1$

No solution

27. $x_1 \quad + 3x_4 = 0$

$x_2 \ - 2x_4 = 0$

$x_3 + 7x_4 = 0$

$0 = 0$

There are an infinite number of solutions of the form (-3k, 2k, -7k, k)

29. $\begin{bmatrix} 1 & 4 & -2 & | & 13 \\ 3 & -1 & 4 & | & 6 \\ 2 & -5 & 6 & | & -4 \end{bmatrix} \quad \begin{bmatrix} 1 & 4 & -2 & | & 13 \\ 0 & -13 & 10 & | & -33 \\ 0 & -13 & 10 & | & -33 \end{bmatrix} \quad \begin{bmatrix} 1 & 0 & 14/13 & | & 37/13 \\ 0 & 1 & -10/13 & | & 33/13 \\ 0 & 0 & 0 & | & 3 \end{bmatrix}$

No solution

31. $\begin{bmatrix} 3 & -2 & 2 & | & 10 \\ 2 & 1 & 3 & | & 3 \\ 1 & 1 & -1 & | & 5 \end{bmatrix} \quad \begin{bmatrix} 1 & 1 & -1 & | & 5 \\ 2 & 1 & 3 & | & 3 \\ 3 & -2 & 2 & | & 10 \end{bmatrix} \quad \begin{bmatrix} 1 & 1 & -1 & | & 5 \\ 0 & -1 & 5 & | & -7 \\ 0 & -5 & 5 & | & -5 \end{bmatrix}$

$\begin{bmatrix} 1 & 0 & 4 & | & -2 \\ 0 & 1 & -5 & | & 7 \\ 0 & 0 & -20 & | & 30 \end{bmatrix} \quad \begin{bmatrix} 1 & 0 & 0 & | & 4 \\ 0 & 1 & 0 & | & -1/2 \\ 0 & 0 & 1 & | & -3/2 \end{bmatrix}$ $\qquad x_1 = 4, \ x_2 = -1/2, \ x_3 = -3/2$

33.
$$\begin{bmatrix} 1 & 1 & 1 & -1 & | & -3 \\ 2 & 3 & 1 & -5 & | & -9 \\ 1 & 3 & -1 & -6 & | & 7 \end{bmatrix} \quad \begin{bmatrix} 1 & 1 & 1 & -1 & | & -3 \\ 0 & 1 & -0 & -3 & | & -3 \\ 0 & 2 & -2 & -5 & | & 10 \end{bmatrix} \quad \begin{bmatrix} 1 & 0 & 2 & 2 & | & 0 \\ 0 & 1 & -1 & -3 & | & -3 \\ 0 & 0 & 0 & 1 & | & 16 \end{bmatrix}$$

$$\begin{bmatrix} 1 & 0 & 2 & 0 & | & -32 \\ 0 & 1 & -1 & 0 & | & 45 \\ 0 & 0 & 0 & 1 & | & 16 \end{bmatrix} \quad x_1 = -32 - 2x_3, \ x_2 = 45 + x_3, \ x_4 = 16 \text{ or } (-32 - 2k, 45 +$$

k, k, 16)

35.
$$\begin{bmatrix} 1 & -1 & 1 & | & 3 \\ -2 & 3 & 1 & | & -8 \\ 4 & -2 & 10 & | & 10 \end{bmatrix} \quad \begin{bmatrix} 1 & -1 & 1 & | & 3 \\ 0 & 1 & 3 & | & -2 \\ 0 & 2 & 6 & | & -2 \end{bmatrix} \quad \begin{bmatrix} 1 & 0 & 4 & | & 1 \\ 0 & 1 & 3 & | & -2 \\ 0 & 0 & 0 & | & 2 \end{bmatrix} \qquad \text{No solution}$$

37.
$$\begin{bmatrix} 1 & 1 & 1 & | & 6 \\ 1 & -3 & 2 & | & 1 \\ 3 & -1 & 4 & | & 5 \end{bmatrix} \quad \begin{bmatrix} 1 & 1 & 1 & | & 6 \\ 0 & -4 & 1 & | & -5 \\ 0 & -4 & 1 & | & -13 \end{bmatrix} \quad \begin{bmatrix} 1 & 1 & 1 & | & 6 \\ 0 & -4 & 1 & | & -5 \\ 0 & 0 & 0 & | & -8 \end{bmatrix} \qquad \text{No solution}$$

39.
$$\begin{bmatrix} 1 & 2 & -1 & | & -13 \\ 2 & 5 & 3 & | & -3 \end{bmatrix} \quad \begin{bmatrix} 1 & 2 & -1 & | & -13 \\ 0 & 1 & 5 & | & 23 \end{bmatrix} \quad \begin{bmatrix} 1 & 0 & -11 & | & -59 \\ 0 & 1 & 5 & | & 23 \end{bmatrix}$$

$x_1 = -59 + 11x_3, \ x_2 = 23 - 5x_3 \text{ or } (-59 + 11k, 23 - 5k, k)$

41.
$$\begin{bmatrix} 1 & -2 & 1 & 1 & -2 & | & -9 \\ 5 & 1 & -6 & -6 & 1 & | & 21 \end{bmatrix} \quad \begin{bmatrix} 1 & -2 & 1 & 1 & -2 & | & -9 \\ 0 & 11 & -11 & -11 & 11 & | & 66 \end{bmatrix}$$

$$\begin{bmatrix} 1 & 0 & -1 & -1 & 0 & | & 3 \\ 0 & 1 & -1 & -1 & 1 & | & 6 \end{bmatrix}$$

$x_1 = 3 + x_3 + x_4, \ x_2 = 6 + x_3 + x_4 - x_5 \text{ or } (3 + k + r, 6 + k + r - s, k, r, s)$

43.
$$\begin{bmatrix} 1 & 1 & -3 & 1 & | & 4 \\ -2 & -2 & 6 & -2 & | & 3 \end{bmatrix} \quad \begin{bmatrix} 1 & 1 & -3 & 1 & | & 4 \\ 0 & 0 & 0 & 0 & | & 11 \end{bmatrix} \qquad \text{No solution}$$

45.
$$\begin{bmatrix} 1 & 1 & -1 & -1 & | & -1 \\ 3 & -2 & -4 & 2 & | & 1 \\ 4 & -1 & -5 & 1 & | & 5 \end{bmatrix} \quad \begin{bmatrix} 1 & 1 & -1 & -1 & | & -1 \\ 0 & -5 & -1 & 5 & | & 4 \\ 0 & -5 & -1 & 5 & | & 9 \end{bmatrix} \quad \begin{bmatrix} 1 & 1 & -1 & -1 & | & -1 \\ 0 & -5 & -1 & 5 & | & 4 \\ 0 & 0 & 0 & 0 & | & 5 \end{bmatrix}$$

No solution

47.
$$\begin{bmatrix} 1 & 4 & | & -10 \\ -2 & 3 & | & -13 \\ 5 & -2 & | & 16 \end{bmatrix} \quad \begin{bmatrix} 1 & 4 & | & -10 \\ 0 & 11 & | & -33 \\ 0 & -22 & | & 66 \end{bmatrix} \quad \begin{bmatrix} 1 & 0 & | & 2 \\ 0 & 1 & | & -3 \\ 0 & 0 & | & 0 \end{bmatrix} \qquad x = 2, y = -3$$

49.
$$\begin{bmatrix} 1 & -1 & | & -7 \\ 1 & 1 & | & -3 \\ 3 & -1 & | & -17 \end{bmatrix} \quad \begin{bmatrix} 1 & -1 & | & -7 \\ 0 & 2 & | & 4 \\ 0 & 2 & | & 4 \end{bmatrix} \quad \begin{bmatrix} 1 & 0 & | & -5 \\ 0 & 1 & | & 2 \\ 0 & 0 & | & 0 \end{bmatrix} \qquad x = -5, y = 2$$

51.
$$\begin{bmatrix} 0 & 1 & 2 & | & 7 \\ 1 & -2 & -6 & | & -18 \\ 1 & -1 & -2 & | & -5 \\ 2 & -5 & -15 & | & -46 \end{bmatrix} \quad \begin{bmatrix} 1 & -2 & -6 & | & -18 \\ 0 & 1 & 2 & | & 7 \\ 1 & -1 & -2 & | & -5 \\ 2 & -5 & -15 & | & -46 \end{bmatrix} \quad \begin{bmatrix} 1 & -2 & -6 & | & -18 \\ 0 & 1 & 2 & | & 7 \\ 0 & 1 & 4 & | & 13 \\ 0 & -1 & -3 & | & -10 \end{bmatrix}$$

$$\begin{bmatrix} 1 & 0 & -2 & | & -4 \\ 0 & 1 & 2 & | & 7 \\ 0 & 0 & 2 & | & 6 \\ 0 & 0 & -1 & | & -3 \end{bmatrix} \quad \begin{bmatrix} 1 & 0 & 0 & | & 2 \\ 0 & 1 & 0 & | & 1 \\ 0 & 0 & 1 & | & 3 \\ 0 & 0 & 0 & | & 0 \end{bmatrix} \qquad x_1 = 2, \; x_2 = 1, \; x_3 = 3$$

53.
$$\begin{bmatrix} 3 & -2 & 4 & | & 4 \\ 2 & 5 & -1 & | & -2 \\ 1 & -7 & 5 & | & 6 \\ 5 & 3 & 3 & | & 3 \end{bmatrix} \quad \begin{bmatrix} 1 & -7 & 5 & | & 6 \\ 3 & -2 & 4 & | & 4 \\ 2 & 5 & -1 & | & -2 \\ 5 & 3 & 3 & | & 3 \end{bmatrix} \quad \begin{bmatrix} 1 & -7 & 5 & | & 6 \\ 0 & 19 & -11 & | & -14 \\ 0 & 19 & -11 & | & -14 \\ 0 & 38 & -22 & | & -27 \end{bmatrix}$$

$$\begin{bmatrix} 1 & 0 & 18/19 & | & 16/19 \\ 0 & 1 & -11/19 & | & -14/19 \\ 0 & 0 & 0 & | & 0 \\ 0 & 0 & 0 & | & 1 \end{bmatrix} \qquad \text{No solution}$$

55.
$$\begin{bmatrix} 2 & -5 & | & 5 \\ 6 & 1 & | & 31 \\ 2 & 11 & | & 18 \end{bmatrix} \quad \begin{bmatrix} 1 & -5/2 & | & 5/2 \\ 0 & 16 & | & 16 \\ 0 & 16 & | & 13 \end{bmatrix} \quad \begin{bmatrix} 1 & -5/2 & | & 5/2 \\ 0 & 16 & | & 16 \\ 0 & 0 & | & -3 \end{bmatrix} \qquad \text{No solution}$$

57.
$$\begin{bmatrix} 1 & -1 & 2 & 0 & | & 7 \\ 3 & -4 & 18 & -13 & | & 17 \\ 2 & -2 & 2 & -4 & | & 12 \\ -1 & 1 & -1 & 2 & | & -6 \\ -3 & 1 & -8 & -10 & | & -21 \end{bmatrix} \quad \begin{bmatrix} 1 & -1 & 2 & 0 & | & 7 \\ 0 & -1 & 12 & -13 & | & -4 \\ 0 & 0 & -2 & -4 & | & -2 \\ 0 & 0 & 1 & 2 & | & 1 \\ 0 & -2 & -2 & -10 & | & 0 \end{bmatrix} \quad \begin{bmatrix} 1 & 0 & -10 & 13 & | & 11 \\ 0 & 1 & -12 & 13 & | & 4 \\ 0 & 0 & -2 & -4 & | & -2 \\ 0 & 0 & 1 & 2 & | & 1 \\ 0 & 0 & -26 & 16 & | & 8 \end{bmatrix}$$

$$\begin{bmatrix} 1 & 0 & 0 & 33 & | & 21 \\ 0 & 1 & 0 & 37 & | & 16 \\ 0 & 0 & 1 & 2 & | & 1 \\ 0 & 0 & 0 & 0 & | & 0 \\ 0 & 0 & 0 & 68 & | & 34 \end{bmatrix} \quad \begin{bmatrix} 1 & 0 & 0 & 0 & | & 9/2 \\ 0 & 1 & 0 & 0 & | & -5/2 \\ 0 & 0 & 1 & 0 & | & 0 \\ 0 & 0 & 0 & 0 & | & 0 \\ 0 & 0 & 0 & 1 & | & 1/2 \end{bmatrix} \qquad x_1 = \frac{9}{2}, \; x_2 = -\frac{5}{2}, \; x_3 = 0, \; x_4 =$$

$\frac{1}{2}$

59.
$$\begin{bmatrix} 1 & 2 & -1 & -1 & | & 0 \\ 1 & 2 & 0 & 1 & | & 4 \\ -1 & -2 & 2 & 4 & | & 5 \\ -1 & -1 & -1 & 0 & | & 1 \end{bmatrix} \quad \begin{bmatrix} 1 & 2 & -1 & -1 & | & 0 \\ 0 & 0 & 1 & 2 & | & 4 \\ 0 & 0 & 1 & 3 & | & 5 \\ 0 & 1 & -2 & -1 & | & 1 \end{bmatrix} \quad \begin{bmatrix} 1 & 0 & 3 & 1 & | & -2 \\ 0 & 1 & -2 & -1 & | & 1 \\ 0 & 0 & 1 & 3 & | & 5 \\ 0 & 0 & 1 & 2 & | & 4 \end{bmatrix}$$

$$\begin{bmatrix} 1 & 0 & 0 & -8 & | & -17 \\ 0 & 1 & 0 & 5 & | & 11 \\ 0 & 0 & 1 & 3 & | & 5 \\ 0 & 0 & 0 & -1 & | & -1 \end{bmatrix} \quad \begin{bmatrix} 1 & 0 & 0 & 0 & | & -9 \\ 0 & 1 & 0 & 0 & | & 6 \\ 0 & 0 & 1 & 0 & | & 2 \\ 0 & 0 & 0 & 1 & | & 1 \end{bmatrix} \qquad (-9, 6, 2, 1)$$

61.
$$\begin{bmatrix} 1 & 2 & -3 & 2 & 5 & -1 & 0 \\ -2 & -4 & 6 & -1 & -4 & 5 & 0 \\ 3 & 6 & -9 & 5 & 13 & -4 & 0 \end{bmatrix} \quad \begin{bmatrix} 1 & 2 & -3 & 2 & 5 & -1 & 0 \\ 0 & 0 & 0 & 3 & 6 & 3 & 0 \\ 0 & 0 & 0 & -1 & -2 & -1 & 0 \end{bmatrix}$$

$$\begin{bmatrix} 1 & 2 & -3 & 0 & 1 & -3 & 0 \\ 0 & 0 & 0 & 1 & 2 & 1 & 0 \\ 0 & 0 & 0 & 0 & 0 & 0 & 0 \end{bmatrix}$$

$x_1 = -2x_2 + 3x_3 - x_5 + 3x_6$, $x_4 = -2x_5 - x_6$ or $(-2k + 3r - s + 3t, k, r, -2s - t, s, t)$

63. Let x_1 = amount in stocks, x_2 = amount in bonds, x_3 = amount in money markets.

$$x_1 + x_2 + x_3 = 45{,}000$$
$$-2x_1 + x_2 + x_3 = 0$$
$$0.10x_1 + 0.07x_2 + 0.075x_3 = 3{,}660$$

$$\begin{bmatrix} 1 & 1 & 1 & 45{,}000 \\ -2 & 1 & 1 & 0 \\ 0.10 & 0.07 & 0.075 & 3660 \end{bmatrix} \quad \begin{bmatrix} 1 & 1 & 1 & 45{,}000 \\ 0 & 3 & 3 & 90{,}000 \\ 0 & -30 & -25 & -840{,}000 \end{bmatrix}$$

$$\begin{bmatrix} 1 & 0 & 0 & 15{,}000 \\ 0 & 1 & 1 & 30{,}000 \\ 0 & 0 & 5 & 60{,}000 \end{bmatrix} \quad \begin{bmatrix} 1 & 0 & 0 & 15{,}000 \\ 0 & 1 & 0 & 18{,}000 \\ 0 & 0 & 1 & 12{,}000 \end{bmatrix}$$

$15,000 in stocks, $18,000 in bonds, $12,000 in money market

65. Let x = minutes jogging, y = minutes playing handball, z = minutes biking.

$$x + y + z = 60$$
$$13x + 11y + 7z = 660$$
$$x \quad\quad - 2z = 0$$

$$\begin{bmatrix} 1 & 1 & 1 & 60 \\ 13 & 11 & 7 & 660 \\ 1 & 0 & -2 & 0 \end{bmatrix} \quad \begin{bmatrix} 1 & 1 & 1 & 60 \\ 0 & -2 & -6 & -120 \\ 0 & -1 & -3 & -60 \end{bmatrix} \quad \begin{bmatrix} 1 & 0 & -2 & 0 \\ 0 & 1 & 3 & 60 \\ 0 & 0 & 0 & 0 \end{bmatrix}$$

$x = 2z$, $y = 60 - 3z$

She may bike from 0 to 20 min., then should jog for twice that time, and play handball for the rest of the 60 minutes.

67. Let x_1 = hours of Math, x_2 = hours of English, x_3 = hours of Chemistry, x_4 = hours of History

$$x_1 + x_2 + x_3 + x_4 = 42$$
$$x_1 + x_2 \quad\quad = 21$$
$$x_1 \quad\quad = 2x_2$$
$$x_2 \quad\quad = 2x_4$$

$$\begin{bmatrix} 1 & 1 & 1 & 1 & | & 42 \\ 1 & 1 & 0 & 0 & | & 21 \\ 1 & -2 & 0 & 0 & | & 0 \\ 0 & 1 & 0 & -2 & | & 0 \end{bmatrix} \quad \begin{bmatrix} 1 & 1 & 1 & 1 & | & 42 \\ 0 & 0 & -1 & -1 & | & -21 \\ 0 & -3 & -1 & -1 & | & -42 \\ 0 & 1 & 0 & -2 & | & 0 \end{bmatrix} \quad \begin{bmatrix} 1 & 1 & 1 & 1 & | & 42 \\ 0 & 1 & 1/3 & 1/3 & | & 14 \\ 0 & 0 & 1 & 1 & | & 21 \\ 0 & 1 & 0 & -2 & | & 0 \end{bmatrix}$$

$$\begin{bmatrix} 1 & 0 & 2/3 & 2/3 & | & 28 \\ 0 & 1 & 1/3 & 1/3 & | & 14 \\ 0 & 0 & 1 & 1 & | & 21 \\ 0 & 0 & 1/3 & 7/3 & | & 14 \end{bmatrix} \quad \begin{bmatrix} 1 & 0 & 0 & 0 & | & 14 \\ 0 & 1 & 0 & 0 & | & 7 \\ 0 & 0 & 1 & 1 & | & 21 \\ 0 & 0 & 0 & 2 & | & 7 \end{bmatrix} \quad \begin{bmatrix} 1 & 0 & 0 & 0 & | & 14 \\ 0 & 1 & 0 & 0 & | & 7 \\ 0 & 0 & 1 & 0 & | & 7.5 \\ 0 & 0 & 0 & 1 & | & 3.5 \end{bmatrix}$$

14 hours for Math, 7 hours for English, and 17.5 hours for Chemistry, and 3.5 hours for History.

69. Let x_1 = federal tax, x_2 = state tax, and x_3 = city tax.

$$x_1 = 0.40(58{,}400 - x_2 - x_3)$$
$$x_2 = 0.20(58{,}400 - x_1 - x_3)$$
$$x_3 = 0.10(58{,}400 - x_1 - x_2)$$

We can reduce these equations to:

$$x_1 + 0.40x_2 + 0.40x_3 = 23{,}360$$
$$0.20x_1 + x_2 + 0.20x_3 = 11{,}680$$
$$0.10x_1 + 0.10x_2 + x_3 = 5{,}840$$

The reduced echelon form of the augmented matrix of this system is

$$\begin{bmatrix} 1 & 0 & 0 & | & 19{,}200 \\ 0 & 1 & 0 & | & 7{,}200 \\ 0 & 0 & 1 & | & 3{,}200 \end{bmatrix}$$

The federal tax is $19,200, the state tax is $7200, and the city tax is $3200.

71. Let x_1 = cost of a CD, x_2 = cost of a DVD movie, and x_3 = cost of a cassette tape.

$$6x_1 + 2x_2 + 4x_3 = 40 \qquad \text{(Amy's purchase)}$$
$$3x_1 + 6x_2 + x_3 = 53 \qquad \text{(Bill's purchase)}$$
$$6x_1 + 7x_2 + 3x_3 = 73 \qquad \text{(Carlton's purchase)}$$

The reduced echelon form of the augmented matrix is $\begin{bmatrix} 1 & 0 & 0.7333 & | & 4.4666 \\ 0 & 1 & -0.2 & | & 6.6 \\ 0 & 0 & 0 & | & 0 \end{bmatrix}$

This indicates an infinite number of solutions where

$x_1 = 4.4666 - 0.7333x_3$ which indicates $x_1 \le 4.466$
$x_2 = 6.6 + 0.2x_3$ which indicates $x_2 = 6.6$

Since x_1 must be greater than or equal to zero,

$4.4666 - 0.7333x_3 = 0$ or $x_3 \le 6.09$

73. (a) Let x_1 = number supplied by Sweats-Plus to Spirit Shop 1

x_2 = number supplied by Sweats-Plus to Spirit Shop 2

x_3 = number supplied by Sweats-Plus to Spirit Shop 3

x_4 = number supplied by Imprint-Sweats to Spirit Shop 1

x_5 = number supplied by Imprint-Sweats to Spirit Shop 2

x_6 = number supplied by Imprint-Sweats to Spirit Shop 3

The system of equations is

$$x_1 + x_4 = 15$$
$$x_2 + x_5 = 20$$
$$x_3 + x_6 = 30$$
$$x_1 + x_2 + x_3 = 40$$
$$x_4 + x_5 + x_6 = 25$$

The solution is

$$x_1 = -10 + x_5 + x_6$$
$$x_2 = 20 - x_5$$
$$x_3 = 30 - x_6$$
$$x_4 = 25 - x_5 - x_6$$

(b) $x_5 = 5$, $x_6 = 5$ yields $x_1 = 0$, $x_2 = 15$, $x_3 = 25$, $x_4 = 15$

$x_5 = 10$, $x_6 = 5$ yields $x_1 = 5$, $x_2 = 10$, $x_3 = 25$, $x_4 = 10$

$x_5 = 0$, $x_6 = 15$ yields $x_1 = 5$, $x_2 = 20$, $x_3 = 15$, $x_4 = 10$

(c) x_5 represents the number of sweatshirts supplied by Imprint-Sweats to Spirit Shop 2. Since $x_2 = 20 - x_5$, x_5 must be no greater than 20 or else x_2 would be a negative number. Thus, $x_5 = 20$.

(d)   If Imprint-Sweats supplies no sweatshirts to Spirit Shop 2 and Spirit Shop 3, then $x_5 = 0$ and $x_6 = 0$ so $x_1 = -10$. Thus, the order cannot be filled.

(e)   Since $x_1 = -10 + x_5 + x_6$, in order for $x_1 = 0$, then $x_5 + x_6 = 10$. Imprint-Sweats must supply a total of 10 or more to Spirit Shops 2 and 3.

(f)   Since $x_2 = 20 - x_5$, and $0 = x_5 = 20$, then $x_2 = 20$. Thus, Sweats-Plus supplies 20 or less to Spirit Shop 2.

75.   The reduced echelon form of the augmented matrix is $\begin{bmatrix} 1 & 0 & 7/5 & | & 0 \\ 0 & 1 & -1/5 & | & 0 \\ 0 & 0 & 0 & | & 1 \end{bmatrix}$

which indicates no solution. In order to have a solution, the 2 in the third equation must be changed so that the bottom row of the reduced matrix is all zeros. Substitute c for 2 in the third equation and reduce the augmented matrix.

The augmented matrix is $\begin{bmatrix} 1 & 2 & 1 & | & 3 \\ 2 & -1 & 3 & | & 2 \\ 3 & 1 & 4 & | & c \end{bmatrix}$ which reduces to $\begin{bmatrix} 1 & 2 & 1 & | & 3 \\ 0 & -5 & 1 & | & -4 \\ 0 & -5 & 1 & | & c-9 \end{bmatrix}$

We need to reduce the matrix further by getting a zero in row 3, column 2.

$\begin{bmatrix} 1 & 2 & 1 & | & 3 \\ 0 & -5 & 1 & | & -4 \\ 0 & 0 & 0 & | & c-5 \end{bmatrix}$

Since c - 5 must be zero, c = 5.

The last equation should be $3x + y + 4z = 5$.

81. The system has infinitely many solutions.

83.

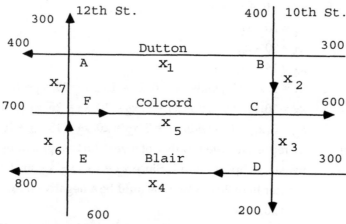

(a) Label the traffic flow on each block by $x_1, x_2, \ldots, x_7$ as shown. Then the condition that incoming traffic equals outgoing traffic at each intersection gives:

At A: $x_1 + x_7 = 700$

At B: $x_1 + x_2 = 700$

At C: $x_2 + x_5 = 600 + x_3$

At D: $x_3 + 300 = 200 + x_4$

At E: $600 + x_4 = 800 + x_6$

At F: $x_6 + 700 = x_5 + x_7$

(b) The solution to the above system is

$x_1 = 700 - x_7$

$x_2 = x_7$

$x_3 = 100 + x_6$

$$x_4 = 200 + x_6$$
$$x_5 = 700 + x_6 - x_7$$

(c) x_3 represents the traffic flow on 10th St. between Colcord and Blair. Since $x_3 = 100 + x_6$, $x_3 = 100$. The minimum traffic flow is 100 vehicles per hour when $x_6 = 0$. The other traffic flows then become

$$x_1 = 700 - x_7$$
$$x_2 = x_7$$
$$x_4 = 200$$
$$x_5 = 700 - x_7$$

85.   c ? 14

87. $x_1 = 0.2 + 1.6x_3$
          $x_2 = -0.2 + 1.4x_3$

89. $(-19, 28, -10)$

91.   (a) $x_1 = 0.5 + 3.5x_4$
                $x_2 = 0.5 - 1.83x_4$
                $x_3 = 2.33x_4$

          (b) $x_2 = 0.5 - 1.83x_4$ implies $0.5 - 1.83x_4 > 0$
                so $x_4 < \dfrac{0.5}{1.83} = 0.273$.
                Thus, $0 < x_4 < 0.273$
                $x_3 = 2.33x_4$ implies $0 < x_3 < 0.636$

Section 2.4 Matrix Operations

$x_1 = 0.5 + 3.5x_4$ implies $x_1 > 0.5$ and
the maximum value of x_1 occurs when x_4 is maximum. Thus, $x_1 < 0.5 +$
$3.5(0.273) = 1.456$. $0.50 < x_1 < 1.456$
$x_2 = 0.5 - 1.83x_4$ implies $x_2 < 0.5$ and x_2 is minimum, 0, when x_4
reaches 0.273 so $0 < x_2 < 0.5$.

Using Your TI-83

1.   (2, -3, 4)

3.   (3, 4, 6)

1. $\begin{bmatrix} 1 & 0 & 0 & 1.3 & 2.5 \\ 0 & 1 & 0 & 0.7 & 0.5 \\ 0 & 0 & 1 & -12 & 0.5 \end{bmatrix}$

3.   (-2, 5, -3)

Using Excel

1.   (-5, 2, 6)

3.   (6, 2, 7)

Section 2.4

1.

	Alpha	Beta
Salv. Army	50	65
Boys' Club	85	32
Girl Scouts	68	94

3.

	Joe	Jane	Judy
Checking	12	11	5
Savings	15	18	8
Boxes	8	9	21

5.   2 by 2

7.   3 by 3

9.   4 by 1

11.   2 by 4

13.   2 by 3

15.   2 by 2

17.   Not equal

19.   Equal

21.   Not equal

23. $\begin{bmatrix} 3 & 3 & 2 \\ 7 & 4 & 3 \end{bmatrix}$

25. $\begin{bmatrix} 2 \\ 58 \end{bmatrix}$

27.   Can <u>not</u> add them

29. $\begin{bmatrix} 6 & 5 & 8 \\ 5 & 1 & 3 \\ 5 & -3 & 4 \end{bmatrix}$

31. $\begin{bmatrix} 12 & 3 \\ 6 & 15 \end{bmatrix}$

33. $\begin{bmatrix} 20 \\ 15 \\ 5 \\ 10 \end{bmatrix}$

35. $\begin{bmatrix} -12 & 6 & -15 \end{bmatrix}$

37. $\begin{bmatrix} 0 & 0 \\ 0 & 0 \end{bmatrix}$

39.   **(a)** $3A = \begin{bmatrix} 3 & 12 \\ -6 & 9 \end{bmatrix},\ -2B = \begin{bmatrix} 0 & -4 \\ -8 & -2 \end{bmatrix},\ 5C = \begin{bmatrix} 5 & -10 \\ 5 & -15 \end{bmatrix}$

      (b) $A + C = \begin{bmatrix} 2 & 2 \\ -1 & 0 \end{bmatrix}$ **(c)** $3A - 2B = \begin{bmatrix} 3 & 8 \\ -14 & 7 \end{bmatrix}$

      (d) $A - 2B + 5C = \begin{bmatrix} 1 & 4 \\ -2 & 3 \end{bmatrix} + \begin{bmatrix} 0 & -4 \\ -8 & -2 \end{bmatrix} + \begin{bmatrix} 5 & -10 \\ 5 & -15 \end{bmatrix} = \begin{bmatrix} 6 & -10 \\ -5 & -14 \end{bmatrix}$

41.

$\begin{array}{c} \\ \text{PC} \\ \text{Print} \\ \text{Disk} \end{array} \begin{array}{ccc} \text{I} & \text{II} & \text{III} \\ \left[\begin{array}{ccc} 23 & 17 & 20 \\ 19 & 22 & 11 \\ 151 & 151 & 105 \end{array}\right. \end{array}$

43. $x = 3$

45. $6x + 4 = 14x - 13$ so $x = \dfrac{17}{8}$

47. $\dfrac{1}{12}A = \begin{array}{c} \\ \\ \\ \end{array}\overset{\begin{array}{ccc} \text{A} & \text{B} & \text{C} \end{array}}{\begin{bmatrix} 65/12 & 55/6 & 20/3 \\ 30/4 & 45/4 & 5 \\ 25/4 & 28/3 & 7 \end{bmatrix}} \begin{array}{l} \text{small} \\ \text{regular} \\ \text{giant} \end{array}$

49.

$12M = \begin{array}{c} \\ \text{Fairfield} \\ \text{Tyler} \end{array} \overset{\begin{array}{ccc} \text{S.D.} & \text{N.O.} & \text{P.M.} \end{array}}{\begin{bmatrix} 2760 & 1080 & 1680 \\ 3120 & 1380 & 1992 \end{bmatrix}}$

51. $\dfrac{1}{3}\left(\begin{bmatrix} 90 \\ 62 \\ 76 \\ 82 \\ 74 \end{bmatrix} + \begin{bmatrix} 88 \\ 69 \\ 78 \\ 80 \\ 76 \end{bmatrix} + \begin{bmatrix} 91 \\ 73 \\ 72 \\ 84 \\ 77 \end{bmatrix}\right) = \dfrac{1}{3}\begin{bmatrix} 269 \\ 204 \\ 226 \\ 246 \\ 227 \end{bmatrix} = \begin{bmatrix} 89.7 \\ 68.0 \\ 75.3 \\ 82.0 \\ 75.7 \end{bmatrix}$

53.   **(a)**   Each row in a matrix represents a school, in the order given, and the columns represent students, tuition, and room and board in that order.

$$Y2002 = \begin{bmatrix} 1609 & 25,345 & 6760 \\ 929 & 20,904 & 6543 \\ 7496 & 17,080 & 6090 \\ 1537 & 19,303 & 5671 \\ 2870 & 10,738 & 4720 \\ 4019 & 15,680 & 6940 \end{bmatrix} \qquad Y1994 = \begin{bmatrix} 1410 & 17,355 & 5855 \\ 862 & 15,160 & 4676 \\ 8409 & 9,900 & 4350 \\ 1414 & 14,916 & 4708 \\ 3194 & 6,074 & 3436 \\ 3996 & 10,970 & 4470 \end{bmatrix}$$

The change in the data is found by subtracting:

$$Y2002 - Y1994 = \begin{bmatrix} 199 & 7990 & 905 \\ 67 & 5744 & 1867 \\ -913 & 7180 & 1740 \\ 123 & 4387 & 963 \\ -324 & 4664 & 1284 \\ 23 & 4710 & 2470 \end{bmatrix}$$

(b)   Bowdoin had the largest increase in tuition, $7990.

(c)   Maraquette had a decrease of 913 students, and Samford had a decrease of 324 students.

57.   **(a)** $\begin{bmatrix} 5 & 1 & 9 \\ 10 & 4 & 10 \\ 3 & 6 & 3 \end{bmatrix}$ **(b)** $\begin{bmatrix} 1 & 1 & -1 \\ -6 & -4 & -6 \\ -1 & -4 & -5 \end{bmatrix}$

    (c) $\begin{bmatrix} 6 & 2 & 8 \\ 4 & 0 & 4 \\ 2 & 2 & -2 \end{bmatrix}$ **(d)** $\begin{bmatrix} 18 & 4 & 31 \\ 32 & 12 & 32 \\ 10 & 19 & 8 \end{bmatrix}$

59.   **(a)** $\begin{bmatrix} 9 & 0 & 8 \\ 6 & 2 & 7 \\ 4 & 8 & 8 \end{bmatrix}$ **(b)** $\begin{bmatrix} -1.4 & 8.3 \\ 7.7 & 6.6 \end{bmatrix}$

    (c) $\begin{bmatrix} 5.4 & 3.0 & 3.1 \\ 3.8 & 4.2 & 1.6 \\ 3.1 & 6.2 & -5.0 \end{bmatrix}$ **(d)** $\begin{bmatrix} 3 & 15 & 3 \\ 6 & 3 & 6 \\ 9 & 0 & 12 \end{bmatrix}$

    (e) $\begin{bmatrix} 16.20 & 11.88 \\ 32.94 & 39.42 \\ -7.56 & -13.50 \end{bmatrix}$ **(f)** $\begin{bmatrix} 6 & 9 & 0 \\ 9 & 2 & 13 \\ 0 & 13 & -8 \end{bmatrix}$

Using Your TI- 83

1.   **(a)** $\begin{bmatrix} 4 & 1 & 3 \\ 10 & 13 & 2 \end{bmatrix}$ **(b)** $\begin{bmatrix} 2 & 6 & 4 \\ 8 & 10 & 14 \end{bmatrix}$

    (c) $\begin{bmatrix} -2 & 5 & 1 \\ -2 & -3 & 12 \end{bmatrix}$ **(d)** $\begin{bmatrix} 9 & 5 & 8 \\ 24 & 31 & 11 \end{bmatrix}$

Using EXCEL

1. $\begin{bmatrix} 9 & 5 & -2 \\ 2 & 14 & 11 \\ -2 & 9 & 7 \end{bmatrix}$

3. $\begin{bmatrix} 4 & 12 & -8 \\ 20 & 36 & 28 \\ -16 & 0 & 24 \end{bmatrix}$

5. $\begin{bmatrix} 52 & 24 & -8 \\ 2 & 66 & 52 \\ -4 & 54 & 30 \end{bmatrix}$

Section 2.5

1. $2 + 12 = 14$

3. $12 + 0 = 12$

5. $6 + 0 + 8 = 14$

7. $[0.90 \quad 1.85 \quad 0.65] \begin{bmatrix} 2 \\ 1 \\ 4 \end{bmatrix} = 1.80 + 1.85 + 2.60 = \$6.25$

9. $\begin{bmatrix} -5 & 11 \\ 0 & 14 \end{bmatrix}$

11. $\begin{bmatrix} 30 & 2 \\ 39 & -3 \end{bmatrix}$

13. **(a)** Not possible
    (b) Possible, a 2 X 3 matrix
    (c) Not possible

15. **(a)** Possible, a 3 X 5 matrix   **(b)** Not possible

17. $\begin{bmatrix} -7 & 7 \\ -1 & 6 \end{bmatrix}$

19. $\begin{bmatrix} 15 \\ -2 \end{bmatrix}$

21. Multiplication is not possible.

23. Multiplication is not possible.

25. $\begin{bmatrix} -4 & 4 \\ 3 & 27 \end{bmatrix}$

27. Multiplication not possible

29. $\begin{bmatrix} 5 & 9 & 12 \\ 5 & 17 & 21 \end{bmatrix}$

31. $\begin{bmatrix} 8 \\ 13 \\ 7 \end{bmatrix}$

33. $AB = \begin{bmatrix} 1 & 8 \\ -1 & 2 \end{bmatrix}$ $BA = \begin{bmatrix} 3 & 10 \\ -1 & 0 \end{bmatrix}$

35. $AB = \begin{bmatrix} -3 & 10 \\ -2 & 5 \end{bmatrix}$ $BA = \begin{bmatrix} -1 & 2 \\ -4 & 3 \end{bmatrix}$

37. $AB = \begin{bmatrix} 6 & 2 & 13 \\ -11 & -6 & -4 \end{bmatrix}$ BA not possible

39. $AB = [27 \quad 38]$ BA not possible

41. $AB = BA = \begin{bmatrix} 10 & 6 \\ -9 & 10 \end{bmatrix}$

43. $\begin{bmatrix} 10 & -3 \\ 10 & 2 \end{bmatrix}\begin{bmatrix} -1 & 1 \\ 3 & -2 \end{bmatrix} = \begin{bmatrix} -19 & 16 \\ -4 & 6 \end{bmatrix}$

45. $\begin{bmatrix} 8 & 1 & -2 & -14 \\ 5 & 3 & 9 & -11 \\ 8 & 2 & 12 & 2 \end{bmatrix}$

47. $\begin{bmatrix} 22 & -8 \\ 7 & 1 \end{bmatrix} + \begin{bmatrix} -18 & 12 \\ 3 & 9 \end{bmatrix} = \begin{bmatrix} 4 & 4 \\ 10 & 10 \end{bmatrix}$

49. $\begin{bmatrix} 3 & 4 \\ 1 & 2 \end{bmatrix}$

51. $\begin{bmatrix} 2 & -10 \\ 3 & 7 \end{bmatrix}$

53. $\begin{bmatrix} 3x + y \\ 2x + 4y \end{bmatrix}$

55. $\begin{bmatrix} x_1 + 2x_2 - x_3 \\ 3x_1 + x_2 + 4x_3 \\ 2x_1 - x_2 - x_3 \end{bmatrix}$

57. $\begin{bmatrix} x_1 + 3x_2 + 5x_3 + 6x_4 \\ -2x_1 + 9x_2 + 6x_3 + x_4 \\ 8x_1 + 17x_3 + 5x_4 \end{bmatrix}$

59. **(a)** $R1 + R2 \to R2$ **(b)** $2R1 \to R1, \ R2 + R3 \to R3$

(c) $R1 - R3 \to R1$ **(d)** $R1 + R2 + R3 \to R2$

(e) $R1 + 4R3 \to R1$

61. **(a)**

$$\begin{bmatrix} 2 & 0 & 0 & 0 \\ 0 & 1 & 0 & 0 \\ 0 & 0 & -3 & 0 \\ 0 & 0 & 0 & 1 \end{bmatrix} \begin{bmatrix} 1 & 5 & 3 & 9 \\ -2 & 7 & 4 & 11 \\ 9 & 0 & 2 & 5 \\ 6 & 3 & 3 & 2 \end{bmatrix} = \begin{bmatrix} 2 & 10 & 6 & 18 \\ -2 & 7 & 4 & 11 \\ -27 & 0 & -6 & -15 \\ 6 & 3 & 3 & 2 \end{bmatrix}$$

(b)

$$\begin{bmatrix} 3 & 0 & 0 & 0 \\ 0 & 1 & 0 & 0 \\ 0 & 2 & 1 & 0 \\ 0 & 0 & 0 & 1 \end{bmatrix} \begin{bmatrix} 1 & 5 & 3 & 9 \\ -2 & 7 & 4 & 11 \\ 9 & 0 & 2 & 5 \\ 6 & 3 & 3 & 2 \end{bmatrix} = \begin{bmatrix} 3 & 15 & 9 & 27 \\ -2 & 7 & 4 & 11 \\ 5 & 14 & 10 & 27 \\ 6 & 3 & 3 & 2 \end{bmatrix}$$

(c)

$$\begin{bmatrix} 1 & 0 & 0 & 0 \\ 2 & 1 & 0 & 0 \\ 0 & 0 & 1 & 0 \\ 0 & 0 & 0 & 1 \end{bmatrix} \begin{bmatrix} 1 & 5 & 3 & 9 \\ -2 & 7 & 4 & 11 \\ 9 & 0 & 2 & 5 \\ 6 & 3 & 3 & 2 \end{bmatrix} = \begin{bmatrix} 1 & 5 & 3 & 9 \\ 0 & 17 & 10 & 29 \\ 9 & 0 & 2 & 5 \\ 6 & 3 & 3 & 2 \end{bmatrix}$$

(d)

$$\begin{bmatrix} 1 & 0 & 0 & 0 \\ 0 & 1 & 0 & 0 \\ -9 & 0 & 1 & 0 \\ 0 & 0 & 0 & 1 \end{bmatrix} \begin{bmatrix} 1 & 5 & 3 & 9 \\ -2 & 7 & 4 & 11 \\ 9 & 0 & 2 & 5 \\ 6 & 3 & 3 & 2 \end{bmatrix} = \begin{bmatrix} 1 & 5 & 3 & 9 \\ -2 & 7 & 4 & 11 \\ 0 & -45 & -25 & -76 \\ 6 & 3 & 3 & 2 \end{bmatrix}$$

(e)

$$\begin{bmatrix} 1 & 0 & 0 & 0 \\ 2 & 1 & 0 & 0 \\ -9 & 0 & 1 & 0 \\ -6 & 0 & 0 & 1 \end{bmatrix} \begin{bmatrix} 1 & 5 & 3 & 9 \\ -2 & 7 & 4 & 11 \\ 9 & 0 & 2 & 5 \\ 6 & 3 & 3 & 2 \end{bmatrix} = \begin{bmatrix} 1 & 5 & 3 & 9 \\ 0 & 17 & 10 & 29 \\ 0 & -45 & -25 & -76 \\ 0 & -27 & -15 & -52 \end{bmatrix}$$

63.

$$\begin{bmatrix} 4 & 5.5 \\ 1 & 2 \end{bmatrix} \begin{bmatrix} 300 \\ 450 \end{bmatrix} = \begin{bmatrix} 3675 \\ 1200 \end{bmatrix}$$

3675 hrs assembly time, 1200 hrs checking

65.

$$\begin{bmatrix} 114 & 85 \\ 118 & 84 \\ 116 & 86 \end{bmatrix} \begin{bmatrix} 60 \\ 140 \end{bmatrix} = \begin{bmatrix} 18740 \\ 18840 \\ 19000 \end{bmatrix}$$

$18,740 on Monday, $18,840 on Wednesday, $19,000 on Friday

67.

$$\begin{bmatrix} 0.5 & 1.5 & 0.5 & 1.0 & 1.0 \\ 0 & 1.0 & 1.0 & 3.0 & 2.0 \end{bmatrix} \begin{bmatrix} 500 & 0.2 & 0 & 129 \\ 0 & 0.2 & 0 & 0 \\ 1560 & 0.32 & 1.7 & 6 \\ 0 & 0 & 0 & 0 \\ 460 & 0 & 0 & 0 \end{bmatrix} = \begin{bmatrix} 1490 & 0.56 & 0.85 & 67.5 \\ 2480 & 0.52 & 1.70 & 6 \end{bmatrix} \begin{matrix} \text{I} \\ \text{II} \end{matrix}$$

$$\begin{matrix} \text{A} & \text{B}_1 & \text{B}_2 & \text{C} \end{matrix}$$

69.

$$\begin{bmatrix} 0.80 & 0.75 & 0.50 \\ 0.20 & 0.20 & 0.40 \\ 0 & 0.05 & 0.10 \end{bmatrix} \begin{bmatrix} 400 \\ 360 \\ 230 \end{bmatrix} = \begin{bmatrix} 705 \\ 244 \\ 41 \end{bmatrix}$$

They need 705 pounds of regular coffee, 244 pounds of High Mountain coffee, and 41 pounds of chocolate.

71.

(a)

$$\begin{bmatrix} 78 & 83 & 81 & 86 \\ 84 & 88 & 79 & 85 \\ 70 & 72 & 77 & 73 \\ 88 & 91 & 94 & 87 \\ 96 & 95 & 98 & 92 \\ 65 & 72 & 74 & 81 \end{bmatrix} \begin{bmatrix} 0.25 \\ 0.25 \\ 0.25 \\ 0.25 \end{bmatrix} = \begin{bmatrix} 82.0 \\ 84.0 \\ 73.0 \\ 90.0 \\ 95.3 \\ 73.0 \end{bmatrix}$$

(b)

$$\begin{bmatrix} 78 & 83 & 81 & 86 \\ 84 & 88 & 79 & 85 \\ 70 & 72 & 77 & 73 \\ 88 & 91 & 94 & 87 \\ 96 & 95 & 98 & 92 \\ 65 & 72 & 74 & 81 \end{bmatrix} \begin{bmatrix} 0.20 \\ 0.20 \\ 0.25 \\ 0.35 \end{bmatrix} = \begin{bmatrix} 82.6 \\ 83.9 \\ 73.2 \\ 89.8 \\ 94.9 \\ 74.3 \end{bmatrix}$$

73.

(a)

$$AB = \begin{matrix} \\ \text{Well} \\ \text{Sick} \\ \text{Carrier} \end{matrix} \begin{matrix} \text{Male} \quad \text{Female} \\ \begin{bmatrix} 104750 & 102000 \\ 42000 & 40000 \\ 13250 & 13000 \end{bmatrix} \end{matrix}$$

This matrix gives the number of males and the number of females who are well, sick, or carriers.

(b)   42,000 sick males   (c)   102,000 well females

75.   (a)   (b)   The desired matrix is A^2.

$$\begin{matrix} & \text{SEA} & \text{LA} & \text{DEN} & \text{KC} & \text{SLC} \\ \text{SEA} & \begin{bmatrix} 0 & 1 & 0 & 0 & 1 \\ \text{LA} & 1 & 0 & 1 & 0 & 0 \\ \text{DEN} & 0 & 1 & 0 & 1 & 1 \\ \text{KC} & 0 & 0 & 1 & 0 & 1 \\ \text{SLC} & 1 & 0 & 1 & 1 & 0 \end{bmatrix} = A \end{matrix}$$

$$A^2 = \begin{matrix} & \text{SEA} & \text{LA} & \text{DEN} & \text{KC} & \text{SLC} \\ \text{SEA} & \begin{bmatrix} 2 & 0 & 2 & 1 & 0 \\ \text{LA} & 0 & 2 & 0 & 1 & 2 \\ \text{DEN} & 2 & 0 & 3 & 1 & 1 \\ \text{KC} & 1 & 1 & 1 & 2 & 1 \\ \text{SLC} & 0 & 2 & 1 & 1 & 3 \end{bmatrix} \end{matrix}$$

77.   The number of columns of A equals the number of rows of B. The number of columns of B equals the number of rows of A.

81. $AB = \begin{bmatrix} 99.9 & 97.2 \\ 71.4 & 65.6 \\ 133.0 & 144.0 \end{bmatrix}$ gives the total shipping cost by department for each company.

Section 2.5 Multiplication of Matrices

83. **(a)** $AB = \begin{bmatrix} 55600 & 9550 \\ 29550 & 5310 \\ 34850 & 5970 \end{bmatrix}$ gives salary and benefits by school

(b) $CD = \begin{bmatrix} 55600 & 29550 & 34850 \\ 9550 & 5310 & 5970 \end{bmatrix}$ gives salary and benefits by level

(c) $FE = \begin{bmatrix} 631.50 & 1967.20 \end{bmatrix}$ gives the total income tax and total FICA withheld for one person at each level.

85. $AB = \begin{bmatrix} 1 & 10 \\ -2 & 22 \end{bmatrix}$

87. $AB = \begin{bmatrix} 4 & -7 & 7 \\ 23 & 10 & 35 \\ 12 & 6 & 18 \end{bmatrix}$ $\qquad BA = \begin{bmatrix} 19 & 9 & 13 \\ -4 & -1 & -12 \\ 14 & 6 & 14 \end{bmatrix}$

89. $AB = \begin{bmatrix} 16 & -13 \\ 6 & -8 \end{bmatrix}$

91. **(a)** $\begin{bmatrix} 1000 \\ 1000 \\ 1000 \end{bmatrix} \begin{bmatrix} 12,000 \\ 500 \\ 200 \end{bmatrix} \begin{bmatrix} 3000 \\ 6000 \\ 125 \end{bmatrix} \begin{bmatrix} 1500 \\ 1500 \\ 1500 \end{bmatrix} \begin{bmatrix} 18,000 \\ 750 \\ 375 \end{bmatrix} \begin{bmatrix} 4500 \\ 9000 \\ 188 \end{bmatrix} \begin{bmatrix} 2250 \\ 2250 \\ 2250 \end{bmatrix} \begin{bmatrix} 27,000 \\ 1125 \\ 563 \end{bmatrix}$

The population increases another 50% every 3 years.

(b) $A^{15}B = \begin{bmatrix} 7593.75 \\ 7593.75 \\ 7593.75 \end{bmatrix}, A^{16}B = \begin{bmatrix} 91,125.0 \\ 3796.875 \\ 1898.4375 \end{bmatrix}, A^{17}B = \begin{bmatrix} 22,781.25 \\ 45,562.50 \\ 949.21875 \end{bmatrix},$

$A^{18}B = \begin{bmatrix} 11,390.625 \\ 11,390.625 \\ 11,390.625 \end{bmatrix}$

93. $\begin{bmatrix} 2000 \\ 1000 \\ 1000 \end{bmatrix} \begin{bmatrix} 4000 \\ 1000 \\ 500 \end{bmatrix} \begin{bmatrix} 2000 \\ 2000 \\ 500 \end{bmatrix} \begin{bmatrix} 2000 \\ 1000 \\ 1000 \end{bmatrix} \begin{bmatrix} 4000 \\ 1000 \\ 500 \end{bmatrix} \begin{bmatrix} 2000 \\ 2000 \\ 500 \end{bmatrix}$

The population follows a cycle of three years.

TI-83

1. $\begin{bmatrix} 2 & 10 \\ 29 & -11 \end{bmatrix}$ **3.** $\begin{bmatrix} 13 & 7 \\ 12 & -10 \end{bmatrix}$

5. $A^2 = \begin{bmatrix} 2 & 3 & 2 \\ 6 & 7 & 6 \\ 8 & 6 & 8 \end{bmatrix}$, $A^3 = \begin{bmatrix} 10 & 9 & 10 \\ 26 & 25 & 26 \\ 28 & 30 & 28 \end{bmatrix}$, $A^4 = \begin{bmatrix} 38 & 39 & 38 \\ 102 & 103 & 102 \\ 116 & 114 & 116 \end{bmatrix}$

Using EXCEL

1. $AB = \begin{bmatrix} 12 & 15 \\ 22 & 29 \end{bmatrix}$

3. $AB = \begin{bmatrix} -3 & 7 & 17 \\ 8 & 7 & 13 \end{bmatrix}$

Section 2.6

1. $25^{-1} = 0.04$, $(2/3)^{-1} = 3/2$, $(-5)^{-1} = -1/5$, $(0.75)^{-1} = 4/3$, $11^{-1} = 1/11$

3. $AB = \begin{bmatrix} 1 & 0 & 0 \\ 0 & 1 & 0 \\ 0 & 0 & 1 \end{bmatrix}$ Yes

5. $AB = \begin{bmatrix} 1 & -3 \\ 0 & 10 \end{bmatrix}$ No

7. $AB = \begin{bmatrix} 1 & 0 & 0 \\ 0 & 1 & 0 \\ 0 & 0 & 1 \end{bmatrix}$ Yes

9. $\begin{bmatrix} 1 & 2 & | & 1 & 0 \\ 3 & 5 & | & 0 & 1 \end{bmatrix}$ $\begin{bmatrix} 1 & 2 & | & 1 & 0 \\ 0 & -1 & | & -3 & 1 \end{bmatrix}$ $\begin{bmatrix} 1 & 0 & | & -5 & 2 \\ 0 & 1 & | & 3 & -1 \end{bmatrix}$ $A^{-1} = \begin{bmatrix} -5 & 2 \\ 3 & -1 \end{bmatrix}$

11. $\begin{bmatrix} 3 & 2 & | & 1 & 0 \\ 4 & 3 & | & 0 & 1 \end{bmatrix}$ $\begin{bmatrix} 1 & 2/3 & | & 1/3 & 0 \\ 0 & 1/3 & | & -4/3 & 1 \end{bmatrix}$ $\begin{bmatrix} 1 & 0 & | & 3 & -2 \\ 0 & 1 & | & -4 & 3 \end{bmatrix}$ $A^{-1} = \begin{bmatrix} 3 & -2 \\ -4 & 3 \end{bmatrix}$

13. $\begin{bmatrix} 1 & 3 & 9 & | & 1 & 0 & 0 \\ 0 & 1 & 4 & | & 0 & 1 & 0 \\ 3 & 2 & 3 & | & 0 & 0 & 1 \end{bmatrix}$ $\begin{bmatrix} 1 & 3 & 9 & | & 1 & 0 & 0 \\ 0 & 1 & 4 & | & 0 & 1 & 0 \\ 0 & -7 & -24 & | & -3 & 0 & 1 \end{bmatrix}$ $\begin{bmatrix} 1 & 0 & -3 & | & 1 & -3 & 0 \\ 0 & 1 & 4 & | & 0 & 1 & 0 \\ 0 & 0 & 4 & | & -3 & 7 & 1 \end{bmatrix}$

$\begin{bmatrix} 1 & 0 & 0 & | & -5/4 & 9/4 & 3/4 \\ 0 & 1 & 0 & | & 3 & -6 & -1 \\ 0 & 0 & 1 & | & -3/4 & 7/4 & 1/4 \end{bmatrix}$ $A^{-1} = \begin{bmatrix} -5/4 & 9/4 & 3/4 \\ 3 & -6 & -1 \\ -3/4 & 7/4 & 1/4 \end{bmatrix}$

15. $\begin{bmatrix} 0 & 4 & -2 & | & 1 & 0 & 0 \\ 1 & 3 & 5 & | & 0 & 1 & 0 \\ 1 & 4 & 2 & | & 0 & 0 & 1 \end{bmatrix}$ $\begin{bmatrix} 1 & 3 & 5 & | & 0 & 1 & 0 \\ 0 & 4 & -2 & | & 1 & 0 & 0 \\ 1 & 4 & 2 & | & 0 & 0 & 1 \end{bmatrix}$ $\begin{bmatrix} 1 & 3 & 5 & | & 0 & 1 & 0 \\ 0 & 4 & -2 & | & 1 & 0 & 0 \\ 0 & 1 & -3 & | & 0 & -1 & 1 \end{bmatrix}$

$$\begin{bmatrix} 1 & 0 & 13/2 & -3/4 & 1 & 0 \\ 0 & 1 & -1/2 & 1/4 & 0 & 0 \\ 0 & 0 & -5/2 & -1/4 & -1 & 1 \end{bmatrix} \qquad \begin{bmatrix} 1 & 0 & 0 & -7/5 & -8/5 & 13/5 \\ 0 & 1 & 0 & 3/10 & 1/5 & -1/5 \\ 0 & 0 & 1 & 1/10 & 2/5 & -2/5 \end{bmatrix}$$

$$A^{-1} = \begin{bmatrix} -7/5 & -8/5 & 13/5 \\ 3/10 & 1/5 & -1/5 \\ 1/10 & 2/5 & -2/5 \end{bmatrix}$$

17. $\begin{bmatrix} 4 & -2 & 1 & 0 \\ -2 & 1 & 0 & 1 \end{bmatrix}$ $\begin{bmatrix} 0 & 0 & 1 & 2 \\ -2 & 1 & 0 & 1 \end{bmatrix}$ No inverse

19. $\begin{bmatrix} 1 & 3 & 1 & 1 & 0 & 0 \\ 2 & 0 & -2 & 0 & 1 & 0 \\ 3 & 3 & -1 & 0 & 0 & 1 \end{bmatrix}$ $\begin{bmatrix} 1 & 3 & 1 & 1 & 0 & 0 \\ 0 & -6 & -4 & -2 & 1 & 0 \\ 0 & -6 & -4 & -3 & 0 & 1 \end{bmatrix}$ $\begin{bmatrix} 1 & 3 & 1 & 1 & 0 & 0 \\ 0 & -6 & -4 & -2 & 1 & 0 \\ 0 & 0 & 0 & -1 & -1 & 1 \end{bmatrix}$

No inverse

21. $\begin{bmatrix} 1 & 2 & 1 & 1 & 0 & 0 \\ 1 & -3 & 2 & 0 & 1 & 0 \\ 2 & -1 & 3 & 0 & 0 & 1 \end{bmatrix}$ $\begin{bmatrix} 1 & 2 & 1 & 1 & 0 & 0 \\ 0 & -5 & 1 & -1 & 1 & 0 \\ 0 & -5 & 1 & -2 & 0 & 1 \end{bmatrix}$ $\begin{bmatrix} 1 & 2 & 1 & 1 & 0 & 0 \\ 0 & -5 & 1 & -1 & 1 & 0 \\ 0 & 0 & 0 & -1 & -1 & 1 \end{bmatrix}$

No inverse

23. $\begin{bmatrix} 2 & 1 & 1 & 0 \\ 4 & 3 & 0 & 1 \end{bmatrix}$ $\begin{bmatrix} 1 & 1/2 & 1/2 & \\ 0 & 1 & -2 & 1 \end{bmatrix}$ $\begin{bmatrix} 1 & 0 & 3/2 & -1/2 \\ 0 & 1 & -2 & 1 \end{bmatrix}$ $A^{-1} = \begin{bmatrix} 3/2 & -1/2 \\ -2 & 1 \end{bmatrix}$

25. $\begin{bmatrix} 1 & 2 & 3 & 1 & 0 & 0 \\ 2 & -1 & 4 & 0 & 1 & 0 \\ 0 & -1 & 1 & 0 & 0 & 1 \end{bmatrix}$ $\begin{bmatrix} 1 & 2 & 3 & 1 & 0 & 0 \\ 0 & -5 & -2 & -2 & 1 & 0 \\ 0 & -1 & 1 & 0 & 0 & 1 \end{bmatrix}$ $\begin{bmatrix} 1 & 2 & 3 & 1 & 0 & 0 \\ 0 & -1 & 1 & 0 & 0 & 1 \\ 0 & -5 & -2 & -2 & 1 & 0 \end{bmatrix}$

$\begin{bmatrix} 1 & 0 & 5 & 1 & 0 & 2 \\ 0 & 1 & -1 & 0 & 0 & -1 \\ 0 & 0 & -7 & -2 & 1 & -5 \end{bmatrix}$ $\begin{bmatrix} 1 & 0 & 0 & -3/7 & 5/7 & -11/7 \\ 0 & 1 & 0 & 2/7 & -1/7 & -2/7 \\ 0 & 0 & 1 & 2/7 & -1/7 & 5/7 \end{bmatrix}$

$$A^{-1} = \begin{bmatrix} -3/7 & 5/7 & -11/7 \\ 2/7 & -1/7 & -2/7 \\ 2/7 & -1/7 & 5/7 \end{bmatrix}$$

27. **(a)** $\begin{bmatrix} 3 & 4 & -5 & 4 \\ 2 & -1 & 3 & -1 \\ 1 & 1 & -1 & 2 \end{bmatrix}$ **(b)** $\begin{bmatrix} 3 & 4 & -5 \\ 2 & -1 & 3 \\ 1 & 1 & -1 \end{bmatrix}$

(c) $\begin{bmatrix} 3 & 4 & -5 \\ 2 & -1 & 3 \\ 1 & 1 & -1 \end{bmatrix} \begin{bmatrix} x_1 \\ x_2 \\ x_3 \end{bmatrix} = \begin{bmatrix} 4 \\ -1 \\ 2 \end{bmatrix}$

29. **(a)** $\begin{bmatrix} 4 & 5 & | & 2 \\ 3 & -2 & | & 7 \end{bmatrix}$ **(b)** $\begin{bmatrix} 4 & 5 \\ 3 & -2 \end{bmatrix}$

(c) $\begin{bmatrix} 4 & 5 \\ 3 & -2 \end{bmatrix} \begin{bmatrix} x \\ y \end{bmatrix} = \begin{bmatrix} 2 \\ 7 \end{bmatrix}$

31. $\begin{bmatrix} 1 & 3 \\ 2 & -1 \end{bmatrix} \begin{bmatrix} x_1 \\ x_2 \end{bmatrix} = \begin{bmatrix} 5 \\ 6 \end{bmatrix}$

33. $\begin{bmatrix} 1 & 2 & -3 & 4 \\ 1 & 1 & 0 & 1 \\ 3 & 2 & 1 & 2 \end{bmatrix} \begin{bmatrix} x_1 \\ x_2 \\ x_3 \\ x_4 \end{bmatrix} = \begin{bmatrix} 0 \\ 5 \\ 4 \end{bmatrix}$

35. $\begin{bmatrix} 1 & 1 & 0 & 0 & | & 1 & 0 & 0 & 0 \\ 0 & 1 & 1 & 0 & | & 0 & 1 & 0 & 0 \\ 1 & 0 & 0 & 1 & | & 0 & 0 & 1 & 0 \\ 0 & 0 & 1 & 1 & | & 0 & 0 & 0 & 1 \end{bmatrix}$ $\begin{bmatrix} 1 & 1 & 0 & 0 & | & 1 & 0 & 0 & 0 \\ 0 & 1 & 1 & 0 & | & 0 & 1 & 0 & 0 \\ 0 & -1 & 0 & 1 & | & -1 & 0 & 1 & 0 \\ 0 & 0 & 1 & 1 & | & 0 & 0 & 0 & 1 \end{bmatrix}$

$\begin{bmatrix} 1 & 0 & -1 & 0 & | & 1 & -1 & 0 & 0 \\ 0 & 1 & 1 & 0 & | & 0 & 1 & 0 & 0 \\ 0 & 0 & 1 & 1 & | & -1 & 1 & 1 & 0 \\ 0 & 0 & 1 & 1 & | & 0 & 0 & 0 & 1 \end{bmatrix}$ No inverse

37. $\begin{bmatrix} 1 & 2 & -1 & | & 1 & 0 & 0 \\ 1 & 1 & 2 & | & 0 & 1 & 0 \\ 1 & -1 & -1 & | & 0 & 0 & 1 \end{bmatrix}$ $\begin{bmatrix} 1 & 2 & -1 & | & 1 & 0 & 0 \\ 0 & -1 & 3 & | & -1 & 1 & 0 \\ 0 & -3 & 0 & | & -1 & 0 & 1 \end{bmatrix}$ $\begin{bmatrix} 1 & 0 & 5 & | & -1 & 2 & 0 \\ 0 & 1 & -3 & | & 1 & -1 & 0 \\ 0 & 0 & -9 & | & 2 & -3 & 1 \end{bmatrix}$

$\begin{bmatrix} 1 & 0 & 5 & | & -1 & 2 & 0 \\ 0 & 1 & -3 & | & 1 & -1 & 0 \\ 0 & 0 & 1 & | & -2/9 & 1/3 & -1/9 \end{bmatrix}$ $\begin{bmatrix} 1 & 0 & 0 & | & 1/9 & 1/3 & 5/9 \\ 0 & 1 & 0 & | & 1/3 & 0 & -1/3 \\ 0 & 0 & 1 & | & -2/9 & 1/3 & -1/9 \end{bmatrix}$

$A^{-1} = \begin{bmatrix} 1/9 & 1/3 & 5/9 \\ 1/3 & 0 & -1/3 \\ -2/9 & 1/3 & -1/9 \end{bmatrix}$ $\begin{bmatrix} 1/9 & 1/3 & 5/9 \\ 1/3 & 0 & -1/3 \\ -2/9 & 1/3 & -1/9 \end{bmatrix} \begin{bmatrix} 2 \\ 0 \\ 1 \end{bmatrix} = \begin{bmatrix} 7/9 \\ 1/3 \\ 5/9 \end{bmatrix}$

$x_1 = \dfrac{7}{9}, \ x_2 = \dfrac{1}{3}, \ x_3 = \dfrac{-5}{9}$

39. $\begin{bmatrix} 1 & 1 & 2 & 1 & | & 1 & 0 & 0 & 0 \\ 2 & 0 & -1 & 1 & | & 0 & 1 & 0 & 0 \\ 0 & 1 & 3 & -1 & | & 0 & 0 & 1 & 0 \\ 3 & 2 & 0 & 1 & | & 0 & 0 & 0 & 1 \end{bmatrix}$ $\begin{bmatrix} 1 & 1 & 2 & 1 & | & 1 & 0 & 0 & 0 \\ 0 & -2 & -5 & -1 & | & -2 & 1 & 0 & 0 \\ 0 & 1 & 3 & -1 & | & 0 & 0 & 1 & 0 \\ 0 & -1 & -6 & -2 & | & -3 & 0 & 0 & 1 \end{bmatrix}$

$\begin{bmatrix} 1 & 1 & 2 & 1 & | & 1 & 0 & 0 & 0 \\ 0 & 1 & 3 & -1 & | & 0 & 0 & 1` & 0 \\ 0 & -2 & -5 & -1 & | & -2 & 1 & 0 & 0 \\ 0 & -1 & -6 & -2 & | & -3 & 0 & 0 & 1 \end{bmatrix}$ $\begin{bmatrix} 1 & 0 & -1 & 2 & | & 1 & 0 & -1 & 0 \\ 0 & 1 & 3 & -1 & | & 0 & 0 & 1 & 0 \\ 0 & 0 & 1 & -3 & | & -2 & 1 & 2 & 0 \\ 0 & 0 & -3 & -3 & | & -3 & 0 & 1 & 1 \end{bmatrix}$

$$\left[\begin{array}{cccc|cccc} 1 & 0 & 0 & -1 & -1 & 1 & 1 & 0 \\ 0 & 1 & 0 & 8 & 6 & -3 & -5 & 0 \\ 0 & 0 & 1 & -3 & -2 & 1 & 2 & 0 \\ 0 & 0 & 0 & -12 & -9 & 3 & 7 & 1 \end{array}\right] \quad \left[\begin{array}{cccc|cccc} 1 & 0 & 0 & 0 & -1/4 & 3/4 & 5/12 & -1/12 \\ 0 & 1 & 0 & 0 & 0 & -1 & -1/3 & 2/3 \\ 0 & 0 & 1 & 0 & 1/4 & 1/4 & 1/4 & -1/4 \\ 0 & 0 & 0 & 1 & 3/4 & -1/4 & -7/12 & -1/12 \end{array}\right]$$

$$A^{-1} = \left[\begin{array}{cccc} -1/4 & 3/4 & 5/12 & -1/12 \\ 0 & -1 & -1/3 & 2/3 \\ 1/4 & 1/4 & 1/4 & -1/4 \\ 3/4 & -1/4 & -7/12 & -1/12 \end{array}\right] \qquad A^{-1}\begin{bmatrix} 4 \\ 6 \\ 3 \\ 9 \end{bmatrix} = \begin{bmatrix} 4 \\ -1 \\ 1 \\ -1 \end{bmatrix}$$

$x_1 = 4,\ x_2 = -1,\ x_3 = 1,\ x_4 = -1$

41.

$$\left[\begin{array}{ccc|ccc} -2 & 1 & 3 & 1 & 0 & 0 \\ 2 & 4 & -1 & 0 & 1 & 0 \\ 3 & 0 & -4 & 0 & 1 & 0 \end{array}\right] \quad \left[\begin{array}{ccc|ccc} 1 & -1/2 & -3/2 & -1/2 & 0 & 0 \\ 0 & 5 & 2 & 1 & 1 & 0 \\ 0 & 3/2 & 1/2 & 3/2 & 0 & 1 \end{array}\right]$$

$$\left[\begin{array}{ccc|ccc} 1 & 0 & -13/10 & -2/5 & 1/10 & 0 \\ 0 & 1 & 2/5 & 1/5 & 1/5 & 0 \\ 0 & 0 & -1/10 & 6/5 & -3/10 & 1 \end{array}\right] \quad \left[\begin{array}{ccc|ccc} 1 & 0 & 0 & -16 & 4 & -13 \\ 0 & 1 & 0 & 5 & -1 & 4 \\ 0 & 0 & 1 & -12 & 3 & -10 \end{array}\right]$$

$$A^{-1} = \left[\begin{array}{ccc} -16 & 4 & -13 \\ 5 & -1 & 4 \\ -12 & 3 & -10 \end{array}\right]$$

$$A^{-1}\begin{bmatrix} 1 \\ 5 \\ 2 \end{bmatrix} = \begin{bmatrix} -22 \\ 8 \\ -17 \end{bmatrix} \quad x_1 = -22,\ x_2 = 8,\ x_3 = -17$$

$$A^{-1}\begin{bmatrix} -1 \\ 3 \\ 1 \end{bmatrix} = \begin{bmatrix} 15 \\ -4 \\ 11 \end{bmatrix} \quad x_1 = 15,\ x_2 = -4,\ x_3 = 11$$

$$A^{-1}\begin{bmatrix} 0 \\ 1 \\ 2 \end{bmatrix} = \begin{bmatrix} -22 \\ 7 \\ -17 \end{bmatrix} \quad x_1 = -22,\ x_2 = 7,\ x_3 = -17$$

43.

$$\left[\begin{array}{cc|cc} 1 & 2 & 1 & 0 \\ 3 & 5 & 0 & 1 \end{array}\right] \quad \left[\begin{array}{cc|cc} 1 & 2 & 1 & 0 \\ 0 & -1 & -3 & 1 \end{array}\right] \quad \left[\begin{array}{cc|cc} 1 & 0 & -5 & 2 \\ 0 & 1 & 3 & -1 \end{array}\right] \quad A^{-1} = \begin{bmatrix} -5 & 2 \\ 3 & -1 \end{bmatrix}$$

(a) $\qquad A^{-1}\begin{bmatrix} 3 \\ 8 \end{bmatrix} = \begin{bmatrix} 1 \\ 1 \end{bmatrix} \qquad\qquad x_1 = 1,\ x_2 = 1$

(b) $\qquad A^{-1}\begin{bmatrix} 4 \\ 9 \end{bmatrix} = \begin{bmatrix} -2 \\ 3 \end{bmatrix} \qquad\qquad x_1 = -2,\ x_2 = 3$

(c) $\qquad A^{-1}\begin{bmatrix} 4 \\ 9 \end{bmatrix} = \begin{bmatrix} -2 \\ 3 \end{bmatrix} \qquad\qquad x_1 = -1,\ x_2 = 2$

45. **(a)** Vitamin C intake $=$ $32x + 24y$
Vitamin A intake $= 900x + 425y$

$$\text{so} \quad \begin{bmatrix} 32 & 24 \\ 900 & 425 \end{bmatrix} \begin{bmatrix} x \\ y \end{bmatrix} = \begin{bmatrix} b_1 \\ b_2 \end{bmatrix}$$

where $b_1 =$ vitamin C intake and $b_2 =$ vitamin A intake

(b) $\begin{bmatrix} 32 & 24 \\ 900 & 425 \end{bmatrix} \begin{bmatrix} 3.2 \\ 2.5 \end{bmatrix} = \begin{bmatrix} 162.4 \\ 3942.5 \end{bmatrix}$ 162.4 mg of C, 3,942.5 iu of A

(c) $\begin{bmatrix} 32 & 24 \\ 900 & 425 \end{bmatrix} \begin{bmatrix} 1.5 \\ 3.0 \end{bmatrix} = \begin{bmatrix} 120 \\ 2625 \end{bmatrix}$ 120 mg of C, 2,625 iu of A

(d) $\begin{bmatrix} -0.053125 & 0.003 \\ 0.1125 & -0.004 \end{bmatrix} \begin{bmatrix} 107.2 \\ 2315.0 \end{bmatrix} = \begin{bmatrix} 1.25 \\ 2.8 \end{bmatrix}$ 1.25 units of A, 2.8 units of B

(e) $\begin{bmatrix} -0.053125 & 0.003 \\ 0.1125 & -0.004 \end{bmatrix} \begin{bmatrix} 104 \\ 2575 \end{bmatrix} = \begin{bmatrix} 2.2 \\ 1.4 \end{bmatrix}$ 2.2 units of A, 1.4 units of B

47. **(a)** $\begin{bmatrix} 2 & -0.25 \\ -1 & 0.25 \end{bmatrix} \begin{bmatrix} 900 \\ 5840 \end{bmatrix} = \begin{bmatrix} 340 \\ 560 \end{bmatrix}$ **(b)** $\begin{bmatrix} 2 & -0.25 \\ -1 & 0.25 \end{bmatrix} \begin{bmatrix} 1000 \\ 6260 \end{bmatrix} = \begin{bmatrix} 435 \\ 565 \end{bmatrix}$

340 children, 560 adults   435 children, 565 adults

(c) $\begin{bmatrix} 2 & -0.25 \\ -1 & 0.25 \end{bmatrix} \begin{bmatrix} 750 \\ 5560 \end{bmatrix} = \begin{bmatrix} 110 \\ 640 \end{bmatrix}$

110 children, 640 adults

49. **(a)**
$$\begin{array}{ccc} \text{Early R.} & \text{After D.} & \text{Deluxe} \end{array}$$
$$A = \begin{bmatrix} 0.80 & 0.75 & 0.50 \\ 0.20 & 0.20 & 0.40 \\ 0 & 0.05 & 0.10 \end{bmatrix} \begin{array}{l} \text{Regular} \\ \text{High Mt.} \\ \text{Chocolate} \end{array}$$

(b) $A^{-1} = \begin{bmatrix} 0 & 5 & -20 \\ 2 & -8 & 22 \\ -1 & 4 & -1 \end{bmatrix}$

(c) Let $x =$ pounds of Early Riser, $y =$ pounds of After Dinner, and $z =$ pounds of Deluxe. Then

$$A \begin{bmatrix} x \\ y \\ z \end{bmatrix} = \begin{bmatrix} \text{Pounds Regular} \\ \text{Pounds High Mt.} \\ \text{Pounds Chocolate} \end{bmatrix} \text{ so}$$

$$A^{-1} \begin{bmatrix} \text{Pounds Regular} \\ \text{Pounds High Mt.} \\ \text{Pounds Chocolate} \end{bmatrix} = \begin{bmatrix} x \\ y \\ z \end{bmatrix}$$

$$\begin{bmatrix} 0 & 5 & -20 \\ 2 & -8 & 22 \\ -1 & 4 & -1 \end{bmatrix} \begin{bmatrix} 505 & 766 & 571 \\ 170 & 244 & 196 \\ 25 & 40 & 33 \end{bmatrix} = \begin{bmatrix} 350 & 420 & 320 \\ 200 & 460 & 300 \\ 150 & 170 & 180 \end{bmatrix}$$

The amount of each blend that could be produced is

$$\begin{array}{c}\text{Day} \\ \begin{array}{ccc}1 & 2 & 3\end{array}\end{array}$$
$$\begin{bmatrix} 350 & 420 & 320 \\ 200 & 460 & 300 \\ 150 & 170 & 180 \end{bmatrix} \begin{array}{l}\text{Early Riser} \\ \text{After Dinner} \\ \text{Deluxe}\end{array}$$

51. **(a)** $\begin{bmatrix} 1/4 & 0 \\ 0 & 1/5 \end{bmatrix}$ **(b)** $\begin{bmatrix} 1/a & 0 \\ 0 & 1/b \end{bmatrix}$

(c) $\begin{bmatrix} 1/a & 0 & 0 \\ 0 & 1/b & 0 \\ 0 & 0 & 1/c \end{bmatrix}$

53. No because a row of zeros in the A part of [A | I] indicates no solution.

55. $A^{-1} = \begin{bmatrix} 1 & -1 \\ -2 & 3 \end{bmatrix}$ **57.** $A^{-1} = \begin{bmatrix} 0.50 & 0 & 0 \\ 0 & 0.25 & 0 \\ 0 & 0 & 0.20 \end{bmatrix}$

59. $A^{-1} = \begin{bmatrix} 1 & -1 & 0 \\ -2 & 5 & 1 \\ -1 & 3 & 1 \end{bmatrix}$

61. $A^{-1} = \begin{bmatrix} 0.4 & 0.8 & -0.2 & -0.8 \\ -0.4 & 0.2 & 0.2 & -0.2 \\ 0 & -0.1429 & 0.1429 & 0 \\ -0.1 & -0.4857 & 0.0857 & 0.7 \end{bmatrix}$

63. **(a)** $AB = \begin{bmatrix} 1 & 0 & 1 \\ 0 & 0 & 1 \\ 1 & -1 & 1 \end{bmatrix}$ $A^{-1} = \begin{bmatrix} 1 & -1 & 0 \\ 0 & 1 & -1 \\ 0 & 0 & 1 \end{bmatrix}$

$B^{-1} = \begin{bmatrix} 1 & 0 & 0 \\ 1 & 1 & 0 \\ 0 & 1 & 1 \end{bmatrix}$ $(AB)^{-1} = \begin{bmatrix} 1 & -1 & 0 \\ 1 & 0 & -1 \\ 0 & 1 & 0 \end{bmatrix}$

$A^{-1}B^{-1} = \begin{bmatrix} 0 & -1 & 0 \\ 1 & 0 & -1 \\ 0 & 1 & 1 \end{bmatrix}$ $B^{-1}A^{-1} = \begin{bmatrix} 1 & -1 & 0 \\ 1 & 0 & -1 \\ 0 & 1 & 0 \end{bmatrix}$

(b) $(AB)^{-1} = B^{-1}A^{-1}$

65. $(AB)^{-1} = B^{-1}A^{-1}$

67. **(a)** $A^{-1} = \begin{bmatrix} 2 & -1.5 & 2.5 \\ 0.5 & -0.25 & 0.75 \\ 0 & -0.25 & 0.25 \end{bmatrix}$ The reduced echelon form of A is $\begin{bmatrix} 1 & 0 & 0 \\ 0 & 1 & 0 \\ 0 & 0 & 1 \end{bmatrix}$

Section 2.7

1. $AX = \begin{bmatrix} 0.15 & 0.08 \\ 0.30 & 0.20 \end{bmatrix} \begin{bmatrix} 8 \\ 12 \end{bmatrix} = \begin{bmatrix} 2.16 \\ 4.8 \end{bmatrix}$

3. $\begin{bmatrix} 0.06 & 0.12 & 0.09 \\ 0.15 & 0.05 & 0.10 \\ 0.08 & 0.04 & 0.02 \end{bmatrix} \begin{bmatrix} 8 \\ 14 \\ 10 \end{bmatrix} = \begin{bmatrix} 3.06 \\ 2.90 \\ 1.40 \end{bmatrix}$

5. $\begin{bmatrix} 0.8 & -0.3 & | & 1 & 0 \\ -0.2 & 0.7 & | & 0 & 1 \end{bmatrix}$ $\begin{bmatrix} 1 & -0.375 & | & 1.25 & 0 \\ 0 & 0.625 & | & 0.25 & 1 \end{bmatrix}$ $\begin{bmatrix} 1 & 0 & | & 1.4 & 0.6 \\ 0 & 1 & | & 0.4 & 1.6 \end{bmatrix}$

    $(I - A)^{-1} = \begin{bmatrix} 1.4 & 0.6 \\ 0.4 & 1.6 \end{bmatrix}$

7. Find $(I-A)^{-1}D$

    $\begin{bmatrix} 0.76 & -0.08 & | & 1 & 0 \\ -0.12 & 0.96 & | & 0 & 1 \end{bmatrix}$ $\begin{bmatrix} 1 & -0.1053 & | & 1.3158 & 0 \\ 0 & 0.9474 & | & 0.15789 & 1 \end{bmatrix}$

    $\begin{bmatrix} 1 & 0 & | & 1.333 & 0.1111 \\ 0 & 1 & | & 0.1667 & 1.0555 \end{bmatrix}$

    $(I-A)^{-1}D = \begin{bmatrix} 1.3333 & 0.1111 \\ 0.1667 & 1.0555 \end{bmatrix} \begin{bmatrix} 5 \\ 12 \end{bmatrix} = \begin{bmatrix} 21.33 \\ 15.17 \end{bmatrix}$

9. $I - A = \begin{bmatrix} 0.8 & -0.4 \\ -0.4 & 0.7 \end{bmatrix}$ and $(I - A)^{-1} = \begin{bmatrix} 1.75 & 1 \\ 1 & 2 \end{bmatrix}$

    The output levels required to meet the demands $\begin{bmatrix} 20 \\ 28 \end{bmatrix}$ and $\begin{bmatrix} 15 \\ 11 \end{bmatrix}$ are obtained

    from

    $X = \begin{bmatrix} 1.75 & 1 \\ 1 & 2 \end{bmatrix} \begin{bmatrix} 20 & 15 \\ 28 & 11 \end{bmatrix} = \begin{bmatrix} 63 & 37.25 \\ 76 & 37.00 \end{bmatrix}$ and are $\begin{bmatrix} 63 \\ 76 \end{bmatrix}$ and $\begin{bmatrix} 37.25 \\ 37.00 \end{bmatrix}$, respectively

11. $I - A = \begin{bmatrix} 0.8 & -0.2 & -0.2 \\ -0.1 & 0.4 & -0.2 \\ -0.1 & -0.1 & 0.6 \end{bmatrix}$ and $(I - A)^{-1} = \begin{bmatrix} 22/15 & 14/15 & 4/5 \\ 8/15 & 46/15 & 6/5 \\ 1/3 & 2/3 & 2 \end{bmatrix}$

    The values of X required to meet the demands $\begin{bmatrix} 30 \\ 24 \\ 42 \end{bmatrix}$ and $\begin{bmatrix} 60 \\ 45 \\ 75 \end{bmatrix}$ come from the

    solution of

5. $\begin{bmatrix} 0.8 & -0.3 & | & 1 & 0 \\ -0.2 & 0.7 & | & 0 & 1 \end{bmatrix}$ $\quad \begin{bmatrix} 1 & -0.375 & | & 1.25 & 0 \\ 0 & 0.625 & | & 0.25 & 1 \end{bmatrix}$ $\quad \begin{bmatrix} 1 & 0 & | & 1.4 & 0.6 \\ 0 & 1 & | & 0.4 & 1.6 \end{bmatrix}$

$(I - A)^{-1} = \begin{bmatrix} 1.4 & 0.6 \\ 0.4 & 1.6 \end{bmatrix}$

7. Find $(I-A)^{-1}D$

$\begin{bmatrix} 0.76 & -0.08 & | & 1 & 0 \\ -0.12 & 0.96 & | & 0 & 1 \end{bmatrix}$ $\quad \begin{bmatrix} 1 & -0.1053 & | & 1.3158 & 0 \\ 0 & 0.9474 & | & 0.15789 & 1 \end{bmatrix}$

$\begin{bmatrix} 1 & 0 & | & 1.333 & 0.1111 \\ 0 & 1 & | & 0.1667 & 1.0555 \end{bmatrix}$

$(I-A)^{-1}D = \begin{bmatrix} 1.3333 & 0.1111 \\ 0.1667 & 1.0555 \end{bmatrix} \begin{bmatrix} 5 \\ 12 \end{bmatrix} = \begin{bmatrix} 21.33 \\ 15.17 \end{bmatrix}$

9. $I - A = \begin{bmatrix} 0.8 & -0.4 \\ -0.4 & 0.7 \end{bmatrix}$ and $(I- A)^{-1} = \begin{bmatrix} 1.75 & 1 \\ 1 & 2 \end{bmatrix}$

The output levels required to meet the demands $\begin{bmatrix} 20 \\ 28 \end{bmatrix}$ and $\begin{bmatrix} 15 \\ 11 \end{bmatrix}$ are obtained from

$X = \begin{bmatrix} 1.75 & 1 \\ 1 & 2 \end{bmatrix} \begin{bmatrix} 20 & 15 \\ 28 & 11 \end{bmatrix} = \begin{bmatrix} 63 & 37.25 \\ 76 & 37.00 \end{bmatrix}$ and are $\begin{bmatrix} 63 \\ 76 \end{bmatrix}$ and $\begin{bmatrix} 37.25 \\ 37.00 \end{bmatrix}$, respectively

11. $I - A = \begin{bmatrix} 0.8 & -0.2 & -0.2 \\ -0.1 & 0.4 & -0.2 \\ -0.1 & -0.1 & 0.6 \end{bmatrix}$ and $(I - A)^{-1} = \begin{bmatrix} 22/15 & 14/15 & 4/5 \\ 8/15 & 46/15 & 6/5 \\ 1/3 & 2/3 & 2 \end{bmatrix}$

The values of X required to meet the demands $\begin{bmatrix} 30 \\ 24 \\ 42 \end{bmatrix}$ and $\begin{bmatrix} 60 \\ 45 \\ 75 \end{bmatrix}$ come from the

solution of

$X = \begin{bmatrix} 22/15 & 14/15 & 4/5 \\ 8/15 & 46/15 & 6/5 \\ 1/3 & 2/3 & 2 \end{bmatrix} \begin{bmatrix} 30 & 60 \\ 24 & 45 \\ 42 & 75 \end{bmatrix} = \begin{bmatrix} 100 & 190 \\ 140 & 260 \\ 110 & 200 \end{bmatrix}$ and are $\begin{bmatrix} 100 \\ 140 \\ 110 \end{bmatrix}$ and $\begin{bmatrix} 190 \\ 260 \\ 200 \end{bmatrix}$,

respectively.

13. **(a)** $AX = \begin{bmatrix} 23.5 \\ 28.5 \end{bmatrix}$ $\qquad$ **(b)** $X - AX = \begin{bmatrix} 16.5 \\ 21.5 \end{bmatrix}$

15. **(a)** $AX = \begin{bmatrix} 21.6 \\ 64.8 \\ 36.0 \end{bmatrix}$ $\qquad$ **(b)** $X - AX = \begin{bmatrix} 14.4 \\ 7.2 \\ 0 \end{bmatrix}$

17.   (a)   The input-output matrix is $A = \begin{matrix} & \\ A \\ N \end{matrix} \begin{matrix} A & N \\ \begin{bmatrix} 0.1 & 0.3 \\ 0.6 & 0.4 \end{bmatrix} \end{matrix}$

    (b)   Internal consumption $= \begin{bmatrix} 0.1 & 0.3 \\ 0.6 & 0.4 \end{bmatrix} \begin{bmatrix} 3.5 \\ 5.2 \end{bmatrix} = \begin{bmatrix} 1.91 \\ 4.18 \end{bmatrix}$

Agriculture internal consumption = \$1.91 million leaving \$1.59 million for export.
Nonagriculture consumption = \$4.18 million leaving \$1.02 million for export.

    (c) $D = \begin{bmatrix} 2 \\ 2 \end{bmatrix}$. The total production required is given by X where

$(I - A)^{-1} D = X$.

$I - A = \begin{bmatrix} 0.9 & -0.3 \\ -0.6 & 0.6 \end{bmatrix}$ and $(I - A)^{-1} = \begin{bmatrix} 5/3 & 5/6 \\ 5/3 & 5/2 \end{bmatrix}$

Total production $= \begin{bmatrix} 5/3 & 5/6 \\ 5/3 & 5/2 \end{bmatrix} \begin{bmatrix} 2 \\ 2 \end{bmatrix} = \begin{bmatrix} 5 \\ 8.3 \end{bmatrix}$

To export \$2 million of each kind of product requires production of \$5 million of agriculture and \$8.3 million of nonagriculture products.

    (d)   For \$2 million of agriculture and \$3 million of nonagriculture exports

Total production $= \begin{bmatrix} 5/3 & 5/6 \\ 5/3 & 5/2 \end{bmatrix} \begin{bmatrix} 2 \\ 3 \end{bmatrix} = \begin{bmatrix} 5.83 \\ 10.83 \end{bmatrix}$

Thus, \$5.83 million of agriculture and \$10.83 million of nonagriculture products are required.

19.   (a) $A = \begin{matrix} & \\ P \\ E \end{matrix} \begin{matrix} P & E \\ \begin{bmatrix} 0.10 & 0.40 \\ 0.20 & 0.20 \end{bmatrix} \end{matrix}$

    (b)   The amounts used internally are

$\begin{matrix} & P & E \\ P & \\ E \end{matrix}$
$\begin{bmatrix} 0.10 & 0.40 \\ 0.20 & 0.20 \end{bmatrix} \begin{bmatrix} 25 \\ 32 \end{bmatrix} = \begin{bmatrix} 15.3 \\ 11.4 \end{bmatrix}$

\$15.3 million worth of plastics and \$11.4 million worth of electronics are used internally.

    (c)   Solve $(I - A)^{-1} \begin{bmatrix} 36 \\ 44 \end{bmatrix} = X$

where X represents total production.

$I - A = \begin{bmatrix} 0.90 & -0.40 \\ -0.20 & 0.80 \end{bmatrix}$ $(I - A)^{-1} = \begin{bmatrix} 1.25 & 0.625 \\ 0.3125 & 1.40625 \end{bmatrix}$

$(I - A)^{-1} \begin{bmatrix} 36 \\ 44 \end{bmatrix} = \begin{bmatrix} 72.5 \\ 73.125 \end{bmatrix}$

The corporation must produce $72.5 million worth of plastics and $73.125 worth of electronics to have $36 million worth of plastics and $44 million worth of electronics available for external sales.

21. **(a)**
$$A = \begin{array}{c} C \\ M \\ US \end{array} \begin{bmatrix} 0.2 & 0.1 & 0.3 \\ 0.2 & 0.4 & 0 \\ 0.4 & 0 & 0.3 \end{bmatrix} \begin{array}{ccc} C & M & US \end{array}$$

(b) The value of components used internally is
$$\begin{bmatrix} 0.2 & 0.1 & 0.3 \\ 0.2 & 0.4 & 0 \\ 0.4 & 0 & 0.3 \end{bmatrix} \begin{bmatrix} 10 \\ 18 \\ 15 \end{bmatrix} = \begin{bmatrix} 8.3 \\ 9.2 \\ 8.5 \end{bmatrix}$$
Canada uses $8.3 million worth of components, Mexico $9.2 million, and US $8.5 million.

(c) Solve $(I - A)^{-1} \begin{bmatrix} 24 \\ 30 \\ 20 \end{bmatrix}$ $I - A = \begin{bmatrix} 0.8 & -0.1 & -0.3 \\ -0.2 & 0.6 & 0 \\ -0.4 & 0 & 0.7 \end{bmatrix}$

$$(I - A)^{-1} = \begin{bmatrix} 1.68 & 0.28 & 0.72 \\ 0.56 & 1.76 & 0.24 \\ 0.96 & 0.16 & 1.84 \end{bmatrix} \text{(rounded to 2 decimals)}$$

$$(I - A)^{-1} \begin{bmatrix} 24 \\ 30 \\ 20 \end{bmatrix} = \begin{bmatrix} 63.12 \\ 71.04 \\ 64.64 \end{bmatrix}$$

Canada must produce $63.12 million worth of vehicles, Mexico $71.04 million, and US $64.64 million.

23. A negative entry indicates that a negative cost is associated with producing a product. An entry greater than 1 indicates it costs more than $1.00 to produce $1.00 worth of a product.

25. Let x_1 = value of services, x_2 = value of retail.
Service used internally $= 0.20x_1 + 0.60x_2$
Retail used internally $= 0.50x_1 + 0.60x_2$
so the input-output matrix is
$$A = \begin{bmatrix} 0.20 & 0.60 \\ 0.50 & 0.60 \end{bmatrix} \text{ and the demand matrix is } D = \begin{bmatrix} 200,000 \\ 100,000 \end{bmatrix}$$
We need to find $(I - A)^{-1}D$
$$(I - A)^{-1}D = \begin{bmatrix} 20 & 30 \\ 25 & 40 \end{bmatrix} \begin{bmatrix} 200,000 \\ 100,000 \end{bmatrix} = \begin{bmatrix} 7,000,000 \\ 9,000,000 \end{bmatrix}$$
The corporation must produce $7,000,000 of service and $9,000,000 of retail in order to have $200,000 of service and $100,000 of retail available to the consumer.

27. **(a)** The demand matrix is of the form $\begin{bmatrix} 0 \\ d \\ d \end{bmatrix}$ so the production is $\begin{bmatrix} 6d \\ 25d \\ 10d \end{bmatrix}$.

The output that produces no grain and equal values of lumber and energy is any scalar multiple of $\begin{bmatrix} 6 \\ 25 \\ 10 \end{bmatrix}$.

(b) The demand matrix is of the form $\begin{bmatrix} d \\ 0 \\ 2d \end{bmatrix}$ so the production is

$$\begin{bmatrix} 2 & 3 & 3 \\ 0 & 15 & 10 \\ 0 & 5 & 5 \end{bmatrix} \begin{bmatrix} d \\ 0 \\ 2d \end{bmatrix} = \begin{bmatrix} 8d \\ 20d \\ 10d \end{bmatrix}.$$

The output that yields no lumber and energy twice the value of grain is any scalar multiple of $\begin{bmatrix} 8 \\ 20 \\ 10 \end{bmatrix}$.

29. **(a)** $AX = \begin{bmatrix} 0.2 & 0.1 & 0.4 \\ 0.3 & 0.4 & 0.6 \\ 0.3 & 0.3 & 0.4 \end{bmatrix} \begin{bmatrix} 10 \\ 30 \\ 20 \end{bmatrix} = \begin{bmatrix} 13 \\ 27 \\ 20 \end{bmatrix}$ is the amount consumed internally.

$\begin{bmatrix} 10 \\ 30 \\ 20 \end{bmatrix} - \begin{bmatrix} 13 \\ 27 \\ 20 \end{bmatrix} = \begin{bmatrix} -3 \\ 3 \\ 0 \end{bmatrix}$ is the amount left for consumers.

More of the first industry is used internally than is produced and all of the third industry is used internally. $AX = \begin{bmatrix} 0.2 & 0.1 & 0.4 \\ 0.3 & 0.4 & 0.6 \\ 0.3 & 0.3 & 0.4 \end{bmatrix} \begin{bmatrix} 20 \\ 40 \\ 30 \end{bmatrix} =$

$\begin{bmatrix} 20 \\ 40 \\ 30 \end{bmatrix}$ i s the amount consumed internally.

$\begin{bmatrix} 20 \\ 40 \\ 30 \end{bmatrix} - \begin{bmatrix} 20 \\ 40 \\ 30 \end{bmatrix} = \begin{bmatrix} 0 \\ 0 \\ 0 \end{bmatrix}$ is the amount left for consumers. All of the production is used internally.

(b) $AX = \begin{bmatrix} 0.2 & 0.1 & 0.4 \\ 0.3 & 0.4 & 0.6 \\ 0.3 & 0.3 & 0.4 \end{bmatrix} \begin{bmatrix} 30 \\ 30 \\ 40 \end{bmatrix} = \begin{bmatrix} 25 \\ 45 \\ 34 \end{bmatrix}$ is the amount consumed internally.

$\begin{bmatrix} 30 \\ 30 \\ 40 \end{bmatrix} - \begin{bmatrix} 25 \\ 45 \\ 34 \end{bmatrix} = \begin{bmatrix} 5 \\ -15 \\ 6 \end{bmatrix}$ is the amount left for consumers.

More of the second industry is used internally than is produced.

$$AX = \begin{bmatrix} 0.2 & 0.1 & 0.4 \\ 0.3 & 0.4 & 0.6 \\ 0.3 & 0.3 & 0.4 \end{bmatrix} \begin{bmatrix} 10 \\ 20 \\ 20 \end{bmatrix} = \begin{bmatrix} 12 \\ 23 \\ 17 \end{bmatrix}$$ is the amount consumed internally.

$$\begin{bmatrix} 10 \\ 20 \\ 20 \end{bmatrix} - \begin{bmatrix} 12 \\ 23 \\ 17 \end{bmatrix} = \begin{bmatrix} -2 \\ -3 \\ 3 \end{bmatrix}$$ is the amount left for consumers.

More of the first and second industry are used internally than is produced.

31.
$$\begin{bmatrix} 1.254 & 0.226 & 0.167 \\ 0.743 & 1.593 & 0.700 \\ 1.087 & 0.883 & 1.604 \end{bmatrix}$$

33.
$$I - A = \begin{bmatrix} 0.60 & -0.10 & -0.20 & -0.15 \\ -0.20 & 0.80 & -0.15 & -0.10 \\ -0.02 & -0.03 & 0.99 & -0.12 \\ -0.25 & -0.04 & -0.01 & 0.90 \end{bmatrix}$$

$$(I - A)^{-1} = \begin{bmatrix} 1.948 & 0.281 & 0.440 & 0.414 \\ 0.582 & 1.349 & 0.325 & 0.290 \\ 0.126 & 0.063 & 1.047 & 0.168 \\ 0.568 & 0.139 & 0.148 & 1.241 \end{bmatrix}$$

$$(I - A)^{-1} \begin{bmatrix} 450,000 \\ 300,000 \\ 620,000 \\ 240,000 \end{bmatrix} = \begin{bmatrix} 1,333,000 \\ 937,560 \\ 764,890 \\ 687,110 \end{bmatrix}$$ production for year 1

$$(I - A)^{-1} \begin{bmatrix} 500,000 \\ 325,000 \\ 600,000 \\ 250,000 \end{bmatrix} = \begin{bmatrix} 1,432,740 \\ 996,770 \\ 753,510 \\ 728,430 \end{bmatrix}$$ production for year 2

$$(I - A)^{-1} \begin{bmatrix} 475,000 \\ 360,000 \\ 590,000 \\ 280,000 \end{bmatrix} = \begin{bmatrix} 1,401,910 \\ 1,034,920 \\ 747,140 \\ 754,830 \end{bmatrix}$$ production for year 3

Section 2.8

1.

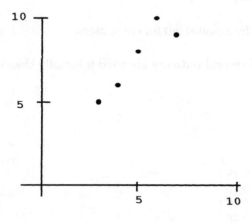

x	y	x²	xy
3	5	9	15
4	6	16	24
5	8	25	40
6	10	36	60
7	9	49	63
25	38	135	202

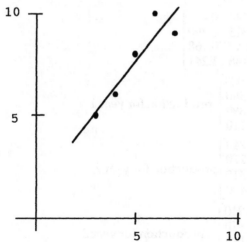

Solve
$135m + 25b = 202$
$25m + 5b = 38$
$m = 1.20, b = 1.60$
Regression line
$y = 1.20x + 1.60$

3.

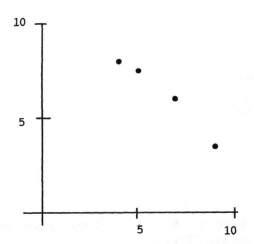

x	y	x^2	xy
4	8	16	32
5	7.5	25	37.5
7	6	49	42
9	3.5	81	31.5
25	25	171	143

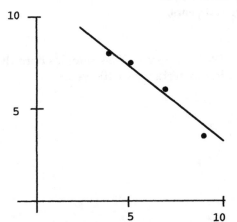

Solve

$171m + 25b = 143$

$25m + 4b = 25$

$m = -0.90, b = 11.86$

Regression line

$y = -0.90x + 11.86$

5.　　**(a)**　　The augmented matrix of the system is

$$\begin{bmatrix} 0 & 1 & 3 & 4 \\ 1 & 1 & 1 & 1 \end{bmatrix} \begin{bmatrix} 0 & 1 & 22.2 \\ 1 & 1 & 32.7 \\ 3 & 1 & 44.4 \\ 4 & 1 & 53.9 \end{bmatrix} = \begin{bmatrix} 26 & 8 & 381.5 \\ 8 & 4 & 153.2 \end{bmatrix}$$

with solution $m = 7.51$ and $b = 23.28$

$y = 7.51x + 23.28$

　　(b)　　$7.51(8) + 23.38 = 83.36\%$

　　(c)　　$100 = 7.51x + 23.28$

$$x = \frac{76.72}{7.51} = 10.21, \text{ in the year 2007.}$$

7.

x	y	x^2	xy
0	65.2	0	0
20	73.1	400	1462
40	77.4	1600	3096
56	79.1	3136	4429.6
60	79.5	3600	4770
176	374.3	8736	13757.6

Solve the system

$$8736m + 176b = 13757.6$$
$$176m + 5b = 374.3$$

m = 0.23, b = 66.79

(a) Least squares line

y = 0.23x + 66.79

(b) For 2010 x = 70 so y = 0.23(70) + 66.79 = 82.89. The life expectancy of females born in 2010 is about 83 years.

(c) For a life expectancy of 100 years,

100 = 0.23x + 66.79

x = 144.39

so about 144 years after 1940, namely in 2084, females born that year are estimated to have a life expectancy of 100 years.

9. (a)

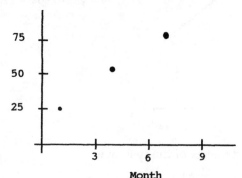

Month

(b)

x	y	x^2	xy
1	24.6	1	24.6
4	53.3	16	213.2
7	76.5	49	535.5
12	154.4	66	773.3

(c)

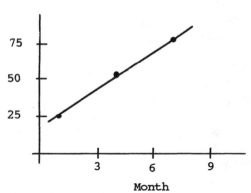

Month

Solve the system
$$66m + 12b = 773.3$$
$$12m + 3b = 154.4$$
$$m = 8.65, b = 16.87$$
The least squares line is
$$y = 8.65x + 16.87$$

Yes

(d) For October, $x = 10$ and $y = 103.4$ estimates the average temperature to be over 103 degrees.

(e) The estimate is not realistic because the average temperature does not rise in a straight line. It peaks in the summer and then declines.

11. **(a)** Israel:

Augmented matrix for the system is

$$\begin{bmatrix} 0 & 5 & 10 & 15 & 23 & 27 \\ 1 & 1 & 1 & 1 & 1 & 1 \end{bmatrix} \begin{bmatrix} 0 & 1 & 28.2 \\ 5 & 1 & 24.1 \\ 10 & 1 & 23.5 \\ 15 & 1 & 22.2 \\ 23 & 1 & 20.0 \\ 27 & 1 & 18.9 \end{bmatrix} = \begin{bmatrix} 1608 & 80 & 1658.8 \\ 80 & 6 & 136.9 \end{bmatrix}$$

$m = -0.31, b = 26.92$
$y = -0.31x + 26.92$

U. S.:

Augmented matrix for the system is

$$\begin{bmatrix} 0 & 5 & 10 & 15 & 23 & 27 \\ 1 & 1 & 1 & 1 & 1 & 1 \end{bmatrix} \begin{bmatrix} 0 & 1 & 14.0 \\ 5 & 1 & 16.2 \\ 10 & 1 & 15.7 \\ 15 & 1 & 16.7 \\ 23 & 1 & 14.4 \\ 27 & 1 & 14.1 \end{bmatrix} = \begin{bmatrix} 1608 & 80 & 1200.4 \\ 80 & 6 & 91.1 \end{bmatrix}$$

$m = -0.026, b = 15.53$
$y = -0.026x + 15.53$

(b) The lines intersect when $x = 40.1$ and $y = 14.5$. The birth rates will be the same, about 14.5, in the year $1975 + 40.1 = 2015.1$, that is, during the year 2015.

13. **(a)** $f(x) = 26.66x + 161.05$

(b) $f(35) = 26.66(35) + 161.05 = 1094.15$

$f(55) = 26.66(55) + 161.05 = 1627.35$

15. **(a)** $y = 0.085x + 1.09$ **(b)**   4.07

(c) $10 = 0.085x + 1.09$

$x = 104.8$ By the year 2080.

17. **(a)** $f(x) = 701.3x + 7499.9$ **(b)** $f(28) = 27,136.3$

(c) $25,000 = 701.3x + 7499.9$

$701.3x = 17,501.1$

$x = 24.95$

It is estimated that the income level will reach \$25,000 in the year 1980 + 24.95, by the year 2005.

19. **(a)** $f(x) = 1697.3x + 38015.2$

(b) $f(-3) = 1697.3(-3) + 38015.2 = 32,923.$

This estimates a median income of about \$32,900 for 1985)

(c) $f(12) = 1697.3(12) + 38015.2 = 58,382.2$

This estimates a median income of about \$58,400 for 2000.

(d) $75,000 = 1697.3x + 38015.2$

$x = 21.8$

This estimates the median income will reach \$75,000 in the year 1988 + 21.8, in the year 2009.

Using Your TI-83

1. $y = 1.64x - 2.71$ **3.** $y = 0.50 + 4.13$

Using EXCEL

1. $y = 1.2x + 0.8$ **3.** $y = 0.99x - 24.14$

Chapter 2 Review

1. $3x + 2y = 5$
    $2x + 4y = 9$

    $x = 9/2 - 2y$
    $3(9/2 - 2y) + 2y = 5$
    $-4y = 5 - 27/2$
    $4y = 17/2$
    $y = 17/8$
    $x = 9/2 - 2(17/8)$
    $= 9/2 - 17/4 = 1/4$

    $(1/4, 17/8)$

3. $5x - y = 34$
    $2x + 3y = 0$

    $15x - 3y = 102$
    $\underline{2x + 3y = \ 0}$
    $17x \quad = 102$
    $x = 6, \ y = -4$

5. $x - 2y + 3z = 3$
    $4x + 7y - 6z = 6$
    $-2x + 4y + 12z = 0$

    Eliminate x using the first two equations
    $4x - 8y + 12z = 12$
    $\underline{4x + 7y - \ 6z = \ 6}$
    $- 15y + 18z = \ 6$

    Eliminate x using the first and third equations
    $2x - 4y + \ 6z = 6$
    $\underline{-2x + 4y + 12z = 0}$
    $18z = 6$
    $z = 1/3$

    $-15y + 6 = 6$
    $y = 0$
    $x - 0 + 1 = 3$
    $x = 2$

    $(2, 0, 1/3)$

7. $\begin{bmatrix} 2 & -4 & -14 & | & 50 \\ 1 & -1 & -5 & | & 17 \\ 2 & -4 & -17 & | & 65 \end{bmatrix}$ $\begin{bmatrix} 1 & -1 & -5 & | & 17 \\ 0 & -2 & -4 & | & 16 \\ 0 & -2 & -7 & | & 31 \end{bmatrix}$ $\begin{bmatrix} 1 & 0 & -3 & | & 9 \\ 0 & 1 & 2 & | & -8 \\ 0 & 0 & -3 & | & 15 \end{bmatrix}$ $\begin{bmatrix} 1 & 0 & 0 & | & -6 \\ 0 & 1 & 0 & | & 2 \\ 0 & 0 & 1 & | & -5 \end{bmatrix}$

    $x_1 = -6, x_2 = 2, x_3 = -5$

9. $\begin{bmatrix} 1 & -1 & | & 3 \\ 4 & 3 & | & 5 \\ 6 & 1 & | & 9 \end{bmatrix}$ $\begin{bmatrix} 1 & -1 & | & 3 \\ 0 & 7 & | & -7 \\ 0 & 7 & | & -9 \end{bmatrix}$ $\begin{bmatrix} 1 & -1 & | & 3 \\ 0 & 7 & | & -7 \\ 0 & 0 & | & -2 \end{bmatrix}$ No solution

11. $\begin{bmatrix} 1 & 0 & 1 & | & 0 \\ 2 & -1 & 1 & | & -1 \\ 1 & -1 & 0 & | & -1 \end{bmatrix}$ $\begin{bmatrix} 1 & 0 & 1 & | & 0 \\ 0 & -1 & -1 & | & -1 \\ 0 & -1 & -1 & | & -1 \end{bmatrix}$ $\begin{bmatrix} 1 & 0 & 1 & | & 0 \\ 0 & 1 & 1 & | & 1 \\ 0 & 0 & 0 & | & 0 \end{bmatrix}$

$x = -z, y = 1 - z$

13. $\begin{bmatrix} 1 & 2 & -1 & 3 & | & 3 \\ 1 & 3 & 1 & -1 & | & 0 \\ 2 & 1 & -6 & 2 & | & -11 \\ 3 & 7 & -1 & 5 & | & 6 \end{bmatrix}$ $\begin{bmatrix} 1 & 2 & -1 & 3 & | & 3 \\ 0 & 1 & 2 & -4 & | & -3 \\ 0 & -3 & -4 & -4 & | & -17 \\ 0 & 1 & 2 & -4 & | & -3 \end{bmatrix}$ $\begin{bmatrix} 1 & 0 & -5 & 11 & | & 9 \\ 0 & 1 & 2 & -4 & | & -3 \\ 0 & 0 & 2 & -16 & | & -26 \\ 0 & 0 & 0 & 0 & | & 0 \end{bmatrix}$

$\begin{bmatrix} 1 & 0 & 0 & -29 & | & -56 \\ 0 & 1 & 0 & 12 & | & 23 \\ 0 & 0 & 1 & -8 & | & -13 \\ 0 & 0 & 0 & 0 & | & 0 \end{bmatrix}$ $x_1 = -56 + 29x_4, x_2 = 23 - 12x_4, x_3 = -13 + 8x_4$

15. $\begin{bmatrix} 2 & 3 & -5 & | & 8 \\ 6 & -3 & 1 & | & 16 \end{bmatrix}$ $\begin{bmatrix} 1 & 3/2 & -5/2 & | & 4 \\ 0 & -12 & 16 & | & -8 \end{bmatrix}$ $\begin{bmatrix} 1 & 0 & -1/2 & | & 3 \\ 0 & 1 & -4/3 & | & 2/3 \end{bmatrix}$

$x_1 = 3 + (1/2)x_3, x_2 = 2/3 + (4/3)x_3$

17. $x + 2 = 5 - x$
$4x = 3$
$x = 3/4$

19. $\begin{bmatrix} -3 & -2 \\ 6 & 7 \end{bmatrix}$

21. $\begin{bmatrix} 11 & -3 \\ 7 & -1 \\ 3 & 0 \end{bmatrix}$

23. [3]   **25.** Cannot multiply them

27. $\begin{bmatrix} 8 & 6 & | & 1 & 0 \\ 7 & 5 & | & 0 & 1 \end{bmatrix}$ $\begin{bmatrix} 1 & 3/4 & | & 1/8 & 0 \\ 0 & -1/4 & | & -7/8 & 1 \end{bmatrix}$ $\begin{bmatrix} 1 & 0 & | & -5/2 & 3 \\ 0 & 1 & | & 7/2 & -4 \end{bmatrix}$

$A^{-1} = \begin{bmatrix} -5/2 & 3 \\ 7/2 & -4 \end{bmatrix}$

29. $\begin{bmatrix} 1 & 0 & 3 & | & 1 & 0 & 0 \\ 2 & -5 & 4 & | & 0 & 1 & 0 \\ 1 & -2 & 2 & | & 0 & 0 & 1 \end{bmatrix}$ $\begin{bmatrix} 1 & 0 & 3 & | & 1 & 0 & 0 \\ 0 & -5 & -2 & | & -2 & 1 & 0 \\ 0 & -2 & -1 & | & -1 & 0 & 1 \end{bmatrix}$

$\begin{bmatrix} 1 & 0 & 3 & | & 1 & 0 & 0 \\ 0 & 1 & 2/5 & | & 2/5 & -1/5 & 0 \\ 0 & 0 & -1/5 & | & -1/5 & -2/5 & 1 \end{bmatrix}$ $\begin{bmatrix} 1 & 0 & 0 & | & -2 & -6 & 15 \\ 0 & 1 & 0 & | & 0 & -1 & 2 \\ 0 & 0 & 1 & | & 1 & 2 & -5 \end{bmatrix}$

$A^{-1} = \begin{bmatrix} -2 & -6 & 15 \\ 0 & -1 & 2 \\ 1 & 2 & -5 \end{bmatrix}$

31. $\begin{bmatrix} 6 & 4 & -5 & | & 10 \\ 3 & -2 & 0 & | & 12 \\ 1 & 1 & -4 & | & -2 \end{bmatrix}$

33.
$$\begin{bmatrix} 2 & 4 & 6 & -2 \\ 3 & 1 & 0 & 5 \\ -2 & 1 & 3 & -11 \end{bmatrix} \quad \begin{bmatrix} 1 & 2 & 3 & -1 \\ 0 & -5 & -9 & 8 \\ 0 & 5 & 9 & -13 \end{bmatrix} \quad \begin{bmatrix} 1 & 0 & -3/5 & 11/5 \\ 0 & 1 & 9/5 & -8/5 \\ 0 & 0 & 0 & -5 \end{bmatrix}$$

$$\begin{bmatrix} 1 & 0 & -3/5 & 0 \\ 0 & 1 & 9/5 & 0 \\ 0 & 0 & 0 & 1 \end{bmatrix}$$

35. Let x_1 = number of free throws, x_2 = number of 2-pointers, and x_3 = number of 3-pointers.

$x_1 + x_2 + x_3 = 19$ $\qquad$ (Number of times scored)
$x_1 + 2x_2 + 3x_3 = 36$ $\qquad$ (Total points scored)
$x_2 = 2x_3 + x_1$ $\qquad$ (Number of 2-pointers)

The augmented matrix of this system is $\begin{bmatrix} 1 & 1 & 1 & 19 \\ 1 & 2 & 3 & 36 \\ 1 & -1 & 2 & 0 \end{bmatrix}$ which reduces to

$$\begin{bmatrix} 1 & 0 & 0 & 5 \\ 0 & 1 & 0 & 11 \\ 0 & 0 & 1 & 3 \end{bmatrix}$$

LaShawn made 5 free throws, 11 2-pointers, and 3 3-pointers.

37. Let x = the amount invested in bonds and y = the amount invested in stocks.
$x + \quad y = 50,000$
$0.04x + 0.065y = 2750$

$\qquad 0.04x + 0.040y = 2000 \quad (2000 = 0.04(50,000))$
$\qquad \underline{0.04x + 0.065y = 2750}$
$\qquad \qquad 0.025y = 750$
$\qquad \qquad \qquad y = 30,000$
$\qquad x + 30,000 = 50,000$
$\qquad x = 20,000$

She invested $20,000 in bonds and $30,000 in stocks.

39. Let x = shares of High-Tech, y = shares of Big Burger
$\qquad \qquad 38x + \quad 16y = 5648$
$\qquad \qquad 40.5x + 15.75y = 5931$

$$\begin{bmatrix} 38 & 16 & 5648 \\ 40.5 & 15.75 & 5931 \end{bmatrix} \quad \begin{bmatrix} 1 & 8/19 & 2824/19 \\ 0 & -99/76 & -1683/19 \end{bmatrix} \quad \begin{bmatrix} 1 & 0 & 120 \\ 0 & 1 & 68 \end{bmatrix}$$

120 of High-Tech, 68 of Big Burger

41. Let x = number at plant A, y = number at plant B.

$$x + y = 900$$
$$3.6x + 1260 + 3.3y + 2637 = 7056$$

We have the system

$$3.6x + 3.3y = 3159$$
$$x + y = 900$$
$$x = 630, y = 270$$

630 at plant A and 270 at plant B.

43. Let t = 0 be year today and P = population.

$P(-6) = 4600$, $P(0) = 5400$ so the slope of the line is

$$m = \frac{5400 - 4600}{0 - (-6)} = \frac{400}{3}$$

The y-intercept is 5400 so the equation is

$$P(t) = \frac{400}{3} t + 5400$$

$$P(15) = \frac{400}{3} (15) + 5400 = 2000 + 5400 = 7400$$

45.

x	y	x^2	xy
0	4	0	0
2	9	4	18
4	13	16	52
6	18	36	108
12	44	56	178

Solve the system

$$56m + 12b = 178$$
$$12m + 4b = 44$$

$$56m + 12b = 178$$
$$36m + 12b = 132$$
$$20m \qquad = 46$$
$$m = 2.3$$

Solve for b.

$$12(2.3) + 4b = 44$$
$$b = 4.1$$

The regression line is $y = 2.3x + 4.1$

Chapter 3
Linear Programming

Section 3.1

1. (1, -1): $5(1) + 2(-1) = 5 - 2 = 3 \le 17$ yes
    (4, 1) : $5(4) + 2(1) = 20 + 2 \nleq 17$ no
    (3, 1) : $5(3) + 2(1) = 15 + 2 = 17 \le 17$ yes
    (4, 4) : $5(4) + 2(4) = 20 + 8 = 28 \nleq 17$ no
    (2, 3) : $5(2) + 2(3) = 10 + 6 = 16 \le 17$ yes

3. $6x + 8y \le 24$
    Find the x and y intercepts.
    $x = 0$: $8y = 24$ $y = 3$
    $y = 0$: $6x = 24$ $x = 4$
    The point (0, 0) is a solution so the
    half plane below the line gives
    the graph.

    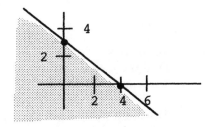

5. $3x - 7y \le 21$
    Find the x and y intercepts
    $x = 0$: $-7y = 21$ $y = -3$
    $y = 0$: $3x = 21$ $x = 7$
    The point (0, 0) is a solution so the
    half plane above the line gives
    the graph.

    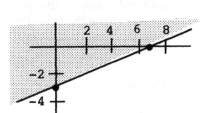

7. $5x + 4y < 20$
    Find the x and y intercepts
    $x = 0$: $4y = 20$ $y = 5$
    $y = 0$: $5x = 20$ $x = 4$
    The point (0, 0) is a solution so the half
    plane below the line gives the graph.
    Since the line is not a part of the
    solution, it is dotted.

    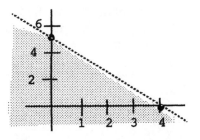

9. $6x + 5y < 30$
Find the x and y intercepts
$x = 0$: $5y = 30$ $y = 6$
$y = 0$: $6x = 30$ $x = 5$
The point $(0, 0)$ is a solution so the half
plane below the line gives the graph.
Since the line is not a part of the
solution, it is dotted.

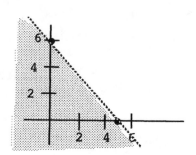

11. $x \le 10$

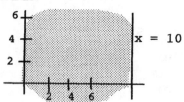

13. $y \ge -3$

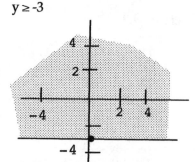

15. $9x - 6y > 30$
Find the x and y intercepts
$x = 0$: $-6y = 30$ $y = -5$
$y = 0$: $9x = 30$ $x = 10/3$
The point $(0, 0)$ is not a solution so the
half plane below the line gives the
graph. Since the line is not a part of
the solution, it is dotted.

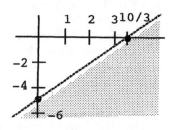

17. $4x - 3y > 12$
Find the x and y intercepts:
$x = 0$: $-3y = 12$ $y = -4$
$y = 0$: $4x = 12$ $x = 3$
The point $(0, 0)$ is not a solution so the
half plane below the line gives the
graph.   The line is not a part of the
solution.

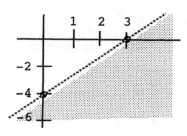

19. $-2x - 5y > 10$
Find the x and y intercepts:
$x = 0$: $-5y = 10$ $y = -2$
$y = 0$: $-2x = 10$ $x = -5$
The point $(0, 0)$ is not a solution so the
half place below the line gives the
graph. The line is not a part of the
solution.

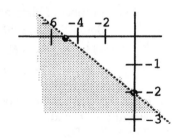

21. Let x = number of air conditioners,
y = number of fans.
$3.2x + 1.8y \le 144$
Find the x and y intercepts
$x = 0$: $1.8y = 144$ $y = 80$
$y = 0$: $3.2x = 144$ $x = 45$
A negative number of items cannot be
produced so $x \ge 0$, $y \ge 0$.

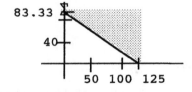

23. Let x = number of members,
y = number of pledges.
(a) $4x + 6y \ge 500$
(b) $x = 0$: $6y = 500$ $y = 83.33$
$y = 0$: $4x = 500$ $x = 125$
A negative number of people
cannot be selected so $x \ge 0$, $y \ge 0$.

25. Let x = number of TV spots, y = number of
newspaper ads.
$900x + 830y \le 75,000$
The number of ads cannot be negative so
$x \ge 0$, $y \ge 0$.

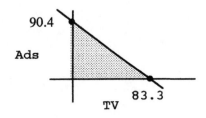

27. Let x = number of acres of strawberries, y = number of acres of tomatoes.
$9x + 6y \le 750$

29. Let x = number of days the Glen Echo plant operates, y = number of days the
Speegleville Road plant operates.
(a) $200x + 300y \ge 2400$ (paperbacks)
(b) $300x + 200y \ge 2100$ (hardbacks)

31.   Let x_1 = number of Ham and Egg, x_2 = number of Roast Beef

$$360x_1 + 300x_2 \leq 2000 \qquad \text{(Calories)}$$
$$14x_1 + 5x_2 \leq 65 \qquad \text{(Total Fat)}$$
$$4x_1 + 2x_2 \leq 20 \qquad \text{(Saturated Fat)}$$

33.   Let x represent the number of servings of milk. Let y represent the number of servings of bread.

$$12x + 15y \geq 50.$$

Section 3.2

1. $x + y = 3$ Intercepts: (3,0), (0,3)
$2x - y = -2$ Intercepts: (-1,0), (0,2)

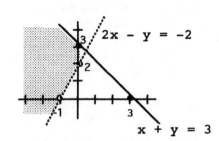

3. $x = 3$
$y = 2$
$3x + 2y = 18$ Intercepts: (6,0), (0,9)

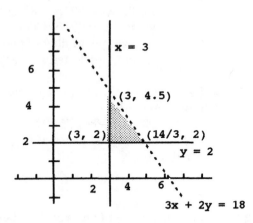

5. $4x + 6y = 18$ Intercepts: (9/2,0), (0,3)
$x + 3y = 6$ Intercepts: (6,0), (0,2)
$x \geq 0$

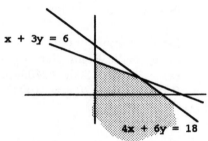

7. $2x + y = 60$ Intercepts: (30,0), (0,60)
       $2x + 3y = 120$ Intercepts: (60,0), (0,40)

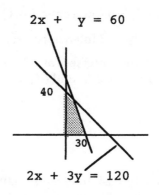

9. $x + y = 4$ Intercepts: (4,0), (0,4)
       $2x - 3y = 8$ Intercepts: (4,0), (0,-8/3)

Corner: (4,0)

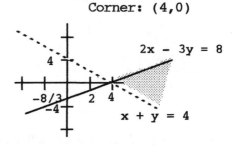

11. $-3x + 10y = 15$ Intercepts: (-5, 0) and (0, 1.5)
       $3x + 5y = 15$ Intercepts: (5, 0) and (0, 3)

       The lines intersect at $(\frac{5}{3}, 2)$.

       Corner (5/3, 2)

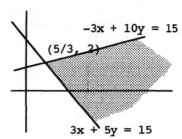

13. $x \le 0$
       $y \ge 2$
       The y-axis and the line $y = 2$ serve as
       boundaries of the feasible region. They
       intersect at (0, 2).
       Corner (0, 2)

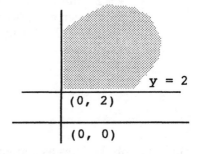

15. $-x + 2y = 6$ Intercepts: $(-6, 0)$ and $(0, 3)$

$3x + 2y = -2$ Intercepts: $(-\frac{2}{3}, 0)$ and $(0, -1)$

The lines intersect at $(-2, 2)$.
Corners: $(-2, 2)$ and $(-6, 0)$

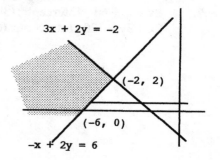

17. $3x + 4y = 24$ Intercepts: $(8,0)$, $(0,6)$
$4x + 5y = 20$ Intercepts: $(5,0)$, $(0,4)$
No feasible region

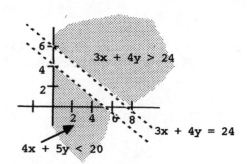

19. $x + y = 2$ Intercepts: $(2,0)$, $(0,2)$
$2x + y = 6$ Intercepts: $(3,0)$, $(0,6)$
No feasible region

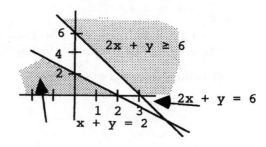

21. $x + y = 1$ Intercepts: $(1,0)$, $(0,1)$
$-x + y = 2$ Intercepts: $(-2,0)$, $(0,2)$
$5x - y = 4$ Intercepts: $(4/5,0)$, $(0,-4)$
Corners: $A(-1/2, 3/2)$, $B(3/2, 7/2)$

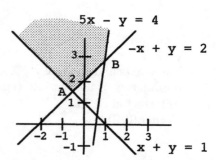

23. Graph the lines as shown.
The lines $-x + y = 5$ and $6x + 7y = 48$ intersect at $(1, 6)$.
The lines $-x + y = 5$ and $5x + 6y = -14$ intersect at $(-4, 1)$.

Section 3.2 Systems of Inequalities

The lines $5x + 6y = -14$ and $-4x + 6y = -32$ intersect at $(2, -4)$.
The lines $-4x + 6y = -32$ and $6x + 7y = 48$ intersect at $(8, 0)$.
Corners: $(-4, 1)$, $(1, 6)$, $(8, 0)$, $(2, -4)$

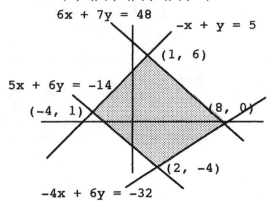

25. $3x + 4y = 12$ Intercepts: $(4, 0)$, $(0, 3)$
$5x + 3y = 15$ Intercepts: $(3, 0)$, $(0, 5)$
Bounded feasible region

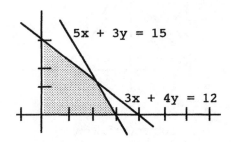

27. $x + y = 6$ $\quad$ Intercepts: $(6, 0)$, $(0, 6)$
$3x + 6y = 24$ $\quad$ Intercepts: $(8, 0)$, $(0, 4)$
Unbounded feasible region

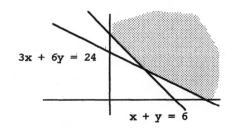

29. $-2x + y = 2$ Intercepts: $(-1,0)$, $(0,2)$
$3x + y = 3$ Intercepts: $(1,0)$, $(0,3)$
$-x + y = -4$ Intercepts: $(4,0)$, $(0,-4)$
$x + y = -3$ Intercepts: $(-3,0)$, $(0,-3)$
Corners: $A(1/2, -7/2)$,
$(B(-5/3, -4/3)$, $C(1/5, 12/5)$,
$D(7/4, -9/4)$

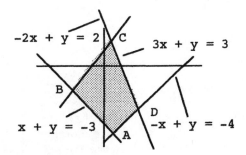

81

31. Let x = ounces of High Fiber,
y = ounces of Corn Bits.
$$0.25x + 0.02y \geq 0.40$$
$$0.04x + 0.10y \geq 0.25$$
$$x \geq 0, y \geq 0$$

33. Let x = number of student tickets,
y = amount of Corn Bits.
$$90x + 120y \geq 600$$
$$160x + 200y < 800$$
$$x \geq 0, y \geq 0$$

35. Let x = number correct,
y = number incorrect.
$$x + y \geq 60$$
$$4x - y \geq 200$$
$$x \geq 0, y \geq 0$$

37. Let x = number of balcony tickets,
y = number of main floor tickets
Total tickets $x + y \geq 3000$
Main floor $y \geq 1200$
Sales $15x + 25y \geq 60{,}000$

39.

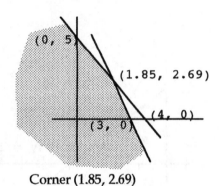

Corner (1.85, 2.69)

41.

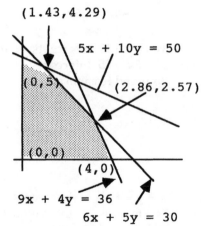

Corners: (0,0), (0, 5) ,(1.43, 4.29),
(2.86, 2.57), (4, 0)

Section 3.2 Systems of Inequalities

43.

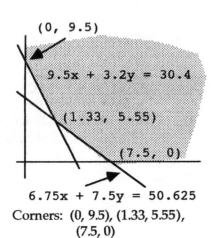

(0, 9.5)

9.5x + 3.2y = 30.4

(1.33, 5.55)

(7.5, 0)

6.75x + 7.5y = 50.625

Corners: (0, 9.5), (1.33, 5.55), (7.5, 0)

45. Corners: (0, 0), (0, 5.47), (1.99, 4.55) ,(4.49, 1.22) , (5.00, 0)

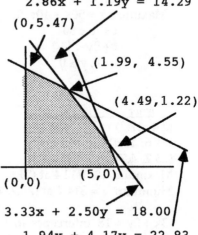

2.86x + 1.19y = 14.29

(0, 5.47)

(1.99, 4.55)

(4.49, 1.22)

(0, 0) (5, 0)

3.33x + 2.50y = 18.00

1.94x + 4.17y = 22.83

TI-83 Exercises

1. Corners: (0, 0), (8, 2), (3, 6), (0, 7), (9.33, 0)

3. Corners: (0, 0), (0, 11), (10, 9), (21, 4), (27, 0)

Excel Exercises

1. The intersection is (7, 5)

3. The first two intersect at (10, 12). The first and third intersect at (14, 4). The second and third intersect at (5, 10)

5. The lines intersect at (2.3, 6.7)

Section 3.3

1. Let x = number of style A,
y = number of style B.
Maximize $z = 50x + 40y$ Subject to
$$x + y \le 80$$
$$x + 2y \le 110$$
$$x \ge 0, y \ge 0$$

3.

Corner	z
(0, 0)	0
(0, 8)	120
(7, 6)	132
(10, 0)	60

Maximum $z = 132$ at $(7, 6)$
Minimum $z = 0$ at $(0, 0)$

5.

Corner	z
(2.4, 17.2)	432
(21.6, 15.4)	542.4
(25.2, 4.7)	314.4

Maximum $z = 542.4$ at $(21.6, 15.4)$
Minimum $z = 314.4$ at $(25.2, 4.7)$

7. $3x + 2y = 18$ Intercepts: $(6, 0), (0, 9)$
$3x + y = 15$ Intercepts: $(5, 0), (0, 15)$

Corner	$z = 20x + 12y$
A(0, 0)	0
B(0, 9)	108
C(4, 3)	116
D(5, 0)	100

Maximum is 116 at $(4, 3)$

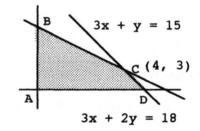

9. $5x + y = 35$ Intercepts: $(7, 0), (0, 35)$
$3x + y = 27$ Intercepts: $(9, 0), (0, 27)$

Corner	$z = 9x + 2y$
A(0, 0)	0
B(0, 27)	54
C(4, 15)	66
D(7, 0)	63

Maximum is 66 at $(4, 15)$

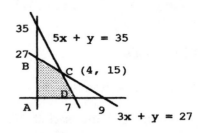

11. $4x + y = 40$ Intercepts: $(10,0), (0,40)$
$4x + 3y = 64$ Intercepts: $(16,0), (0,64/3)$

Corner	$z = 2x + 3y$
A(0,40)	120
B(7,12)	50
C(16,0)	32

Minimum is 32 at $(16, 0)$

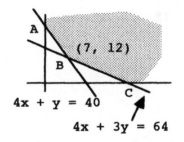

13. $10x + 11y = 330$ Intercepts: $(33, 0), (0, 30)$
$4x + 6y = 156$ Intercepts: $(39, 0), (0, 26)$

	$z = 9x + 13y$
A(0, 0)	0
B(0, 26)	338
C(33/2, 15)	343.5
D(33, 0)	297

Maximum is 343.5 at (16.5, 15)

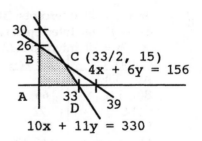

15. $-x + 2y = 40$ Intercepts: $(-40, 0), (0, 20)$
$x + 4y = 54$ Intercepts: $(54, 0), (0, 27/2)$
$3x + y = 63$ Intercepts: $(21, 0), (0, 63)$

Corner	$z = 20x + 30y$
A(0, 0)	0
B(0, 27/2)	405
C(18, 9)	630
D(21, 0)	420

Maximum is 630 at (18, 9)

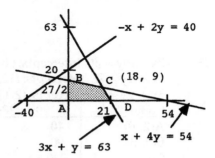

17. $x + 5y = 250$ Intercepts: $(250, 0), (0, 50)$
$2x + 5y = 300$ Intercepts: $(150, 0), (0, 60)$
$x = 75$

Corner	$z = 320x + 140y$
A(0, 0)	0
B(0, 50)	7,000
C(50, 40)	21,600
D(75, 30)	28,200
E(75, 0)	24,000

Maximum is 28,200 at (75, 30)

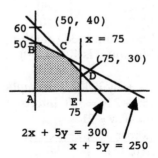

19. $6x + y = 52$ Intercepts: $(26/3, 0), (0, 52)$
$2x + y = 20$ Intercepts: $(10, 0), (0, 20)$
$x + 4y = 24$ Intercepts: $(24, 0), (0, 6)$

Corner	$z = 5x + 3y$
A(0, 52)	156
B(8, 4)	52
C(24, 0)	120

Minimum is 52 at (8, 4)

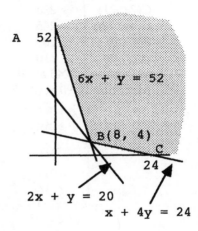

21. $3x + y = 30$ Intercepts: $(10, 0)$, $(0, 30)$
$4x + 3y = 60$ Intercepts: $(15, 0)$, $(0, 20)$
$x + 2y = 20$ Intercepts: $(20, 0)$, $(0, 10)$

Corner	$z = 5x + 3y$
A(0, 30)	90
B(6, 12)	66
C(12, 4)	72
D(20, 0)	100

Minimum is 66 at (6, 12)

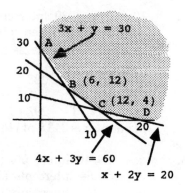

23. $2x + 10y = 80$ Intercepts: $(40, 0)$, $(0, 8)$
$6x + 2y = 72$ Intercepts: $(12, 0)$, $(0, 36)$
$3x + 2y = 6$ Intercepts: $(2, 0)$, $(0, 3)$

Corner	$z = 20x + 30y$
A(2, 0)	40
B(0, 3)	90
C(0, 8)	240
D(10, 6)	380
E(12, 0)	240

Maximum is 380 at (10, 6)
Minimum is 40 at (2, 0)

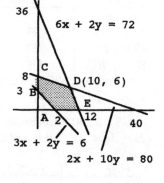

25. $6x + 8y = 300$ Intercepts: $(50, 0)$, $(0, 37.5)$
$15x + 22y = 330$ Intercepts: $(22, 0)$, $(0, 15)$
$x = 40, y = 21$

Corner	$z = 5x + 6y$
A(0, 15)	90
B(0, 21)	126
C(22, 21)	236
D(40, 7.5)	245
E(40, 0)	200
F(22 ,0)	110

Maximum is 245 at (40, 7.5).
Minimum is 90 at (0, 15)

27. $5x + 3y = 30$ Intercepts: $(6, 0)$, $(0, 10)$
$5x + y = 20$ Intercepts: $(4, 0)$, $(0, 20)$

Corners	$z = 15x + 9y$
A(0, 0)	0
B(0, 10)	90
C(3, 5)	90
D(4, 0)	60

Maximum is 90 at any point on line
$5x + 3y = 30$ from $(0, 10)$ to $(3, 5)$

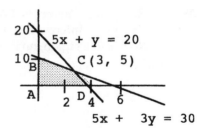

29. $3x + 2y = 60$ Intercepts: $(20, 0)$, $(0, 30)$
$10x + 3y = 180$ Intercepts: $(18, 0)$, $(0, 60)$
$y = 24$

	$z = 9x + 6y$
A(4, 24)	180
B(10.8, 24)	241.2
$C(\frac{180}{11}, \frac{60}{11})$	180

Minimum is 180 at $(4, 24)$, $(\frac{180}{11}, \frac{60}{11})$

and points between

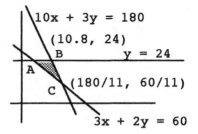

31. $3x + y = 150$ Intercepts: $(50, 0)$, $(0, 150)$
$x + 2y = 100$ Intercepts: $(100, 0)$, $(0, 50)$
$y = 20$ Intercepts: $(0, 20)$

Corner	$z = 5x + 10y$
A(0, 150)	1500
B(40, 30)	500
C(60, 20)	500

Minimum is 500 at $(40, 30)$, $(60, 20)$ and
points between

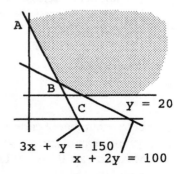

33. $2x - 3y = 10$ $(5, 0)$, $(0, -10/3)$
$x = 8$ $(8, 0)$
Unbounded feasible region,
no maximum

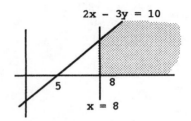

35. $-3x + y = 4$ Intercepts: $(-4/3, 0)$, $(0, 4)$
$-2x - y = 1$ Intercepts: $(-1/2, 0)$, $(0, -1)$
No feasible region, no minimum

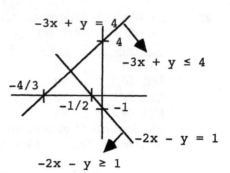

37. $3x - y = 8$ Intercepts: $(8/3, 0)$, $(0, -8)$
$x - 2y = 5$ Intercepts: $(5, 0)$, $(0, -5/2)$
No feasible region, no maximum

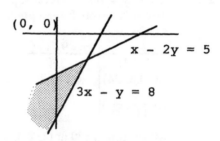

39. Let x = number of standard VCR's,
y = number of deluxe VCR's.
Maximize $z = 39x + 26y$ subject to
$8x + 9y \leq 2200$
$115x + 136y \leq 18,000$
$x \geq 35, y \geq 0$

41. Let x = number shipped to A,
y = number shipped to B.
Minimize $z = 13x + 11y$
subject to
$x + y \geq 250$
$x + 30 \leq 140$
$y + 18 \leq 165$
$x \geq 0, y \geq 0$

43. Let x = number of cartons of regular,
y = number of cartons of diet.
Maximize $z = 0.15x + 0.17y$ subject to
$x + 1.20y \leq 5400$
$x + y \leq 5000$
$x \geq 0, y \geq 0$

45. Let x = number of desk lamps
y = number of floor lamps.
Maximize $z = 2.65x + 3.15y$
subject to
$0.8x + y \leq 1200$ (1500, 0) (0, 1200)
$4x + 3y \leq 4200$ (1050, 0) (0, 1400)
$x \geq 0, y \geq 0$

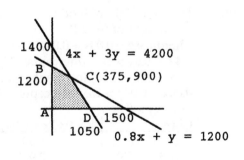

Corner	$z = 2.65x + 3.15y$
A (0, 0)	0
B (0, 1200)	3780
C (375, 900)	3838.75
D (1050, 0)	2782.50

Maximum profit is $3828.75 at 375 desk lamps, 900 floor lamps.

47. Let x = number of sandwiches and
y = number of bowls of soup.
$400x + 200y \leq 2000$ (Calories)
$12x + 8.5y \leq 65$ (Fat)
$z = 9x + 5y$ (Fiber)
We want to solve
Maximize $z = 9x + 5y$
subject to
$400x + 200y \leq 2000$
$12x + 8.5y \leq 65$
$x \geq 0, y \geq 0$

Corner	$z = 9x + 5y$
(0, 0)	0
(0, 65/8.5)	38.24
(4, 2)	46
(5, 0)	45

Maximum fiber is 46 grams with
4 sandwiches and 2 bowls of soup.

49.   Let x = weight of food I,
y = weight of food II.
Minimize z = 0.03x + 0.04y   subject to
0.4x + 0.6y ≥ 10   Intercepts: (25, 0), (0, 50/3)
0.5x + 0.2y ≥ 7.5 Intercepts: (15, 0), (0, 37.5)
0.06x + 0.04y ≥ 1.2 Intercepts: (20, 0), (0, 30)
x ≥ 0, y ≥ 0

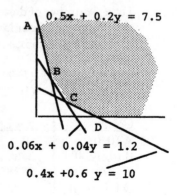

Corner	z = 0.03x + 0.04y
A (0, 37.5)	1.50
B (7.5, 18.75)	0.975
C (16, 6)	0.72
D (25, 0)	0.75

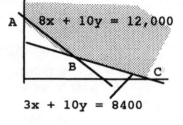

Minimum cost is $0.72 using 16 g of food I and
6 g of food II.

51.   Let x = number of standard gears, y = number
of heavy duty gears.
Minimize z = 15x + 22y subject to
8x + 10y ≥ 12,000 Intercepts: (1500, 0), (0, 1200)
3x + 10y ≥ 8,400 Intercepts: (2800, 0), (0, 840)
x ≥ 0, y ≥ 0

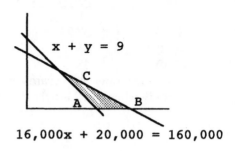

Corner	z =15x + 22y
A (0, 1200)	26,400
B (720, 624)	24,528
C (2800, 0)	42,000

Minimum cost is $24,528 producing 720 standard
gears and 624 heavy duty gears.

53.   Let x = number of SE vans,
y = number of LE vans
Minimize z = 2700x + 2400y
subject to
  x + y ≥ 9   Intercepts: (9, 0), (0, 9)
16,000x + 20,000y ≤ 160,000
        Intercepts: (10, 0), (0, 8)
x ≥ 0, y ≥ 0

Corner	z = 2700x + 2400y
A (9, 0)	24,300
B (10, 0)	27,000
C (5, 4)	23,100

Minimum maintenance
cost = $23,100 with 5 SE vans and
4 LE vans

55. **(a)** Let x = number cartons of regular,
y = number cartons of diet.
Maximize $z = 0.15x + 0.17y$ subject to
$$x + 1.2y \le 5400$$
$$x + y \le 5000$$
$$x \ge 0, y \ge 0$$

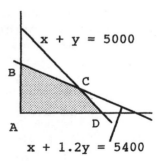

Corner	$z = 0.15x + 0.17y$
A (0, 0)	0
B (0, 4500)	765
C (3000, 2000)	790
D (5000, 0)	750

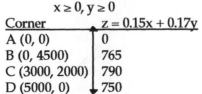

Maximum profit is $790 with 3,000
cartons of regular and 2,000 cartons
of diet.

(b) **(i)** The objective function becomes $z = 0.20x + 0.17y$, the
constraints and corners remain the same.

Corners	$z = 0.20x + 0.17y$
A(0,0)	0
B(0,4500)	765
C(3000,2000)	940
D(5000,0)	1000

The optimal solution changes to the corner (5000, 0) and the
maximum profit increases to $1000.

(ii) The objective function becomes $z = 0.15x + 0.19y$, the
constraints and corners remain the same.

Corners	$0.15x + 0.19y$
A(0,0)	0
B(0,4500)	855
C(3000,2000)	830
D(5000,0)	750

The optimal solution changes to the corner (0, 4500) and the
maximum profit increases to $855.

(iii) The objective function becomes $z = 0.15x + 0.16y$, the
constraints and corners remain the same.

Corners	$0.15x + 0.16y$
A(0,0)	0
B(0,4500)	720
C(3000,2000)	770
D(5000,0)	750

The optimal solution remains at the corner (3000, 2000) but
the maximum profit decreases to $770.

57. Let x = number of square feet of type A glass,
y = number of square feet of type B glass.

(a) Minimize $z = x + 0.25y$ subject to
$x + y \geq 4000$ Intercepts: (4000, 0), (0, 4000)
$0.8x + 1.2y \leq 4500$ Intercepts: (5625, 0), (0, 3750)
$x \geq 0, y \geq 0$

Corner	$z = x + 0.25y$
A(4000, 0)	4000
B(750, 3250)	1562.5
C(5625, 0)	5625

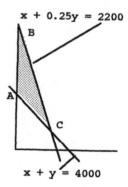

Minimum conductance is 1562.5 BTU using 750 square feet of type A glass and 3250 square feet of type B.

(b) Minimize $z = 0.8x + 1.2y$ subject to
$x + 0.25y \leq 2200$ Intercepts: (2200, 0), (0,8800)
$x + y \geq 4000$ Intercepts: (4000, 0), (0, 4000)

Corner	$z = 0.8x + 1.2y$
A(0, 4000)	4,800
B(0, 8800)	10,560
C(1600, 2400)	4,160

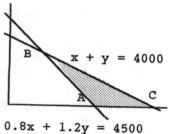

Minimum cost is $4,160 using 1600 square feet of type A glass and 2400 square feet of type B glass.

59. Let x = number of days the Glen Echo plant operates,
y = number of days the Speegleville Road plant operates.
Minimize $z = x + y$ subject to
$200x + 300y \geq 2400$ Intercepts: (12, 0), (0, 8)
$300x + 200y \geq 2100$ Intercepts: (7, 0), (0,10.5)
$x \geq 0, y \geq 0$

Corner	$z = x + y$
A(0, 10.5)	10.5
B(3, 6)	9
C(12, 0)	12

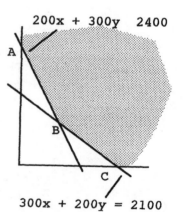

Minimum number of days is 9 with 3 at Glen Echo and 6 at Speegleville Road.

Section 3.3 Linear Programming

61.   Let x = number of computers from Supplier A to Raleigh.
Let y = number of computers from Supplier A to Greensboro.
Then 165 – x = number of computers from Supplier B to Raleigh.
190 – y = number of computers from Supplier B to Greensboro.
We summarize this information in the following table.

	Raleigh	Greensboro	
A cost (ea.)	$35	$30	200
Number	x	y	available
B cost (ea.)	$45	$50	230
Number	165–x	190–y	available
	165 needed	190 needed	

The total shipping cost
$$C = 35x + 30y + 45(165 - x) + 50(190 - y) = -10x - 20y + 16925$$
The constraints are
$$x + y \leq 200 \text{ (available from A)}$$
$$165 - x + 190 - y \leq 230 \text{ (available from B) which reduces to } 125 \leq x + y$$
$$x \geq 0, y \geq 0, x \leq 165, y \leq 190$$
The feasible region and corner points for this problem are the following
Compute C = 16925 – 10x – 20y at each corner.

Corner	C = 16,925 - 10x - 20y
(0, 125)	14425
(0, 190)	13125
(10, 190)	13025
(165, 35)	14575
(165, 0)	15275
(125, 0)	15675

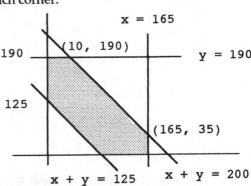

The minimum cost is $13,025 and occurs when x = 10 and y = 190. Thus, Supplier A should ship 10 to Raleigh and 190 to Greensboro. Supplier B should ship 155 to Raleigh and none to Greensboro.

63.   Let x = number of lamps shipped from A to Emporia.
Let y = number of lamps shipped from A to Ardmore.
25 – x = number of lamps shipped from B to Emporia
20 – y = number of lamps shipped from B to Ardmore
We use the following table to summarize this information.

		To	
	Emporia	Ardmore	
A cost (ea.)	$5	$6	10
Number	x	y	available
B cost (ea.)	$7	$4	40
Number	25–x	20–y	available
	25 needed	20 needed	

The total shipping costs are

$$C = 5x + 6y + 7(25 - x) + 4(20 - y)$$
$$= -2x + 2y + 255$$

The constraints are

$x + y \le 10$ (available from A)

$25 - x + 20 - y \le 40$ (available from B) which reduces to

$5 \le x + y$

$x \ge 0, y \ge 0, x \le 25, y \le 20$

Here is the feasible region and its corners.

Calculate $C = 255 - 2x + 2y$ at the corners.

Corner	$C = 255 - 2x + 2y$
(0, 5)	265
(0, 10)	275
(10, 0)	235
(5, 0)	245

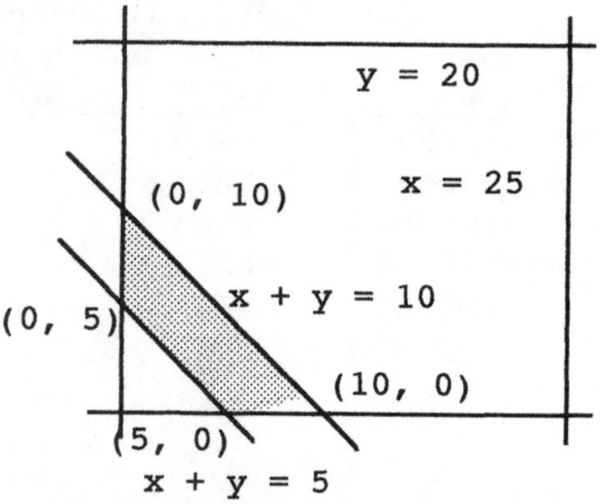

The minimum cost is $235 and it occurs when x = 10 and y = 0. Thus, Supplier A should ship 10 to Emporia and none to Ardmore. Supplier B should ship 15 to Emporia and 20 to Ardmore.

65. In a maximization problem the profit line is moved away from the origin until a corner point is the only point it touches in the feasible region. To find the optimal integer solution move the profit line outward until it touches the last point in the feasible region with integer coordinates. In this case that is the point (7, 0) so the optimal profit occurs when x = 7 and y = 0. If neither x nor y can be zero, the last point is (6, 1).

67. Corner B

69. Corner C

Corner | C = 16,925 - 10x - 20y

Corner	
(0, 125)	14425
(0, 190)	13125
(10, 190)	13025
(165, 35)	14575
(165, 0)	15275
(125, 0)	15675

The minimum cost is \$13,025 and occurs when x = 10 and y = 190. Thus, Supplier A should ship 10 to Raleigh and 190 to Greensboro. Supplier B should ship 155 to Raleigh and none to Greensboro.

63. Let x = number of lamps shipped from A to Emporia.
Let y = number of lamps shipped from A to Ardmore.
$\quad$ 25 – x = number of lamps shipped from B to Emporia
$\quad$ 20 – y = number of lamps shipped from B to Ardmore
$\quad$ We use the following table to summarize this information.

		To Emporia	Ardmore	
A	cost (ea.)	\$5	\$6	10
	Number	x	y	available
B	cost (ea.)	\$7	\$4	40
	Number	25–x	20–y	available
		25 needed	20 needed	

The total shipping costs are
$$C = 5x + 6y + 7(25 – x) + 4(20 – y)$$
$$= -2x + 2y + 255$$
The constraints are
$\quad$ x + y = 10 (available from A)
$\quad$ 25 – x + 20 - y = 40 (available from B) which reduces to
$\quad$ 5 = x + y
$\quad$ x = 0, y = 0, x = 25, y = 20
Here is the feasible region and its corners.
Calculate C = 255 – 2x + 2y at the corners.

Corner	C = 255 - 2x + 2y
(0, 5)	265
(0, 10)	275
(10, 0)	235
(5, 0)	245

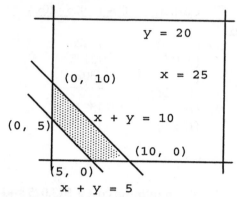

The minimum cost is $235 and it occurs when x = 10 and y = 0. Thus, Supplier A should ship 10 to Emporia and none to Ardmore. Supplier B should ship 15 to Emporia and 20 to Ardmore.

65.   In a maximization problem the profit line is moved away from the origin until a corner point is the only point it touches in the feasible region. To find the optimal integer solution move the profit line outward until it touches the last point in the feasible region with integer coordinates. In this case that is the point (7, 0) so the optimal profit occurs when x = 7 and y = 0. If neither x nor y can be zero, the last point is (6, 1).

67.   Corner B   69.   Corner C

9.   The following chart summarizes the given information.

Plantation

	P-1	P-2	P-3	
Plant A	Cost = 50 No. = x_1	Cost = 65 No. = x_2	Cost = 58 No. = x_3	Process at least 540
Plant B	Cost = 40 No. = x_4	Cost = 55 No. = x_5	Cost = 69 No. = x_6	Process at least 450
	Produces 250	Produces 275	Produces 310	

The linear programming problem is the following.
Minimize shipping costs
$z = 50x_1 + 65x_2 + 58x_3 + 40x_4 + 55x_5 + 69x_6$ subject to

$$x_1 + x_2 + x_3 \geq 540 \quad \text{(Plant A)}$$
$$x_4 + x_5 + x_6 \geq 450 \quad \text{(Plant B)}$$
$$x_1 + x_4 \geq 250 \quad \text{(P-1 production)}$$
$$x_2 + x_5 \geq 275 \quad \text{(P-2 production)}$$
$$x_3 + x_6 \geq 310 \quad \text{(P-3 production)}$$
$$x_1 \geq 0, x_2 \geq 0, x_3 \geq 0, x_4 \geq 0, x_5 \geq 0, x_6 \geq 0$$

11.   Let x_1 = number of days Red Mountain operates, x_2 = number of days Cahaba operates, x_3 = number of days Clear Creek operates
The objective function is the cost function.
Minimize $z = 8000x_1 + 14000x_2 + 12000x_3$ subject to

$$105x_1 + 295x_2 + 270x_3 \geq 1800 \quad \text{(amount of low-grade)}$$
$$90x_1 + 200x_2 + 85x_3 \geq 1350 \quad \text{(amount of high grade)}$$
$$x_1 \geq 0, x_2 \geq 0, x_3 \geq 0$$

13.   Let x_1 = number of Mini Packets, x_2 = number of Mid Packets, x_3 = number of Maxi Packets
Minimize costs
$$z = 19x_1 + 30x_2 + 45x_3 \quad \text{subject to}$$
$$8x_1 + 7x_2 + 6x_3 \geq 260 \quad \text{(Number of "A" tickets)}$$
$$2x_1 + 7x_2 + 14x_3 \geq 175 \quad \text{(Number of "B" tickets)}$$
$$x_1 \geq 0, x_2 \geq 0, x_3 \geq 0$$

15. We set up the linear programming problem in the following way.
Let x_1 = number of servings of milk, x_2 = number of servings of vegetables, x_3 = number of servings of fruit, x_4 = number of servings of bread, x_5 = number of servings of meat.
Maximize carbohydrates $z = 12x_1 + 7x_2 + 11x_3 + 15x_4$ subject to

$$8x_1 + 2x_2 + 2x_4 + 7x_5 \leq 35 \quad \text{(amount of protein)}$$
$$10x_1 + x_4 + 5x_5 \leq 40 \quad \text{(amount of fat)}$$
$$x_1 \geq 0, x_2 \geq 0, x_3 \geq 0, x_4 \geq 0, x_5 \geq 0$$

17. Let x_1 = percent invested in A bonds, x_2 = percent invested in AA bonds, x_3 = percent invested in AAA bonds
Maximize return

$$z = 0.072x_1 + 0.068x_2 + 0.065x_3 \quad \text{subject to}$$
$$x_1 + x_2 + x_3 \leq 100 \quad \text{(Investment Total)}$$
$$x_1 + x_2 \leq 65 \quad \text{(Invested in A and AA)}$$
$$x_2 + x_3 \geq 50 \quad \text{(Invested in AA and AAA)}$$
$$x_1 \geq 0, x_2 \geq 0, x_3 \geq 0$$

19. Let x_1 = number of sheets of cutting plan 1, x_2 = number of sheets of cutting plan 2, etc. The amount to be minimized is

$$\text{waste} = x_2 + 3x_3 + 2x_4 + 12x_5 + 14x_7 \quad \text{subject too}$$
$$3x_2 + 9x_3 + 6x_4 + 6x_6 \geq 165 \quad \text{(no. of 15 inch doors)}$$
$$9x_1 + 6x_2 + 3x_4 + 3x_7 \geq 200 \quad \text{(no. of 16 inch doors)}$$
$$6x_5 + 3x_6 + 3x_7 \geq 85 \quad \text{(no. of 18 inch doors)}$$
$$x_1 \geq 0, x_2 \geq 0, x_3 \geq 0, x_4 \geq 0, x_5 \geq 0, x_6 \geq 0$$

Chapter 3 Review

1. **(a)** $5x + 7y < 70$
Intercepts: (14, 0), (0, 10)

(b) $2x - 3y > 18$
Intercepts: (9, 0), (0, -6)

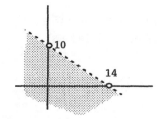

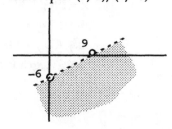

(c) $x + 9y \leq 21$
Intercepts: $(21, 0)$, $(0, 7/3)$

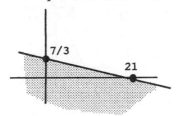

(d) $-2x + 12y \geq 26$
Intercepts: $(-13, 0)$, $(0, 13/6)$

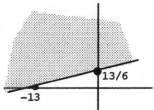

(e) $y \geq -6$

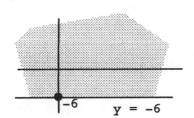

(f) $x \leq 3$

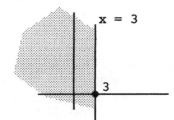

3. $x - 3y = 6$ Intercepts: $(6, 0)$, $(0, -2)$
     $x - y = 4$ Intercepts: $(4, 0)$, $(0, -4)$
     $y = -5$ Intercepts: $(0, -5)$
         Corners: A$(-1, -5)$
                  B$(-9, -5)$
                  C$(3, -1)$

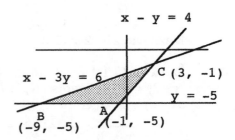

5. $-3x + 4y = 20$ Intercepts: $(-20/3, 0)$, $(0, 5)$
     $x + y = -2$ Intercepts: $(-2, 0)$, $(0, -2)$
     $8x + y = 40$ Intercepts: $(5, 0)$, $(0, 40)$
           Corners: A$(-4, 2)$
                    B$(4, 8)$
                    C$(5, 0)$
                    D$(-2, 0)$

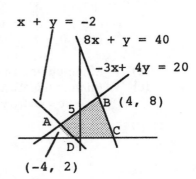

7. $x - 2y = 0$ Intercepts: (0, 0), (2, 1)
$-2x + y = 2$ Intercepts: (-1, 0), (0, 2)
$x = 2$ Intercepts: (2, 0)
$y = 2$ Intercepts: (0, 2)
Corners: A(-4/3, -2/3)
B(0, 2)
C(2, 2)
D(2, 1)

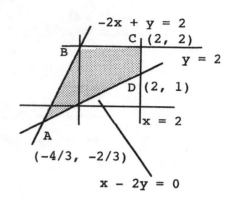

9. $3x + 2y = 12$ Intercepts: (4, 0), (0, 6)
$x + y = 5$ Intercepts: (5, 0), (0, 5)

	$5x + 4y$
A(0, 0)	0
B(0, 5)	20
C(2, 3)	22
D(4, 0)	20

Maximum z is 22 at (2, 3)

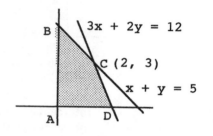

11. $3x + 2y = 18$ Intercepts: (6, 0), (0, 9)
$x + 2y = 10$ Intercepts: (10, 0), (0, 5)
$5x + 6y = 46$ Intercepts: (9.2, 0), (0, 23/3)

	$5x + 4y$	$10x + 12y$
A(0, 9)	36	108
B(2, 6)	34	92
C(8, 1)	44	92
D(10, 0)	50	100

(a)   Minimum is 34 at (2, 6)
(b)   Minimum is 92, at any point on line
$5x + 6y = 46$ from (2, 6) to (8, 1)

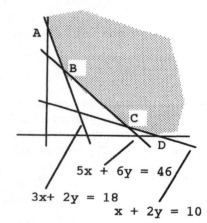

100

13. $2x + y = 90$ Intercepts: $(45, 0)$, $(0, 90)$
$x + 2y = 80$ Intercepts: $(80, 0)$, $(0, 40)$
$x + y = 50$ Intercepts: $(50, 0)$, $(0, 50)$

	$4x + 7y$
A(0, 0)	0
B(0, 40)	280
C(20, 30)	290
D(40, 10)	230
E(45, 0)	180

Maximum z is 290 at $(20, 30)$

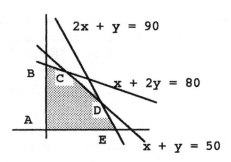

15. Let x = number of hours Line A is used, y = number of hours Line B is used.

   (a) $65x + 105y \le 700$ **(b)**
            $x \ge 0, y \ge 0$

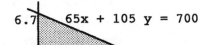

17. Let x = number of adult tickets, y = number of children tickets.
            $x + y \le 275$
            $4.5x + 3y \ge 1100$
            $x \ge 0, y \ge 0$

19. Let x = number of standard bars, y = number of premium bars.
Maximize $r = 90x + 100y$ subject to
$0.9(100)x + 0.8(100)y \le 80,000$ Intercepts: $(888.9, 0)$, $(0, 1000)$
$0.1(100)x + 0.2(100)y \le 12,000$ Intercepts: $(1200, 0)$, $(0, 600)$
$x \ge 0, y \ge 0$

Corners	$r = 90x + 100y$
A(0, 0)	0
B(0, 600)	60,000
C(640, 280)	85,600
D(888.9, 0)	80,000

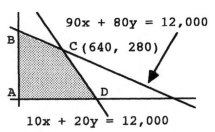

Maximum revenue is $85,600 when 640 bars of standard and 280 of premium are produced.

21. Let x_1 = number of A-teams, x_2 = number of B-teams, x_3 = number of C-teams
Maximize number of inoculations
$$z = 175x_1 + 110x_2 + 85x_3 \text{ subject to}$$
$$x_1 + x_2 + x_3 \leq 75 \text{ (Number of doctors)}$$
$$3x_1 + 2x_2 + x_3 \leq 200 \text{ (Number of nurses)}$$
$$x_1 \geq 0, x_2 \geq 0, x_3 \geq 0$$

23. Let x_1 = number of Type I pattern, x_2 = number of Type II pattern, x_3 = number of Type III pattern, x_4 = number of Type IV pattern
Maximize profit $z = 48x_1 + 45x_2 + 55x_3 + 65x_4$ subject to
$$40x_1 + 25x_2 + 30x_3 + 45x_4 \leq 1250 \text{ (Number tulips)}$$
$$25x_1 + 50x_2 + 40x_3 + 45x_4 \leq 1600 \text{ (Number daffodils)}$$
$$6x_1 + 4x_2 + 8x_3 + 2x_4 \leq 195 \text{ (Number boxwood)}$$
$$x_1 \geq 0, x_2 \geq 0, x_3 \geq 0, x_4 \geq 0$$

Chapter 4
Linear Programming: The Simplex Method

Section 4.1

1.
$$2x_1 + 3x_2 + s_1 = 9$$
$$x_1 + 5x_2 + s_2 = 16$$

3.
$$x_1 + 7x_2 - 4x_3 + s_1 = 150$$
$$5x_1 + 9x_2 + 2x_3 + s_2 = 435$$

5.
$$2x_1 + 6x_2 + s_1 = 9$$
$$x_1 - 5x_2 + s_2 = 14$$
$$-3x_1 + x_2 + s_3 = 8$$
$$-3x_1 - 7x_2 + z = 0$$

7.
$$6x_1 + 7x_2 + 12x_3 + s_1 = 50$$
$$4x_1 + 18x_2 + 9x_3 + s_2 = 85$$
$$x_1 - 2x_2 + 14x_3 + s_3 = 66$$
$$-420x_1 - 260x_2 - 50x_3 + z = 0$$

9.
$$\begin{bmatrix} 4 & 5 & 1 & 0 & 0 & | & 10 \\ 3 & 1 & 0 & 1 & 0 & | & 25 \\ -3 & -17 & 0 & 0 & 1 & | & 0 \end{bmatrix}$$

11.
$$\begin{bmatrix} 16 & -4 & 9 & 1 & 0 & 0 & 0 & | & 128 \\ 8 & 13 & 22 & 0 & 1 & 0 & 0 & | & 144 \\ 5 & 6 & -15 & 0 & 0 & 1 & 0 & | & 225 \\ -20 & -45 & -40 & 0 & 0 & 0 & 1 & | & 0 \end{bmatrix}$$

13. Let x_1 = number of cartons of screwdrivers,
x_2 = number of cartons of chisels
x_3 = number of cartons of putty knives.
Maximize $z = 5x_1 + 6x_2 + 5x_3$ subject to
$$3x_1 + 4x_2 + 5x_3 \le 2200$$
$$15x_1 + 12x_2 + 11x_3 \le 8500$$
$$x_1 \ge 0, x_2 \ge 0, x_3 \ge 0$$

$$\begin{bmatrix} 3 & 4 & 5 & 1 & 0 & 0 & | & 2200 \\ 5 & 12 & 11 & 0 & 1 & 0 & | & 8500 \\ -5 & -6 & -5 & 0 & 0 & 1 & | & 0 \end{bmatrix}$$

15. Let x_1 = amount of salad,

x_2 = amount of potatoes,

x_3 = amount of steak.

Maximize $z = 0.5x_1 + 1x_2 + 9x_3$

subject to

$$20x_1 + 50x_2 + 56x_3 \leq 1000$$
$$1.5x_1 + 3x_2 + 2x_3 \leq 35$$
$$x_1 \geq 0, x_2 \geq 0, x_3 \geq 0$$

$$\begin{bmatrix} 20 & 50 & 56 & 1 & 0 & 0 & | & 1000 \\ 1.5 & 3 & 2 & 0 & 1 & 0 & | & 35 \\ -0.5 & -1 & -9 & 0 & 0 & 1 & | & 0 \end{bmatrix}$$

17. Let x_1 = lbs of Lite, x_2 = lbs. of Trim, x_3 = lbs. of Health Fare.

Maximize $z = 0.25x_1 + 0.25x_2 + 0.32x_3$ subject to

$$0.75x_1 + 0.50x_2 + 0.15x_3 \leq 2320$$
$$0.25x_1 + 0.25x_2 + 0.60x_3 \leq 1380$$
$$0.25x_2 + 0.25x_3 \leq 700$$
$$x_1 \geq 0, x_2 \geq 0, x_3 \geq 0$$

$$\begin{bmatrix} 0.75 & 0.50 & 0.15 & 1 & 0 & 0 & 0 & | & 2320 \\ 0.25 & 0.25 & 0.60 & 0 & 1 & 0 & 0 & | & 1380 \\ 0 & 0.25 & 0.25 & 0 & 0 & 1 & 0 & | & 700 \\ -0.25 & -0.25 & -0.32 & 0 & 0 & 0 & 1 & | & 0 \end{bmatrix}$$

19. Let x_1 = number of military trunks, x_2 = number of commercial trunks, x_3 = number of decorative trunks.

Maximize $z = 6x_1 + 7x_2 + 9x_3$ subject to

$$4x_1 + 3x_2 + 2x_3 \leq 4900 \text{ (Assembly)}$$
$$x_1 + 2x_2 + 4x_3 \leq 2200 \text{ (Finishing)}$$
$$0.1x_1 + 0.2x_2 + 0.3x_3 \leq 210 \text{ (Packing)}$$
$$x_1 \geq 0, x_2 \geq 0, x_3 \geq 0$$

$$\begin{bmatrix} 4 & 3 & 2 & 1 & 0 & 0 & 0 & | & 4900 \\ 1 & 2 & 4 & 0 & 1 & 0 & 0 & | & 2200 \\ 0.1 & 0.2 & 0.3 & 0 & 0 & 1 & 0 & | & 210 \\ -6 & -7 & -9 & 0 & 0 & 0 & 1 & | & 0 \end{bmatrix}$$

21. Let x_1 = number of Majestic, x_2 = number of Traditional, x_3 = number of Wall clocks.

Maximize $z = 400x_1 + 250x_2 + 160x_3$ subject to

$$4x_1 + 2x_2 + x_3 \leq 120 \text{ (Cutting)}$$

$$3x_1 + 2x_2 + x_3 \le 80 \quad \text{(Sanding)}$$
$$x_1 + x_2 + 0.5x_3 \le 40 \quad \text{(Packing)}$$
$$x_1 \ge 0, x_2 \ge 0, x_3 \ge 0$$

$$\begin{bmatrix} 4 & 2 & 1 & 1 & 0 & 0 & 0 & | & 120 \\ 3 & 2 & 1 & 0 & 1 & 0 & 0 & | & 80 \\ 1 & 1 & 0.5 & 0 & 0 & 1 & 0 & | & 40 \\ \hline -400 & -250 & -160 & 0 & 0 & 0 & 1 & | & 0 \end{bmatrix}$$

23. Let x_1 = number of pounds of Early Riser, x_2 = number of pounds of Coffee Time, x_3 = number of pounds of After Dinner, x_4 = number of pounds of Deluxe

Maximize profit $z = 1.00x_1 + 1.10x_2 + 1.15x_3 + 1.20x_4$ subject to

$$0.80x_1 + 0.75x_2 + 0.75x_3 + 0.50x_4 \le 260 \text{ (regular)}$$
$$0.20x_1 + 0.23x_2 + 0.20x_3 + 0.40x_4 \le 90 \text{ (High Mountain)}$$
$$0.02x_2 + 0.05x_3 + 0.10x_4 \le 20 \text{ (Chocolate)}$$

$$x_1 \ge 0, x_2 \ge 0, x_3 \ge 0, x_4 \ge 0$$

$$\begin{bmatrix} 0.80 & 0.75 & 0.75 & 0.50 & 1 & 0 & 0 & 0 & | & 260 \\ 0.20 & 0.23 & 0.20 & 0.40 & 0 & 1 & 0 & 0 & | & 90 \\ 0 & 0.02 & 0.05 & 0.10 & 0 & 0 & 1 & 0 & | & 20 \\ \hline -1.00 & -1.10 & -1.15 & -1.20 & 0 & 0 & 0 & 1 & | & 0 \end{bmatrix}$$

25. Let x_1 = amount invested in stocks, x_2 = amount invested in treasury bonds, x_3 = amount invested in municipal bonds, x_4 = amount invested in corporate bonds

The linear programming problem is stated as follows:
Maximize return on investment

$$z = 0.10x_1 + 0.06x_2 + 0.07x_3 + 0.08x_4 \text{ subject to}$$
$$x_1 + x_2 + x_3 + x_4 \le 10,000 \text{ (total investment)}$$
$$x_1 \le 0.40(x_1 + x_2 + x_3 + x_4) \text{ (stocks less than 40\%)}$$
$$x_2 \le 0.15(x_1 + x_2 + x_3 + x_4) \text{ (treasury bonds no more than 15\%)}$$
$$x_3 \le 0.30(x_1 + x_2 + x_3 + x_4) \text{ (municipal bonds less than 30\%)}$$
$$x_4 \le 0.25(x_1 + x_2 + x_3 + x_4) \text{ (corporate bonds less than 25\%)}$$

Simplifying the last four constraints the problem is
Maximize $z = 0.10x_1 + 0.06x_2 + 0.07x_3 + 0.08x_4$ subject to

$$x_1 + x_2 + x_3 + x_4 \le 10,000$$
$$0.60x_1 - 0.40x_2 - 0.40x_3 - 0.40x_4 \le 0$$
$$-0.15x_1 + 0.85x_2 - 0.15x_3 - 0.15x_4 \le 0$$
$$-0.30x_1 - 0.30x_2 + 0.70x_3 - 0.30x_4 \le 0$$

$$-0.25x_1 - 0.25x_2 - 0.25x_3 + 0.75x_4 \leq 0$$

$$x_1 \geq 0, x_2 \geq 0, x_3 \geq 0, x_4 \geq 0$$

$$\begin{bmatrix} 1 & 1 & 1 & 1 & 1 & 0 & 0 & 0 & 0 & 0 & 10000 \\ 0.60 & -0.40 & -0.40 & -0.40 & 0 & 1 & 0 & 0 & 0 & 0 & 0 \\ -0.15 & 0.85 & -0.15 & -0.15 & 0 & 0 & 1 & 0 & 0 & 0 & 0 \\ -0.30 & -0.30 & 0.70 & -0.30 & 0 & 0 & 0 & 1 & 0 & 0 & 0 \\ -0.25 & -0.25 & -0.25 & 0.75 & 0 & 0 & 0 & 0 & 1 & 0 & 0 \\ -0.10 & -0.06 & -0.07 & -0.08 & 0 & 0 & 0 & 0 & 0 & 1 & 0 \end{bmatrix}$$

27. **(a)** $12x + 10y + s_1 = 120$

$3x + 10y + s_2 = 60$

$7x + 12y + s_3 = 84$

(b) The point (x, y, s_1, s_2, s_3) may not lie in the feasible region. The point (x, y) satisfies the constraint $12x + 10y \leq 120$ so it lies in the half-plane that is the solution to this constraint.

(c) The point (x, y) lies on the boundary line $12x + 10y = 120$. It may or may not be in the feasible region depending on where it is on the line.

(d) The point (x, y) is not in the feasible region when s_1 is negative.

(e) The point $(x, y, 0, s_2, 0)$ satisfies the first equation and $s_1 = 0$ so the point is on the "line" $12x + 10y + s_1 = 120$. Similarly, $(x, y, 0, s_2, 0)$ satisfies the third line and $s_3 = 0$ so the point lies on the "line" $7x + 12y + s_3 = 84$. Thus, the point (x, y) is the intersection of the lines $12x + 10y = 120$ and $7x + 12y = 84$.

29. **(a)** Enter the matrix $\begin{bmatrix} 6 & 3 & 1 & 0 & 0 & 18 \\ 5 & 2 & 0 & 1 & 0 & 27 \\ -40 & -22 & 0 & 0 & 1 & 0 \end{bmatrix}$

(b) Enter the matrix $\begin{bmatrix} 10 & 14 & 1 & 0 & 0 & 0 & 73 \\ 6 & 21 & 0 & 1 & 0 & 0 & 67 \\ 15 & 8 & 0 & 0 & 1 & 0 & 48 \\ -134 & -109 & 0 & 0 & 0 & 1 & 0 \end{bmatrix}$

(c) Enter the matrix $\begin{bmatrix} 1 & 1 & 1 & 1 & 0 & 0 & 0 & 24 \\ 3 & 1 & 4 & 0 & 1 & 0 & 0 & 37 \\ 2 & 5 & 3 & 0 & 0 & 1 & 0 & 41 \\ -15 & -23 & -34 & 0 & 0 & 0 & 1 & 0 \end{bmatrix}$

(d) Enter the matrix $\begin{bmatrix} 7 & 4 & 1 & 2 & 1 & 0 & 0 & 0 & 435 \\ 5 & 3 & 6 & 1 & 0 & 1 & 0 & 0 & 384 \\ 2 & 8 & 4 & 5 & 0 & 0 & 1 & 0 & 562 \\ -24 & -19 & -15 & -33 & 0 & 0 & 0 & 1 & 0 \end{bmatrix}$

(e)   Enter the matrix
$$\begin{bmatrix} 4.7 & 3.2 & 1.58 & 1 & 0 & 0 & 0 & 40.6 \\ 2.14 & 1.82 & 5.09 & 0 & 1 & 0 & 0 & 61.7 \\ 1.63 & 3.44 & 2.84 & 0 & 0 & 1 & 0 & 54.8 \\ -12.9 & -11.27 & -23.85 & 0 & 0 & 0 & 1 & 0 \end{bmatrix}$$

Section 4.2

1.   Basic: $x_1 = 8$, $s_2 = 10$; nonbasic: $x_2 = 0$, $s_1 = 0$

3.   Basic: $x_2 = 86$, $s_1 = 54$, $s_3 = 39$; nonbasic: $x_1 = x_3 = s_2 = 0$

5.   (a)   (i)   7 variables   (ii)   3 basic variables
             (iii)   4 nonbasic variables.
      (b)   (i)   7 variables   (ii)   4 basic variables
             (iii)   3 nonbasic variables.
      (c)   (i)   9 variables   (ii)   4 basic variables
             (iii)   5 nonbasic variables.
      (d)   (i)   4 variables   (ii)   2 basic variables
             (iii)   2 nonbasic variables.

7.
$$\begin{bmatrix} 5 & 4 & 3 & 1 & 0 & 0 & 0 & 8 \\ 2 & 7 & 1 & 0 & 1 & 0 & 0 & 15 \\ 6 & 8 & 5 & 0 & 0 & 1 & 0 & 24 \\ -8 & -10 & -4 & 0 & 0 & 0 & 1 & 0 \end{bmatrix} \begin{matrix} 8/4=2 \\ 15/7 = 2.14 \\ 24/8=3 \\ {} \end{matrix}$$

4 in row 1, column 2

9.
$$\begin{bmatrix} 2 & 5 & 3 & 1 & 0 & 0 & 0 & 15 \\ 4 & 1 & 4 & 0 & 1 & 0 & 0 & 12 \\ 7 & 3 & -5 & 0 & 0 & 1 & 0 & 10 \\ -25 & -30 & -50 & 0 & 0 & 0 & 1 & 0 \end{bmatrix} \begin{matrix} 15/3=5 \\ 12/4=3 \\ 10/-5=-2 \\ {} \end{matrix}$$

4 in row 2, column 3

11.
$$\begin{bmatrix} 2 & 1 & 1 & 0 & 0 & 0 & 7 \\ 3 & 4 & 0 & 1 & 0 & 0 & 12 \\ 2 & 5 & 0 & 0 & 1 & 0 & 15 \\ -5 & -8 & 0 & 0 & 0 & 1 & 0 \end{bmatrix} \begin{matrix} 7/1=7 \\ 12/4=3 \\ 15/5=3 \\ {} \end{matrix}$$

Either the 4 in row 2, column 2 or the 5 in row 3, column 2.

13.
$$\begin{bmatrix} 3 & 5 & 6 & 1 & 0 & 0 & 0 & 9 \\ 2 & 8 & 2 & 0 & 1 & 0 & 0 & 6 \\ 5 & 4 & 3 & 0 & 0 & 1 & 0 & 15 \\ -6 & -12 & -12 & 0 & 0 & 0 & 1 & 0 \end{bmatrix} \begin{matrix} 9/5 \text{ or } 9/6 \\ 6/8 \text{ or } 6/2 \\ 15/4 \text{ or } 15/3 \\ {} \end{matrix}$$

Either the 8 in row 2, column 2 or the 6 in row 1, column 3.

15.
$$\left[\begin{array}{cccccc|c} 6 & 2 & 1 & 0 & 0 & 0 & 3 \\ 4 & 3 & 0 & 1 & 0 & 0 & 0 \\ 3 & 5 & 0 & 0 & 1 & 0 & 8 \\ -12 & -3 & 0 & 0 & 0 & 1 & 0 \end{array}\right] \begin{array}{l} 3/6 = 1/2 \\ 0/4 = 0 \\ 8/3 \\ \, \end{array}$$

4 in row 2, column 1

17.
$$\left[\begin{array}{cccccc|c} 2 & 3 & 1 & 0 & 0 & 0 & 12 \\ 1 & 2 & 0 & 1 & 0 & 0 & 6 \\ 2 & 5 & 0 & 0 & 1 & 0 & 20 \\ -4 & -3 & 0 & 0 & 0 & 1 & 0 \end{array}\right] \quad \left[\begin{array}{cccccc|c} 0 & -1 & 1 & -2 & 0 & 0 & 0 \\ 1 & 2 & 0 & 1 & 0 & 0 & 6 \\ 0 & 1 & 0 & -2 & 1 & 0 & 8 \\ 0 & 5 & 0 & 4 & 0 & 1 & 24 \end{array}\right]$$

19.
$$\left[\begin{array}{ccccccc|c} 6 & 11 & 4 & 1 & 0 & 0 & 0 & 250 \\ -5 & -14 & -8 & 0 & 1 & 0 & 0 & -460 \\ -1 & -1 & -3 & 0 & 0 & 1 & 0 & -390 \\ -10 & -50 & -30 & 0 & 0 & 0 & 1 & 0 \end{array}\right] \left[\begin{array}{ccccccc|c} 14/3 & 29/3 & 0 & 1 & 0 & 4/3 & 0 & -270 \\ -7/3 & -34/3 & 0 & 0 & 1 & -8/3 & 0 & 580 \\ 1/3 & 1/3 & 1 & 0 & 0 & -1/3 & 0 & 130 \\ 0 & -40 & 0 & 0 & 0 & -10 & 1 & 3900 \end{array}\right]$$

21.
$$\left[\begin{array}{ccccc|c} 3 & 1 & 1 & 0 & 0 & 22 \\ 3 & 4 & 0 & 1 & 0 & 34 \\ -2 & -1 & 0 & 0 & 1 & 0 \end{array}\right] \begin{array}{l} 22/3 \\ 34/3 \\ \, \end{array} \quad \left[\begin{array}{ccccc|c} 1 & 1/3 & 1/3 & 0 & 0 & 22/3 \\ 0 & 3 & -1 & 1 & 0 & 12 \\ 0 & -1/3 & 2/3 & 0 & 1 & 44/3 \end{array}\right] \begin{array}{l} 22 \\ 4 \\ \, \end{array}$$

$$\left[\begin{array}{ccccc|c} 1 & 0 & 4/9 & -1/9 & 0 & 6 \\ 0 & 1 & -1/3 & 1/3 & 0 & 4 \\ 0 & 0 & 5/9 & 1/9 & 1 & 16 \end{array}\right] \qquad x_1 = 6,\ x_2 = 4,\ z = 16$$

23.
$$\left[\begin{array}{ccccc|c} 1 & 4 & 1 & 0 & 0 & 9 \\ 4 & 1 & 0 & 1 & 0 & 6 \\ -4 & -5 & 0 & 0 & 1 & 0 \end{array}\right] \begin{array}{l} 9/4 = 2.25 \\ 6/1 = 6 \\ \, \end{array} \quad \left[\begin{array}{ccccc|c} 1/4 & 1 & 1/4 & 0 & 0 & 9/4 \\ 15/4 & 0 & -1/4 & 1 & 0 & 15/4 \\ -11/4 & 0 & 5/4 & 0 & 1 & 45/4 \end{array}\right]$$

$$\left[\begin{array}{ccccc|c} 1 & 0 & 4/15 & -1/15 & 0 & 2 \\ 1 & 0 & -1/15 & 4/15 & 0 & 1 \\ 0 & 0 & 16/15 & 11/15 & 1 & 14 \end{array}\right] \qquad x_1 = 1,\ x_2 = 2,\ z = 14$$

25.
$$\left[\begin{array}{ccccc|c} 1 & 1 & 1 & 0 & 0 & 240 \\ 4 & 3 & 0 & 1 & 0 & 720 \\ -8 & -4 & 0 & 0 & 1 & 0 \end{array}\right] \begin{array}{l} 240 \\ 180 \\ \, \end{array} \quad \left[\begin{array}{ccccc|c} 0 & 1/4 & 1 & -1/4 & 0 & 60 \\ 1 & 3/4 & 0 & 1/4 & 0 & 180 \\ 0 & 2 & 0 & 2 & 1 & 1440 \end{array}\right]$$

$x_1 = 180,\ x_2 = 0,\ z = 1440$

27.
$$\left[\begin{array}{ccccccc|c} 5 & 5 & 10 & 1 & 0 & 0 & 0 & 1000 \\ 10 & 8 & 5 & 0 & 1 & 0 & 0 & 2000 \\ 10 & 5 & 0 & 0 & 0 & 1 & 0 & 500 \\ -100 & -200 & -50 & 0 & 0 & 0 & 1 & 0 \end{array}\right] \left[\begin{array}{ccccccc|c} -5 & 0 & 10 & 1 & 0 & -1 & 0 & 500 \\ -6 & 0 & 5 & 0 & 1 & -8/5 & 0 & 1200 \\ 2 & 1 & 0 & 0 & 0 & 1/5 & 0 & 100 \\ 300 & 0 & -50 & 0 & 0 & 40 & 1 & 20000 \end{array}\right]$$

$$\left[\begin{array}{cccccc|c} -1/2 & 0 & 1 & 1/10 & 0 & -1/10 & 0 & 50 \\ -7/2 & 0 & 0 & -1/2 & 1 & -11/10 & 0 & 950 \\ 2 & 1 & 0 & 0 & 0 & 1/5 & 0 & 100 \\ 275 & 0 & 0 & 5 & 0 & 35 & 1 & 22500 \end{array}\right] \qquad x_1 = 0,\ x_2 = 100,\ x_3 = 50,\ z = 22{,}500$$

29.

$$\begin{bmatrix} 1 & 1 & 1 & 1 & 0 & 0 & 0 & 100 \\ 3 & 2 & 4 & 0 & 1 & 0 & 0 & 210 \\ 1 & 2 & 0 & 0 & 0 & 1 & 0 & 150 \\ \hline -3 & -5 & -5 & 0 & 0 & 0 & 1 & 0 \end{bmatrix} \qquad \begin{bmatrix} 1/4 & 1/2 & 0 & 1 & -1/4 & 0 & 0 & 47.5 \\ 3/4 & 1/2 & 1 & 0 & 1/4 & 0 & 0 & 52.5 \\ 1 & 2 & 0 & 0 & 0 & 1 & 0 & 150 \\ \hline 3/4 & -5/2 & 0 & 0 & 5/4 & 0 & 1 & 262.5 \end{bmatrix}$$

$$\begin{bmatrix} 0 & 0 & 0 & 1 & -1/4 & -1/4 & 0 & 10 \\ 1/2 & 0 & 1 & 0 & 1/4 & -1/4 & 0 & 15 \\ 1/2 & 1 & 0 & 0 & 0 & 1/2 & 0 & 75 \\ \hline 2 & 0 & 0 & 0 & 5/4 & 5/4 & 1 & 450 \end{bmatrix}$$

$x_1 = 0,\ x_2 = 75,\ x_3 = 15,\ z = 450$

31.

$$\begin{bmatrix} 2 & 1 & 4 & 1 & 0 & 0 & 0 & 360 \\ 2 & 5 & 10 & 0 & 1 & 0 & 0 & 850 \\ 3 & 3 & 1 & 0 & 0 & 1 & 0 & 510 \\ \hline -15 & -9 & -15 & 0 & 0 & 0 & 1 & 0 \end{bmatrix} \qquad \begin{bmatrix} 0 & -1 & 10/3 & 1 & 0 & -2/3 & 0 & 20 \\ 0 & 3 & 28/3 & 0 & 1 & -2/3 & 0 & 510 \\ 1 & 1 & 1/3 & 0 & 0 & 1/3 & 0 & 170 \\ \hline 0 & 6 & -10 & 0 & 0 & 5 & 1 & 2250 \end{bmatrix}$$

$$\begin{bmatrix} 0 & -3/10 & 1 & 3/10 & 0 & -1/5 & 0 & 6 \\ 0 & 29/5 & 0 & -14/5 & 1 & 6/5 & 0 & 454 \\ 1 & 11/10 & 0 & -1/10 & 0 & 2/5 & 0 & 168 \\ \hline 0 & 3 & 0 & 3 & 0 & 3 & 1 & 2610 \end{bmatrix}$$

$x_1 = 168,\ x_2 = 0,\ x_3 = 6,\ z = 2610$

33.

$$\begin{bmatrix} 1 & 8 & 1 & 0 & 0 & 0 & 66 \\ 3 & 9 & 0 & 1 & 0 & 0 & 72 \\ 2 & 6 & 0 & 0 & 1 & 0 & 48 \\ \hline -33 & -9 & 0 & 0 & 0 & 1 & 0 \end{bmatrix} \qquad \begin{bmatrix} 0 & 5 & 1 & 0 & -1/2 & 0 & 42 \\ 0 & 0 & 0 & 1 & -3/2 & 0 & 0 \\ 1 & 3 & 0 & 0 & 1/2 & 0 & 24 \\ \hline 0 & 90 & 0 & 0 & 33/2 & 1 & 792 \end{bmatrix}$$

$x_1 = 24,\ x_2 = 0,\ z = 792$

35.

$$\begin{bmatrix} 2 & 1 & 2 & 1 & 0 & 0 & 0 & 100 \\ 1 & 2 & 2 & 0 & 1 & 0 & 0 & 100 \\ 2 & 2 & 1 & 0 & 0 & 1 & 0 & 100 \\ \hline -22 & -20 & -18 & 0 & 0 & 0 & 1 & 0 \end{bmatrix} \qquad \begin{bmatrix} 1 & 1/2 & 1 & 1/2 & 0 & 0 & 0 & 50 \\ 0 & 3/2 & 1 & -1/2 & 1 & 0 & 0 & 50 \\ 0 & 1 & -1 & -1 & 0 & 1 & 0 & 0 \\ \hline 0 & -9 & 4 & 11 & 0 & 0 & 1 & 1100 \end{bmatrix}$$

$$\begin{bmatrix} 1 & 0 & 3/2 & 1 & 0 & -1/2 & 0 & 50 \\ 0 & 0 & 5/2 & 1 & 1 & -3/2 & 0 & 50 \\ 0 & 1 & -1 & -1 & 0 & 1 & 0 & 0 \\ \hline 0 & 0 & -5 & 2 & 0 & 9 & 1 & 1100 \end{bmatrix} \qquad \begin{bmatrix} 1 & 0 & 0 & 2/5 & -3/5 & 2/5 & 0 & 20 \\ 0 & 0 & 1 & 2/5 & 2/5 & -3/5 & 0 & 20 \\ 0 & 1 & 0 & -3/5 & 2/5 & 2/5 & 0 & 20 \\ \hline 0 & 0 & 0 & 4 & 2 & 6 & 1 & 1200 \end{bmatrix}$$

$x_1 = 20,\ x_2 = 20,\ x_3 = 20,\ z = 1200$

37. Let x_1 = number cartons of screwdrivers, x_2 = number cartons of chisels,
x_3 = number cartons of putty knives.

Maximize $z = 5x_1 + 6x_2 + 5x_3$ subject to

$$3x_1 + 4x_2 + 5x_3 \le 2200 \quad \text{(Labor)}$$
$$15x_1 + 12x_2 + 11x_3 \le 8500 \quad \text{(Cost)}$$
$$x_1 \ge 0,\ x_2 \ge 0,\ x_3 \ge 0$$

$$\begin{bmatrix} 3 & 4 & 5 & 1 & 0 & 0 & 2200 \\ 15 & 12 & 11 & 0 & 1 & 0 & 8500 \\ -5 & -6 & -5 & 0 & 0 & 1 & 0 \end{bmatrix}$$

$$\begin{bmatrix} 3/4 & 1 & 5/4 & 1/4 & 0 & 0 & 550 \\ 6 & 0 & -4 & -3 & 1 & 0 & 1900 \\ -1/2 & 0 & 5/2 & 3/2 & 0 & 1 & 3300 \end{bmatrix}$$

$$\begin{bmatrix} 0 & 1 & 7/4 & 5/8 & -1/8 & 0 & 312.5 \\ 1 & 0 & -2/3 & -1/2 & 1/6 & 0 & 950/3 \\ 0 & 0 & 13/6 & 5/4 & 1/12 & 1 & 10375/3 \end{bmatrix}$$

316.7 cartons of screwdrivers, 312.5 of chisels, and no putty knives, for profit of \$3458.33. It is reasonable to round this to 317 cartons of screwdrivers, 312 cartons of chisels, and no putty knives.

39.   Let x_1 = number of Pack I, x_2 = number of Pack II, and x_3 = number of Pack III.
The limitations imposed by the number of books available are:
Short story: $3x_1 + 2x_2 + x_3 \le 660$

Science: $x_1 + 4x_2 + 2x_3 \le 740$

History: $2x_1 + x_2 + 3x_3 \le 853$

The linear programming problem to be solved is
Maximize $z = 1.25_1 + 2.00x_2 + 1.60x_3$ subject to
$3x_1 + 2x_2 + x_3 \le 660$

$x_1 + 4x_2 + 2x_3 \le 740$

$2x_1 + x_2 + 3x_3 \le 853$

$x_1 \ge 0, x_2 \ge 0, x_3 \ge 0$

The initial simplex tableau is

$$\begin{bmatrix} 3 & 2 & 1 & 1 & 0 & 0 & 0 & 660 \\ 1 & 4 & 2 & 0 & 1 & 0 & 0 & 740 \\ 2 & 1 & 3 & 0 & 0 & 1 & 0 & 853 \\ -1.25 & -2.00 & -1.60 & 0 & 0 & 0 & 1 & 0 \end{bmatrix}$$

Here is the sequence of pivots that lead to the solution.

$$\begin{bmatrix} 5/2 & 0 & 0 & 1 & -1/2 & 0 & 0 & 290 \\ 1/4 & 1 & 1/2 & 0 & 1/4 & 0 & 0 & 185 \\ 7/4 & 0 & 5/2 & 0 & -1/4 & 1 & 0 & 668 \\ -3/4 & 0 & -3/5 & 0 & 1/2 & 0 & 1 & 370 \end{bmatrix} \begin{bmatrix} 1 & 0 & 0 & 2/5 & -1/5 & 0 & 0 & 116 \\ 0 & 1 & 1/2 & -1/10 & 3/10 & 0 & 0 & 156 \\ 0 & 0 & 5/2 & 7/10 & 1/10 & 1 & 0 & 465 \\ 0 & 0 & -3/5 & 3/10 & 7/20 & 0 & 1 & 457 \end{bmatrix}$$

$$\begin{bmatrix} 1 & 0 & 0 & 2/5 & -1/5 & 0 & 0 & 116 \\ 0 & 1 & 0 & 1/25 & 7/25 & -1/5 & 0 & 63 \\ 0 & 0 & 1 & -7/25 & 1/25 & 2/5 & 0 & 186 \\ 0 & 0 & 0 & 33/250 & 187/500 & 6/25 & 1 & 568.6 \end{bmatrix}$$

The maximum profit is \$568.60 using 116 of Pack I, 63 of Pack II, and 186 of Pack III.

41. Let x_1 = lbs of Early Riser, x_2 = lbs of After Dinner, x_3 = lbs of Deluxe

Maximize $z = x_1 + 1.1x_2 + 1.2x_3$, subject to

$$0.80x_1 + 0.75x_2 + 0.50x_3 \le 255 \text{ (Regular)}$$
$$0.20x_1 + 0.20x_2 + 0.40x_3 \le 80 \text{ (High Mountain)}$$
$$0.05x_2 + 0.10x_3 \le 15 \text{ (Chocolate)}$$
$$x_1 \ge 0,\ x_2 \ge 0,\ x_3 \ge 0$$

$$\begin{bmatrix} 0.80 & 0.75 & 0.50 & 1 & 0 & 0 & 0 & 255 \\ 0.20 & 0.20 & 0.40 & 0 & 1 & 0 & 0 & 80 \\ 0 & 0.05 & 0.10 & 0 & 0 & 1 & 0 & 15 \\ \hline -1.0 & -1.1 & -1.2 & 0 & 0 & 0 & 1 & 0 \end{bmatrix}
\begin{bmatrix} 4/5 & 1/2 & 0 & 1 & 0 & -5 & 0 & 180 \\ 1/5 & 0 & 0 & 0 & 1 & -4 & 0 & 20 \\ 0 & 1/2 & 1 & 0 & 0 & 10 & 0 & 150 \\ \hline -1 & -1/2 & 0 & 0 & 0 & 12 & 1 & 180 \end{bmatrix}$$

$$\begin{bmatrix} 0 & 1/2 & 0 & 1 & -4 & 11 & 0 & 100 \\ 1 & 0 & 0 & 0 & 5 & -20 & 0 & 100 \\ 0 & 1/2 & 1 & 0 & 0 & 10 & 0 & 150 \\ \hline 0 & -1/2 & 0 & 0 & 5 & -8 & 1 & 280 \end{bmatrix}$$

$$\begin{bmatrix} 0 & 1/22 & 0 & 1/11 & -4/11 & 1 & 0 & 100/11 \\ 1 & 10/11 & 0 & 20/11 & -25/11 & 0 & 0 & 3100/11 \\ 0 & 1/22 & 1 & -10/11 & 40/11 & 0 & 0 & 650/11 \\ \hline 0 & -3/22 & 0 & 8/11 & 23/11 & 0 & 1 & 3880/11 \end{bmatrix}$$

$$\begin{bmatrix} 0 & 1 & 0 & 2 & -8 & 22 & 0 & 200 \\ 1 & 0 & 0 & 0 & 5 & -20 & 0 & 100 \\ 0 & 0 & 1 & -1 & 4 & -1 & 0 & 50 \\ \hline 0 & 0 & 0 & 1 & 1 & 3 & 1 & 380 \end{bmatrix}$$

Maximum profit is \$380 using 100 lbs Early Riser, 200 lbs After Dinner, and 50 lbs Deluxe.

43. Let x_1 = number of packages of TV mix, x_2 = number of packages of Party Mix, x_3 = number of packages of Dinner Mix.

Maximize $z = 4.4x_1 + 4.8x_2 + 5.2x_3$, subject to

$$600x_1 + 500x_2 + 400x_3 \le 39{,}500 \text{ (Peanuts)}$$
$$300x_1 + 300x_2 + 200x_3 \le 22{,}500 \text{ (Cashews)}$$
$$100x_1 + 200x_2 + 400x_3 \le 18{,}000 \text{ (Pecans)}$$
$$x_1 \ge 0,\ x_2 \ge 0,\ x_3 \ge 0$$

$$\begin{bmatrix} 600 & 500 & 400 & 1 & 0 & 0 & 0 & 39500 \\ 300 & 300 & 200 & 0 & 1 & 0 & 0 & 22500 \\ 100 & 200 & 400 & 0 & 0 & 1 & 0 & 18000 \\ \hline -4.4 & -4.8 & -5.2 & 0 & 0 & 0 & 1 & 0 \end{bmatrix}$$

$$\begin{bmatrix} 500 & 300 & 0 & 1 & 0 & -1 & 0 & | & 21500 \\ 250 & 200 & 0 & 0 & 1 & -0.5 & 0 & | & 13500 \\ 0.25 & 0.5 & 1 & 0 & 0 & 0.0025 & 0 & | & 45 \\ \hline -3.1 & -2.2 & 0 & 0 & 0 & 0.013 & 1 & | & 234 \end{bmatrix}$$

$$\begin{bmatrix} 1 & 0.6 & 0 & 0.002 & 0 & -0.002 & 0 & | & 43 \\ 0 & 50 & 0 & -0.5 & 1 & 0 & 0 & | & 2750 \\ 0 & 0.35 & 1 & -0.0005 & 0 & 0.003 & 0 & | & 34.25 \\ \hline 0 & -0.34 & 0 & 0.0062 & 0 & 0.0068 & 1 & | & 367.3 \end{bmatrix}$$

$$\begin{bmatrix} 1 & 0 & 0 & 0.008 & -0.012 & -.002 & 0 & | & 10 \\ 0 & 1 & 0 & -0.01 & 0.02 & 0 & 0 & | & 55 \\ 0 & 0 & 1 & 0.003 & -0.007 & 0.003 & 0 & | & 15 \\ \hline 0 & 0 & 0 & 0.0028 & 0.0068 & 0.0068 & 1 & | & 386 \end{bmatrix}$$

Maximum revenue is \$386 making 10 packages of TV Mix, 55 packages of Party Mix, and 15 packages of Dinner Mix.

45. Let x = number of Majestic, y = number of Traditional, z = number of Wall clocks.

Maximize R = 400x + 250y + 160z, subject to

$$4x + 2y + \;z \le 124 \;\text{(Cutting)}$$
$$3x + \;y + \;z \le 81 \;\text{(Sanding)}$$
$$x + \;y + 0.5z \le 46 \;\text{(Packing)}$$

$$\begin{bmatrix} 4 & 2 & 1 & 1 & 0 & 0 & 0 & | & 124 \\ 3 & 1 & 1 & 0 & 1 & 0 & 0 & | & 81 \\ 1 & 1 & 0.5 & 0 & 0 & 1 & 0 & | & 46 \\ \hline -400 & -250 & -160 & 0 & 0 & 0 & 1 & | & 0 \end{bmatrix}$$

$$\begin{bmatrix} 0 & 2/3 & -1/3 & 1 & -4/3 & 0 & 0 & | & 16 \\ 1 & 1/3 & 1/3 & 0 & 1/3 & 0 & 0 & | & 27 \\ 0 & 2/3 & 1/6 & 0 & -1/3 & 1 & 0 & | & 19 \\ \hline 0 & -350/3 & -80/3 & 0 & 400/3 & 0 & 1 & | & 10800 \end{bmatrix}$$

$$\begin{bmatrix} 0 & 1 & -1/2 & 3/2 & -2 & 0 & 0 & | & 24 \\ 1 & 0 & 1/2 & -1/2 & 1 & 0 & 0 & | & 19 \\ 0 & 0 & 1/2 & -1 & 1 & 1 & 0 & | & 3 \\ \hline 0 & 0 & -85 & 175 & -100 & 0 & 1 & | & 13600 \end{bmatrix}$$

$$\begin{bmatrix} 0 & 1 & 1/2 & -1/2 & 0 & 2 & 0 & | & 30 \\ 1 & 0 & 0 & 1/2 & 0 & -1 & 0 & | & 16 \\ 0 & 0 & 1/2 & -1 & 1 & 1 & 0 & | & 3 \\ \hline 0 & 0 & -35 & 75 & 0 & 100 & 1 & | & 13900 \end{bmatrix}$$

$$\begin{bmatrix} 0 & 1 & 0 & 1/2 & -1 & 1 & 0 & | & 27 \\ 1 & 0 & 0 & 1/2 & 0 & -1 & 0 & | & 16 \\ 0 & 0 & 1 & -2 & 2 & 2 & 0 & | & 6 \\ \hline 0 & 0 & 0 & 5 & 70 & 170 & 1 & | & 14110 \end{bmatrix}$$

Maximum revenue is \$14,110 for 16 Majestic, 27 Traditional, and 6 Wall clocks.

47. The initial tableau and ratios using the pivot row, row 1 are

Ratio

$$\begin{bmatrix} 2 & 1 & 1 & 0 & 0 & 0 & | & 80 \\ 2 & 3 & 0 & 1 & 0 & 0 & | & 120 \\ 4 & 1 & 0 & 0 & 1 & 0 & | & 160 \\ -5 & -4 & 0 & 0 & 0 & 1 & | & 0 \end{bmatrix} \begin{matrix} 40 \\ 60 \\ 40 \\ \\ \end{matrix}$$

$x = 0, y = 0$

Either row 1 or row 3 may be used as the pivot row. Let's use row 1 and we obtain the next tableau

Ratio

$$\begin{bmatrix} 1 & 1/2 & 1/2 & 0 & 0 & 0 & | & 40 \\ 0 & 2 & -1 & 1 & 0 & 0 & | & 40 \\ 0 & -1 & -2 & 0 & 1 & 0 & | & 0 \\ 0 & -3/2 & 5/2 & 0 & 0 & 1 & | & 200 \end{bmatrix} \begin{matrix} 80 \\ 20 \\ -0 \\ \\ \end{matrix}$$

$x = 40, y = 0$

This is not optimal so we use column 2 as the next pivot column. The smallest ratio 0/(-1) has a negative divisor so we choose the least positive divisor 20, to determine that row 2 is the pivot row. Pivoting on 2 in row 2, column 2, we obtain

$$\begin{bmatrix} 1 & 0 & 3/4 & -1/4 & 0 & 0 & | & 30 \\ 0 & 1 & -1/2 & 1/2 & 0 & 0 & | & 20 \\ 0 & 0 & -5/2 & 1/2 & 1 & 0 & | & 20 \\ 0 & 0 & 7/4 & 3/4 & 0 & 1 & | & 230 \end{bmatrix}$$

$x = 30, y = 20$

This yields maximum z = 230 at (30, 20). The sequence of corner points was (0, 0) (40, 0), and (30, 20).

Now go back to the initial tableau and pivot on row 3, column 1. We obtain

Ratio

$$\begin{bmatrix} 0 & 1/2 & 1 & 0 & -1/2 & 0 & | & 0 \\ 0 & 5/2 & 0 & 1 & -1/2 & 0 & | & 40 \\ 1 & 1/4 & 0 & 0 & 1/4 & 0 & | & 40 \\ 0 & -11/4 & 0 & 0 & 5/4 & 1 & | & 200 \end{bmatrix} \begin{matrix} +0 \\ 16 \\ 160 \\ \\ \end{matrix}$$

$x = 40, y = 0$

In this case the zero ratio has a positive divisor so it determines the pivot row, row 1. A pivot on 1/2 in row 1. Column 2 gives the tableau

Ratio

$$\begin{bmatrix} 0 & 1 & 2 & 0 & -1 & 0 & | & 0 \\ 0 & 0 & -5 & 1 & 2 & 0 & | & 40 \\ 1 & 0 & -1/2 & 0 & 1/2 & 0 & | & 40 \\ 0 & 0 & 11/2 & 0 & -3/2 & 1 & | & 200 \end{bmatrix} \begin{matrix} -0 \\ 20 \\ 80 \\ \\ \end{matrix}$$

$x = 40, y = 0$

We still need to pivot using column 5 and row 2 since the zero ratio has a negative divisor. That pivot yields

$$\begin{bmatrix} 0 & 1 & -1/2 & 1/2 & 0 & 0 & | & 20 \\ 0 & 0 & -5/2 & 1/2 & 1 & 0 & | & 20 \\ 1 & 0 & 3/4 & -1/4 & 0 & 0 & | & 30 \\ 0 & 0 & 7/4 & 3/4 & 0 & 1 & | & 230 \end{bmatrix}$$

$x = 30, y = 20$

This tableau gives the optimal solution, maximum z = 230 at (30, 20). The sequence of corner points was (0, 0), (40, 0), (40,0), and (30, 20).
Note that the corner (40, 0) occured twice, once for each constraint through that point.

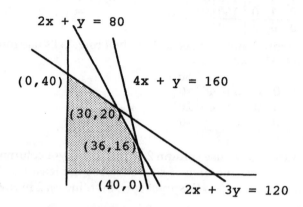

49.   The initial tableau and ratios for the first pivot row are

$$\begin{bmatrix} 1 & 2 & 1 & 0 & 0 & 0 & 0 & | & 100 \\ 1 & 1 & 0 & 1 & 0 & 0 & 0 & | & 60 \\ 4 & 5 & 0 & 0 & 1 & 0 & 0 & | & 265 \\ 3 & 5 & 0 & 0 & 0 & 1 & 0 & | & 255 \\ \hline -7 & -8 & 0 & 0 & 0 & 0 & 1 & | & 0 \end{bmatrix} \begin{matrix} \text{Ratios} \\ 50 \\ 60 \\ 53 \\ 51 \\ \end{matrix}$$

The sequence of tableaux are:

$$\begin{bmatrix} 1/2 & 1 & 1/2 & 0 & 0 & 0 & 0 & | & 50 \\ 1/2 & 0 & -1/2 & 1 & 0 & 0 & 0 & | & 10 \\ 3/2 & 0 & -5/2 & 0 & 1 & 0 & 0 & | & 15 \\ 1/2 & 0 & -5/2 & 0 & 0 & 1 & 0 & | & 5 \\ \hline -3 & 0 & 4 & 0 & 0 & 0 & 1 & | & 400 \end{bmatrix} \begin{matrix} \text{Ratios} \\ 100 \\ 20 \\ 10 \\ 10 \\ \end{matrix}$$

Use row 3 for the pivot row.

$$\begin{bmatrix} 0 & 1 & 4/3 & 0 & -1/3 & 0 & 0 & | & 45 \\ 0 & 0 & 1/3 & 1 & -1/3 & 0 & 0 & | & 5 \\ 1 & 0 & -5/3 & 0 & 2/3 & 0 & 0 & | & 10 \\ 0 & 0 & -5/3 & 0 & -1/3 & 1 & 0 & | & 0 \\ \hline 0 & 0 & -1 & 0 & 2 & 0 & 1 & | & 430 \end{bmatrix} \begin{matrix} \text{Ratios} \\ 33.75 \\ 15 \\ -6 \\ -0 \\ \end{matrix}$$

$$\begin{bmatrix} 0 & 1 & 0 & -4 & 1 & 0 & 0 & | & 25 \\ 0 & 0 & 1 & 3 & -1 & 0 & 0 & | & 15 \\ 1 & 0 & 0 & 5 & -1 & 0 & 0 & | & 35 \\ 0 & 0 & 0 & 5 & -2 & 1 & 0 & | & 25 \\ \hline 0 & 0 & 0 & 3 & 1 & 0 & 1 & | & 445 \end{bmatrix}$$

Maximum z = 445 at (35, 25). The sequence of corner points used was (0, 0), (0, 50), (10, 45), and (35, 25).

Now go back to the second tableau and use row 4 instead of row 3 for the pivot row. We obtain the following sequence.

$$
\begin{array}{c}
\text{Ratios} \\
\left[\begin{array}{ccccccc|c}
0 & 1 & 3 & 0 & 0 & -1 & 0 & 45 \\
0 & 0 & 2 & 1 & 0 & -1 & 0 & 5 \\
0 & 0 & 5 & 0 & 1 & -3 & 0 & 0 \\
1 & 0 & -5 & 0 & 0 & 2 & 0 & 10 \\
\hline
0 & 0 & -11 & 0 & 6 & 6 & 1 & 430
\end{array}\right]
\begin{array}{c}
15 \\
2.5 \\
0 \\
-2 \\
\;
\end{array}
\end{array}
$$

$$
\begin{array}{c}
\text{Ratios} \\
\left[\begin{array}{ccccccc|c}
0 & 1 & 0 & 0 & -3/5 & 4/5 & 0 & 45 \\
0 & 0 & 0 & 1 & -2/5 & 1/5 & 0 & 5 \\
0 & 0 & 1 & 0 & 1/5 & -3/5 & 0 & 0 \\
1 & 0 & 0 & 0 & 1 & -1 & 0 & 10 \\
\hline
0 & 0 & 0 & 0 & 11/5 & -3/5 & 1 & 430
\end{array}\right]
\begin{array}{c}
56.25 \\
25 \\
-0 \\
-10 \\
\;
\end{array}
\end{array}
\qquad
\left[\begin{array}{ccccccc|c}
0 & 1 & 0 & -4 & 1 & 0 & 0 & 25 \\
0 & 0 & 0 & 5 & -2 & 1 & 0 & 25 \\
0 & 0 & 1 & 3 & -1 & 0 & 0 & 15 \\
1 & 0 & 0 & 5 & -1 & 0 & 0 & 35 \\
\hline
0 & 0 & 0 & 3 & 1 & 0 & 1 & 445
\end{array}\right]
$$

Maximum z = 445 at (35, 25). The sequence of corner points used was (0, 0), (0, 50), (10, 45), (10, 45) and (35, 25).

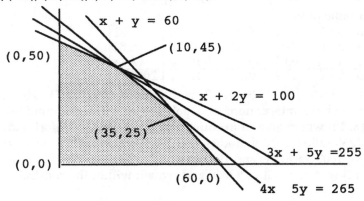

51. **(a)**
$$
\left[\begin{array}{ccccccc|c}
15 & 1.25 & 1 & 0.025 & 0 & 0 & 0 & 9.875 \\
0 & 5 & 0 & -0.5 & 1 & 0 & 0 & 32.5 \\
-50 & -30 & 0 & -1 & 0 & 1 & 0 & -195 \\
\hline
5 & 2.5 & 0 & 0.15 & 0 & 0 & 1 & 59.25
\end{array}\right]
$$

A negative appears in the last column indicating that you are outside the feasible region.

(b)
$$
\left[\begin{array}{ccccccc|c}
1 & 0.83 & 0.67 & 0.02 & 0 & 0 & 0 & 6.58 \\
0 & 5 & 0 & -0.5 & 1 & 0 & 0 & 32.5 \\
0 & 11.67 & 33.33 & -0.17 & 0 & 1 & 0 & 134.17 \\
\hline
0 & -1.67 & -3.33 & 0.07 & 0 & 0 & 1 & 26.33
\end{array}\right]
$$

It goes to the corner (6.58, 0, 0) instead of (0, 0, 5). If the pivoting continues, it will eventually reach the optimal solution.

53. Pivot on 1.5 in the (1, 3) position.

$$\begin{bmatrix} 0.67 & 0 & 1 & 0.67 & 0 & -0.33 & 0 & 33.33 \\ -1.67 & 0 & 0 & -0.67 & 1 & -0.67 & 0 & -8.33 \\ 0.67 & 1 & 0 & -0.33 & 0 & 0.67 & 0 & 58.33 \\ \hline 3.33 & 0 & 0 & 5.33 & 0 & 7.33 & 1 & 1266.67 \end{bmatrix}$$

55. Pivot on the (2, 3) entry.

$$\begin{bmatrix} 0.75 & 1 & 0 & -0.13 & -0.13 & 0 & 0 & 9 \\ -0.17 & 0 & 1 & 0.21 & 0.13 & 0 & 0 & 17 \\ 3.83 & 0 & 0 & -1.79 & -0.88 & 1 & 0 & -55 \\ \hline 3.67 & 0 & 0 & 2.42 & 0.25 & 0 & 1 & 658 \end{bmatrix}$$

Note that this takes you outside the feasible region.

57. The initial tableau is

$$\begin{bmatrix} 40 & 25 & 30 & 45 & 1 & 0 & 0 & 0 & 1250 \\ 25 & 50 & 40 & 45 & 0 & 1 & 0 & 0 & 1600 \\ 6 & 4 & 8 & 2 & 0 & 0 & 1 & 0 & 195 \\ \hline -48 & -45 & -55 & -65 & 0 & 0 & 0 & 1 & 0 \end{bmatrix}$$

The final tableau is

$$\begin{bmatrix} 0.74 & 0 & 0 & 1 & 0.04 & -0.01 & -0.08 & 0 & 12.02 \\ -1.03 & 1 & 0 & 0 & -0.05 & 0.05 & -0.07 & 0 & 6.81 \\ 1.08 & 0 & 1 & 0 & 0.01 & -0.02 & 0.18 & 0 & 17.96 \\ \hline 13.21 & 0 & 0 & 0 & 1.19 & 0.19 & 1.47 & 1 & 2075.71 \end{bmatrix}$$

The solution is (0, 6.81, 17.96, 12.02) with a profit of $2075.71. However, the solution must be an integer number of patterns, so we might round the solution to (0, 7, 18, 12) which gives a maximum profit of $2085. Unfortunately, this requires 1255 tulips, 1610 daffodils, and 196 boxwood while there are only 1250 tulips, 1600 daffodils and 195 boxwood available. We can round to (0, 6, 18, 12) with a profit of $2040 and remain within the constraints.

Using EXCEL

1. Maximum $z = 117$ at $(6,5)$

3. Maximum $z = 696$ at $(12, 12, 24)$

Section 4.3

1. $\begin{bmatrix} 2 & 4 \\ 1 & 0 \\ 3 & 2 \end{bmatrix}$

3. $\begin{bmatrix} 4 & 1 & 6 & 2 \\ 3 & 8 & -7 & 4 \\ 2 & -2 & 1 & 6 \end{bmatrix}$

5. (a) $\left[\begin{array}{cc|c} 6 & 5 & 30 \\ 8 & 3 & 42 \\ 25 & 30 & 1 \end{array}\right]$
   (b) $\left[\begin{array}{cc|c} 6 & 8 & 25 \\ 5 & 3 & 30 \\ 30 & 42 & 1 \end{array}\right]$

   (c) $\left[\begin{array}{ccccc|c} 6 & 8 & 1 & 0 & 0 & 25 \\ 5 & 3 & 0 & 1 & 0 & 30 \\ -30 & -42 & 0 & 0 & 1 & 0 \end{array}\right]$

7. (a) $\left[\begin{array}{cc|c} 22 & 30 & 110 \\ 15 & 40 & 95 \\ 20 & 35 & 68 \\ \hline 500 & 700 & 1 \end{array}\right]$
   (b) $\left[\begin{array}{ccc|c} 22 & 15 & 20 & 500 \\ 30 & 40 & 35 & 700 \\ \hline 110 & 95 & 68 & 1 \end{array}\right]$

   (c) $\left[\begin{array}{cccccc|c} 22 & 15 & 20 & 1 & 0 & 0 & 500 \\ 30 & 40 & 35 & 0 & 1 & 0 & 700 \\ \hline -110 & -95 & -68 & 0 & 0 & 1 & 0 \end{array}\right]$

9. $\begin{array}{ccccc} y_1 & y_2 & x_1 & x_2 & w \end{array}$
$\left[\begin{array}{ccccc|c} 0 & 1 & 1 & -1 & 0 & 1 \\ 1 & 0 & -1 & 2 & 0 & 2 \\ \hline 0 & 0 & 6 & 4 & 1 & 40 \end{array}\right]$

11. $\begin{array}{ccccccc} y_1 & y_2 & y_3 & x_1 & x_2 & x_3 & w \end{array}$
$\left[\begin{array}{ccccccc|c} 1/6 & 0 & 1 & 1/5 & -1/8 & 0 & 0 & 7/6 \\ 1/3 & 1 & 0 & -1/5 & 1/4 & 0 & 0 & 4/3 \\ -1 & 0 & 0 & -2/5 & -1 & 1 & 0 & 1 \\ \hline 45 & 0 & 0 & 12 & 10 & 0 & 1 & 510 \end{array}\right]$

From the last row $x_1 = 6$, $x_2 = 4$, $z = 40$.   Minimum $z = 510$ at $(12,10,0)$

13. $\begin{array}{ccccc} y_1 & y_2 & x_1 & x_2 & w \end{array}$
$\left[\begin{array}{ccccc|c} 1 & 2 & 1 & 0 & 0 & 4 \\ 1 & 1 & 0 & 1 & 0 & 3 \\ \hline -8 & -14 & 0 & 0 & 1 & 0 \end{array}\right]$

$\begin{array}{ccccc} y_1 & y_2 & x_1 & x_2 & w \end{array}$
$\left[\begin{array}{ccccc|c} 1/2 & 1 & 1/2 & 0 & 0 & 2 \\ 1/2 & 0 & -1/2 & 1 & 0 & 1 \\ \hline -1 & 0 & 7 & 0 & 1 & 28 \end{array}\right]$

$\begin{array}{ccccc} y_1 & y_2 & x_1 & x_2 & w \end{array}$
$\left[\begin{array}{ccccc|c} 0 & 1 & 1 & -1 & 0 & 1 \\ 1 & 0 & -1 & 2 & 0 & 2 \\ \hline 0 & 0 & 6 & 2 & 1 & 30 \end{array}\right]$

Minimum $z = 30$ at $(6, 2)$

15.

$$\begin{array}{ccccccc|c} y_1 & y_2 & y_3 & x_1 & x_2 & x_3 & w & \\ 3 & 1 & 0 & 1 & 0 & 0 & 0 & 10 \\ 1 & 1 & 4 & 0 & 1 & 0 & 0 & 16 \\ 6 & 0 & 1 & 0 & 0 & 1 & 0 & 20 \\ \hline -9 & -9 & -12 & 0 & 0 & 0 & 1 & 0 \end{array}$$

$$\begin{array}{ccccccc|c} y_1 & y_2 & y_3 & x_1 & x_2 & x_3 & w & \\ 3 & 1 & 0 & 1 & 0 & 0 & 0 & 10 \\ 1/4 & 1/4 & 1 & 0 & 1/4 & 0 & 0 & 4 \\ 23/4 & -1/4 & 0 & 0 & -1/4 & 1 & 0 & 16 \\ \hline -6 & -6 & 0 & 0 & 3 & 0 & 1 & 48 \end{array}$$

$$\begin{array}{ccccccc|c} y_1 & y_2 & y_3 & x_1 & x_2 & x_3 & w & \\ 3 & 1 & 0 & 1 & 0 & 0 & 0 & 10 \\ -1/2 & 0 & 1 & -1/4 & 1/4 & 0 & 0 & 3/2 \\ 13/2 & 0 & 0 & 1/4 & -1/4 & 1 & 0 & 37/2 \\ \hline 12 & 0 & 0 & 6 & 3 & 0 & 1 & 108 \end{array}$$

Minimum z = 108 at (6, 3, 0)

17.

$$\begin{array}{ccccccc|c} y_1 & y_2 & y_3 & x_1 & x_2 & x_3 & w & \\ 1 & 3 & 3 & 1 & 0 & 0 & 0 & 8 \\ 1 & 1 & 6 & 0 & 1 & 0 & 0 & 5 \\ 1 & 3 & 8 & 0 & 0 & 1 & 0 & 12 \\ \hline -37 & -81 & -216 & 0 & 0 & 0 & 1 & 0 \end{array}$$

$$\begin{array}{ccccccc|c} y_1 & y_2 & y_3 & x_1 & x_2 & x_3 & w & \\ 1/2 & 5/2 & 0 & 1 & -1/2 & 0 & 0 & 11/2 \\ 1/6 & 1/6 & 1 & 0 & 1/6 & 0 & 0 & 5/6 \\ -1/3 & 5/3 & 0 & 0 & -4/3 & 1 & 0 & 16/3 \\ \hline -1 & -45 & 0 & 0 & 36 & 0 & 1 & 180 \end{array}$$

$$\begin{array}{ccccccc|c} y_1 & y_2 & y_3 & x_1 & x_2 & x_3 & w & \\ 1/5 & 1 & 0 & 2/5 & -1/5 & 0 & 0 & 11/5 \\ 2/15 & 0 & 1 & -1/15 & 1/5 & 0 & 0 & 7/15 \\ -2/3 & 0 & 0 & -2/3 & -1 & 1 & 0 & 5/3 \\ \hline 8 & 0 & 0 & 18 & 27 & 0 & 1 & 279 \end{array}$$

Minimum z = 279 at (18, 27, 0)

19. Let x_1 = number of days at Dallas, x_2 = number of days at New Orleans.

Minimize $z = 22{,}000x_1 + 12{,}000x_2$ subject to

$$800x_1 + 500x_2 \geq 28{,}000 \quad \text{(Radial)}$$
$$280x_1 + 150x_2 \geq 9{,}000 \quad \text{(Standard)}$$
$$x_1 \geq 0, x_2 \geq 0$$

$$\begin{array}{ccccc|c} 800 & 280 & 1 & 0 & 0 & 22000 \\ 500 & 150 & 0 & 1 & 0 & 12000 \\ \hline -28000 & -9000 & 0 & 0 & 1 & 0 \end{array}$$

$$\begin{array}{ccccc|c} 0 & 40 & 1 & -8/5 & 0 & 2800 \\ 1 & 3/10 & 0 & 1/500 & 0 & 24 \\ \hline 0 & -600 & 0 & 56 & 1 & 672000 \end{array}$$

$$\begin{array}{ccccc|c} 0 & 1 & 1/40 & -1/25 & 0 & 70 \\ 1 & 0 & -3/400 & 7/500 & 0 & 3 \\ \hline 0 & 0 & 15 & 32 & 1 & 714000 \end{array}$$

15 at Dallas, 32 at New Orleans for cost of $714,000

21. Let x_1 = number of days the Chicago plant operates, x_2 = number of days the Detroit plant operates.

Minimize $z = 20{,}000x_1 + 15{,}000x_2$ subject to

$$600x_1 + 300x_2 \geq 24{,}000 \quad \text{(Radial)}$$
$$100x_1 + 100x_2 \geq 5{,}000 \quad \text{(Standard)}$$
$$x_1 \geq 0, x_2 \geq 0.$$

This is a standard minimization problem so we solve it using its dual problem. The augmented matrix of the problem is

$$\begin{bmatrix} 600 & 300 & | & 24000 \\ 100 & 100 & | & 5000 \\ 20000 & 15000 & | & 1 \end{bmatrix} \qquad \begin{bmatrix} 600 & 100 & | & 20000 \\ 300 & 100 & | & 15000 \\ 24000 & 5000 & | & 1 \end{bmatrix}$$

The second matrix represents the dual problem.

The tableaux that solve this dual problem are:

$$\begin{bmatrix} 600 & 100 & 1 & 0 & 0 & | & 20000 \\ 300 & 100 & 0 & 1 & 0 & | & 15000 \\ \hline -24000 & -5000 & 0 & 0 & 1 & | & 0 \end{bmatrix} \qquad \begin{bmatrix} 1 & 1/6 & 1/600 & 0 & 0 & | & 200/6 \\ 300 & 100 & 0 & 1 & 0 & | & 15000 \\ \hline -24000 & -5000 & 0 & 0 & 1 & | & 0 \end{bmatrix}$$

$$\begin{bmatrix} 1 & 1/6 & 1/600 & 0 & 0 & | & 200/6 \\ 0 & 50 & -1/2 & 1 & 0 & | & 5000 \\ \hline 0 & -1000 & 40 & 0 & 1 & | & 800000 \end{bmatrix} \qquad \begin{bmatrix} 1 & 1/6 & 1/600 & 0 & 0 & | & 200/6 \\ 0 & 1 & -1/100 & 1/50 & 0 & | & 100 \\ \hline 0 & -1000 & 40 & 0 & 1 & | & 800000 \end{bmatrix}$$

$$\begin{bmatrix} 1 & 0 & 1/300 & -1/300 & 0 & | & 100/6 \\ 0 & 1 & -1/100 & 1/50 & 0 & | & 100 \\ \hline 0 & 0 & 30 & 20 & 1 & | & 900000 \end{bmatrix}$$

The last row gives the solution to the original problem $x_1 = 30$, $x_2 = 20$, $z = 900{,}000$.

The minimum operating costs are \$900,000 when the Chicago plant operates 30 days and the Detroit plant 20 days.

23. Let x_1 = number of cereal, x_2 = number of cheeseburgers, and x_3 = number of fries. Then we want to minimize calories,

$$z = 250x_1 + 350x_2 + 500x_3 \quad \text{subject to}$$
$$30x_1 + 45x_2 + 60x_3 \geq 210 \quad \text{(Carbohydrates)}$$
$$8x_1 + 3x_2 + 4x_3 \geq 20 \quad \text{(Fiber)}$$
$$20x_1 + 10x_2 \qquad \geq 80 \quad \text{(Vitamin A)}$$
$$x_1 \geq 0, x_2 \geq 0, x_3 \geq 0$$

Write the matrix of the minimum problem and its transpose.

$$A = \begin{bmatrix} 30 & 45 & 60 & 210 \\ 8 & 3 & 4 & 20 \\ 20 & 10 & 0 & 80 \\ 250 & 350 & 500 & 0 \end{bmatrix} \qquad \text{Transpose} = \begin{bmatrix} 30 & 8 & 20 & 250 \\ 45 & 3 & 10 & 350 \\ 60 & 4 & 0 & 500 \\ 210 & 20 & 80 & 0 \end{bmatrix}$$

Form the initial tableau from the transpose and solve the maximum problem.

$$\left[\begin{array}{cccccc|c} 30 & 8 & 20 & 1 & 0 & 0 & 250 \\ 45 & 3 & 10 & 0 & 1 & 0 & 350 \\ 60 & 4 & 0 & 0 & 0 & 1 & 500 \\ \hline -210 & -20 & -80 & 0 & 0 & 0 & 0 \end{array}\right] \quad \left[\begin{array}{cccccc|c} 0 & 6 & 40/3 & 1 & -2/3 & 0 & 50/3 \\ 1 & 1/15 & 2/9 & 0 & 1/45 & 0 & 70/9 \\ 0 & 0 & -40/3 & 0 & -4/3 & 1 & 100/3 \\ \hline 0 & -6 & -100/3 & 0 & 14/3 & 0 & 4900/3 \end{array}\right]$$

$$\left[\begin{array}{cccccc|c} 0 & 9/20 & 1 & 3/40 & -1/20 & 0 & 5/4 \\ 1 & -1/30 & 0 & -1/60 & 1/30 & 0 & 15/2 \\ 0 & 6 & 0 & 1 & -2 & 1 & 50 \\ \hline 0 & 9 & 0 & 5/2 & 3 & 0 & 1675 \end{array}\right]$$

In the bottom row of the final tableau the numbers 2.5, 3, 0, 1675 give the solution to the original problem: minimum $z = 1675$ at $x_1 = 2.5$, $x_2 = 3$, $x_3 = 0$. Elizabeth should select 2.5 cereals, 3 cheeseburgers, and no fries for a total of 1675 calories.

Section 4.4

1. We change minimize $z = 2x_1 - 5x_2$ to maximize $w = -2x_1 + 5x_2$. The initial simplex tableau is

$$\left[\begin{array}{ccccc|c} 4 & 3 & 1 & 0 & 0 & 120 \\ 2 & 1 & 0 & 1 & 0 & 50 \\ 2 & -5 & 0 & 0 & 1 & 0 \end{array}\right] \quad \left[\begin{array}{ccccc|c} 4/3 & 1 & 1/3 & 0 & 0 & 40 \\ 2/3 & 0 & -1/3 & 1 & 0 & 10 \\ 26/3 & 0 & 5/3 & 0 & 1 & 200 \end{array}\right]$$

Minimum z is -200 at $x_1 = 0$, $x_2 = 40$

3.

$$\left[\begin{array}{ccccccc|c} 3 & 2 & -12 & 1 & 0 & 0 & 0 & 120 \\ 2 & 4 & 6 & 0 & 1 & 0 & 0 & 120 \\ 1 & -2 & 3 & 0 & 0 & 1 & 0 & 52 \\ \hline 4 & 5 & -9 & 0 & 0 & 0 & 1 & 0 \end{array}\right] \quad \left[\begin{array}{ccccccc|c} 7 & -6 & 0 & 1 & 0 & 4 & 0 & 328 \\ 0 & 8 & 0 & 0 & 1 & -2 & 0 & 16 \\ 1/3 & -2/3 & 1 & 0 & 0 & 1/3 & 0 & 52/3 \\ \hline 7 & -1 & 0 & 0 & 0 & 3 & 1 & 156 \end{array}\right]$$

$$\left[\begin{array}{ccccccc|c} 7 & 0 & 0 & 1 & 3/4 & 5/2 & 0 & 340 \\ 0 & 1 & 0 & 0 & 1/8 & -1/4 & 0 & 2 \\ 1/3 & 0 & 1 & 0 & 1/12 & 1/6 & 0 & 56/3 \\ \hline 7 & 0 & 0 & 0 & 1/8 & 11/4 & 1 & 158 \end{array}\right]$$

Minimum $z = -158$ at $x_1 = 0$, $x_2 = 2$, $x_3 = 56/3$

5. Modify the last two constraints to
$$-3x_1 - 5x_2 - 8x_3 \le -106$$
$$-6x_1 - 12x_2 - x_3 \le -98$$

The initial simplex tableau then is

$$\left[\begin{array}{ccccccc|c} 9 & 7 & 10 & 1 & 0 & 0 & 0 & 154 \\ -3 & -5 & -8 & 0 & 1 & 0 & 0 & -106 \\ -6 & -12 & -1 & 0 & 0 & 1 & 0 & -98 \\ \hline -5 & -3 & -8 & 0 & 0 & 0 & 1 & 0 \end{array}\right]$$

120

7. Change the objective function to maximize
$w = -7x_1 - 5x_2 - 8x_3$ and change the second
and third constraints to
$$-7x_1 - 9x_2 - 15x_3 \le -48$$
$$x_1 + 3x_2 + 5x_3 \le 27$$
$$-x_1 - 3x_2 - 5x_3 \le -27$$

The initial simplex tableau then is
$$\begin{bmatrix} 15 & 23 & 9 & 1 & 0 & 0 & 0 & 0 & 85 \\ -7 & -9 & -15 & 0 & 1 & 0 & 0 & 0 & -48 \\ 1 & 3 & 5 & 0 & 0 & 1 & 0 & 0 & 27 \\ -1 & -3 & -5 & 0 & 0 & 0 & 1 & 0 & -27 \\ \hline 7 & 5 & 8 & 0 & 0 & 0 & 0 & 1 & 0 \end{bmatrix}$$

9. $\begin{bmatrix} 3 & 2 & 1 & 0 & 0 & 36 \\ -2 & 1 & 0 & 1 & 0 & -3 \\ \hline -5 & -2 & 0 & 0 & 1 & 0 \end{bmatrix}$ $\begin{bmatrix} 0 & 7/2 & 1 & 3/2 & 0 & 63/2 \\ 1 & -1/2 & 0 & -1/2 & 0 & 3/2 \\ \hline 0 & -9/2 & 0 & -5/2 & 1 & 15/2 \end{bmatrix}$

$\begin{bmatrix} 0 & 1 & 2/7 & 3/7 & 0 & 9 \\ 1 & 0 & 1/7 & -2/7 & 0 & 6 \\ \hline 0 & 0 & 9/7 & -4/7 & 1 & 48 \end{bmatrix}$ $\begin{bmatrix} 0 & 7/3 & 2/3 & 1 & 0 & 21 \\ 1 & 2/3 & 1/3 & 0 & 0 & 12 \\ \hline 0 & 4/3 & 5/3 & 0 & 1 & 60 \end{bmatrix}$

Maximum $z = 60$ at $(12, 0)$

11. $\begin{bmatrix} 5 & 11 & 1 & 0 & 0 & 350 \\ -15 & -8 & 0 & 1 & 0 & -300 \\ \hline -15 & -22 & 0 & 0 & 1 & 0 \end{bmatrix}$ $\begin{bmatrix} 0 & 25/3 & 1 & 1/3 & 0 & 250 \\ 1 & 8/15 & 0 & -1/15 & 0 & 20 \\ \hline 0 & -14 & 0 & -1 & 1 & 300 \end{bmatrix}$

$\begin{bmatrix} 0 & 1 & 3/25 & 1/25 & 0 & 30 \\ 1 & 0 & -8/125 & -11/125 & 0 & 4 \\ \hline 0 & 0 & 42/25 & -11/25 & 1 & 720 \end{bmatrix}$ $\begin{bmatrix} 0 & 25 & 3 & 1 & 0 & 750 \\ 1 & 11/5 & 1/5 & 0 & 0 & 70 \\ \hline 0 & 11 & 3 & 0 & 1 & 1050 \end{bmatrix}$

Maximum is $z = 1050$ at $x_1 = 70$, $x_2 = 0$

13. Replace the second constraint with $-3x_1 - 6x_2 \le -120$ and we obtain the
initial simplex tableau
$$\begin{bmatrix} 5 & 8 & 1 & 0 & 0 & 180 \\ -3 & -6 & 0 & 1 & 0 & -120 \\ \hline -11 & -20 & 0 & 0 & 1 & 0 \end{bmatrix}$$
We pivot on -6 because of the -120 in the last column. The next tableau is
$$\begin{bmatrix} 1 & 0 & 1 & 4/3 & 0 & 20 \\ 1/2 & 1 & 0 & -1/6 & 0 & 20 \\ \hline -1 & 0 & 0 & -10/3 & 1 & 400 \end{bmatrix}$$
This is not optimal so we choose column 4 as the pivot column, row 1 as the
pivot row, and pivot on $4/3$ to obtain

121

$$\begin{bmatrix} 3/4 & 0 & 3/4 & 1 & 0 & 15 \\ 5/8 & 1 & 1/8 & 0 & 0 & 45/2 \\ 3/2 & 0 & 5/2 & 0 & 1 & 450 \end{bmatrix}$$

This is optimal with $z = 450$ at $(0, 22.5)$.

15.

$$\begin{bmatrix} 6 & 12 & 4 & 1 & 0 & 0 & 0 & 900 \\ -5 & -16 & -8 & 0 & 1 & 0 & 0 & -120 \\ -3 & -1 & -1 & 0 & 0 & 1 & 0 & -300 \\ -10 & -50 & -30 & 0 & 0 & 0 & 1 & 0 \end{bmatrix}$$

$$\begin{bmatrix} 0 & 10 & 2 & 1 & 0 & 2 & 0 & 300 \\ 0 & -43/3 & -19/3 & 0 & 1 & -5/3 & 0 & 380 \\ 1 & 1/3 & 1/3 & 0 & 0 & -1/3 & 0 & 100 \\ 0 & -140/3 & -80/3 & 0 & 0 & -10/3 & 1 & 1000 \end{bmatrix}$$

$$\begin{bmatrix} 0 & 1 & 1/5 & 1/10 & 0 & 1/5 & 0 & 30 \\ 0 & 0 & -52/15 & 43/30 & 1 & 6/5 & 0 & 810 \\ 1 & 0 & 4/15 & -1/30 & 0 & -2/5 & 0 & 90 \\ 0 & 0 & -52/3 & 14/3 & 0 & 6 & 1 & 2400 \end{bmatrix}$$

$$\begin{bmatrix} 0 & 5 & 1 & 1/2 & 0 & 1 & 0 & 150 \\ 0 & 52/3 & 0 & 19/6 & 1 & 14/3 & 0 & 1330 \\ 1 & -4/3 & 0 & -1/6 & 0 & -2/3 & 0 & 50 \\ 0 & 260/3 & 0 & 40/3 & 0 & 70/3 & 1 & 5000 \end{bmatrix}$$

Maximum $z = 5000$ at $(50, 0, 150)$

17. To set up the initial simplex tableau we replace the objective function with maximize $w = -15x_1 - 8x_2$ and replace the second constraint with

$$-3x_1 - 2x_2 \leq -36.$$

This gives the initial tableau

$$\begin{bmatrix} 1 & 2 & 1 & 0 & 0 & 20 \\ -3 & -2 & 0 & 1 & 0 & -36 \\ 15 & 8 & 0 & 0 & 1 & 0 \end{bmatrix}$$

To obtain a tableau with a basic feasible solution we pivot on -3 to obtain

$$\begin{bmatrix} 0 & 4/3 & 1 & 1/3 & 0 & 8 \\ 1 & 2/3 & 0 & -1/3 & 0 & 12 \\ 0 & -2 & 0 & 5 & 1 & -180 \end{bmatrix}$$

Now the pivot column is column 2, and the pivot row is row 1. Thus, we pivot on $4/3$ and obtain

$$\begin{bmatrix} 0 & 1 & 3/4 & 1/4 & 0 & 6 \\ 1 & 0 & -1/2 & -1/2 & 0 & 8 \\ 0 & 0 & 3/2 & 11/2 & 1 & -168 \end{bmatrix}$$

This gives the optimal solution maximum $w = -168$ at $(8, 6)$ so the original problem has the minimum value $z = 168$ at $(8, 6)$.

19.

$$\begin{bmatrix} -10 & -12 & -5 & 1 & 0 & 0 & | & -100 \\ 5 & 7 & 5 & 0 & 1 & 0 & | & 75 \\ 4 & 5 & 1 & 0 & 0 & 1 & | & 0 \end{bmatrix}$$

$$\begin{bmatrix} 5/6 & 1 & 5/12 & -1/12 & 0 & 0 & | & 25/3 \\ -5/6 & 0 & 25/12 & 7/12 & 1 & 0 & | & 50/3 \\ -1/6 & 0 & -13/12 & 5/12 & 0 & 1 & | & -125/3 \end{bmatrix}$$

$$\begin{bmatrix} 1 & 1 & 0 & -1/5 & -1/5 & 0 & | & 5 \\ -2/5 & 0 & 1 & 7/25 & 12/25 & 0 & | & 8 \\ -3/5 & 0 & 0 & 18/25 & 13/25 & 1 & | & -33 \end{bmatrix} \qquad \begin{bmatrix} 1 & 1 & 0 & -1/5 & -1/5 & 0 & | & 5 \\ 0 & 2/5 & 1 & 1/5 & 2/5 & 0 & | & 10 \\ 0 & 3/5 & 0 & 3/5 & 2/5 & 1 & | & -30 \end{bmatrix}$$

Minimum $z = 30$ at $(5, 0, 10)$

21.

$$\begin{bmatrix} 3 & 2 & 1 & 0 & 0 & 0 & | & 48 \\ 2 & 4 & 0 & 1 & 0 & 0 & | & 64 \\ -4 & -6 & 0 & 0 & 1 & 0 & | & -84 \\ -8 & -4 & 0 & 0 & 0 & 1 & | & 0 \end{bmatrix} \qquad \begin{bmatrix} 5/3 & 0 & 1 & 0 & 1/3 & 0 & | & 20 \\ -2/3 & 0 & 0 & 1 & 2/3 & 0 & | & 8 \\ 2/3 & 1 & 0 & 0 & -1/6 & 0 & | & 14 \\ -16/3 & 0 & 0 & 0 & -2/3 & 1 & | & 56 \end{bmatrix}$$

$$\begin{bmatrix} 1 & 0 & 3/5 & 0 & 1/5 & 0 & | & 12 \\ 0 & 0 & 2/5 & 1 & 4/5 & 0 & | & 16 \\ 0 & 1 & -2/5 & 0 & -3/10 & 0 & | & 6 \\ 0 & 0 & 16/5 & 0 & 2/5 & 1 & | & 120 \end{bmatrix}$$

Maximum $z = 120$ at $(12, 6)$

23.

$$\begin{bmatrix} 3 & 2 & 1 & 0 & 0 & 0 & 0 & | & 60 \\ -2 & -3 & 0 & 1 & 0 & 0 & 0 & | & -24 \\ 1 & 1 & 0 & 0 & 1 & 0 & 0 & | & 25 \\ -1 & -1 & 0 & 0 & 0 & 1 & 0 & | & -25 \\ -6 & -4 & 0 & 0 & 0 & 0 & 1 & | & 0 \end{bmatrix} \qquad \begin{bmatrix} 0 & -1 & 1 & 0 & 0 & 3 & 0 & | & -15 \\ 0 & -1 & 0 & 1 & 0 & -2 & 0 & | & 26 \\ 0 & 0 & 0 & 0 & 1 & 1 & 0 & | & 0 \\ 1 & 1 & 0 & 0 & 0 & -1 & 0 & | & 25 \\ 0 & 2 & 0 & 0 & 0 & -6 & 1 & | & 150 \end{bmatrix}$$

$$\begin{bmatrix} 0 & 1 & -1 & 0 & 0 & -3 & 0 & | & 15 \\ 0 & 0 & -1 & 1 & 0 & -5 & 0 & | & 41 \\ 0 & 0 & 0 & 0 & 1 & 1 & 0 & | & 0 \\ 1 & 0 & 1 & 0 & 0 & 2 & 0 & | & 10 \\ 0 & 0 & 2 & 0 & 0 & 0 & 1 & | & 120 \end{bmatrix}$$

Maximum $z = 120$ at $x_1 = 10$, $x_2 = 15$

25.

$$\begin{bmatrix} 7 & 12 & 12 & 1 & 0 & 0 & 0 & 0 & | & 312 \\ -13 & -20 & -12 & 0 & 1 & 0 & 0 & 0 & | & -384 \\ 5 & 4 & 12 & 0 & 0 & 1 & 0 & 0 & | & 168 \\ -5 & -4 & -12 & 0 & 0 & 0 & 1 & 0 & | & -168 \\ -10 & -24 & -26 & 0 & 0 & 0 & 0 & 1 & | & 0 \end{bmatrix}$$

$$\begin{bmatrix} -4/5 & 0 & 24/5 & 1 & 3/5 & 0 & 0 & 0 & | & 408/5 \\ 13/20 & 1 & 3/5 & 0 & -1/20 & 0 & 0 & 0 & | & 96/5 \\ 12/5 & 0 & 48/5 & 0 & 1/5 & 1 & 0 & 0 & | & 456/5 \\ -12/5 & 0 & -48/5 & 0 & -1/5 & 0 & 1 & 0 & | & -456/5 \\ 28/5 & 0 & -58/5 & 0 & -6/5 & 0 & 0 & 1 & | & 2304/5 \end{bmatrix}$$

$$\left[\begin{array}{cccccccc|c}
-2 & 0 & 0 & 1 & 1/2 & 0 & 1/2 & 0 & 36 \\
1/2 & 1 & 0 & 0 & -1/16 & 0 & 1/16 & 0 & 27/2 \\
0 & 0 & 0 & 0 & 0 & 1 & 1 & 0 & 0 \\
1/4 & 0 & 1 & 0 & 1/48 & 0 & -5/48 & 0 & 19/2 \\
\hline
17/2 & 0 & 0 & 0 & -23/24 & 0 & -29/24 & 1 & 571
\end{array}\right]$$

$$\left[\begin{array}{cccccccc|c}
-2 & 0 & 0 & 1 & 1/2 & 0 & -1/2 & 0 & 36 \\
1/2 & 1 & 0 & 0 & -1/16 & 0 & -1/16 & 0 & 27/2 \\
0 & 0 & 0 & 0 & 0 & 1 & 1 & 0 & 0 \\
1/4 & 0 & 1 & 0 & 1/48 & 0 & 5/48 & 0 & 19/2 \\
\hline
17/2 & 0 & 0 & 0 & -23/24 & 0 & 29/24 & 1 & 571
\end{array}\right]$$

$$\left[\begin{array}{cccccccc|c}
-4 & 0 & 0 & 2 & 1 & -1 & 0 & 0 & 72 \\
1/4 & 1 & 0 & 1/8 & 0 & -1/8 & 0 & 0 & 18 \\
0 & 0 & 0 & 0 & 0 & 1 & 1 & 0 & 0 \\
1/3 & 0 & 1 & -1/24 & 0 & 1/8 & 0 & 0 & 8 \\
\hline
14/3 & 0 & 0 & 23/12 & 0 & 1/4 & 0 & 1 & 640
\end{array}\right]$$
Maximum z = 640 at (0,18,8)

27.

$$\left[\begin{array}{ccccccc|c}
-2 & 5 & 1 & 0 & 0 & 0 & 0 & 90 \\
4 & 3 & 0 & 1 & 0 & 0 & 0 & 80 \\
-4 & -3 & 0 & 0 & 1 & 0 & 0 & -80 \\
-2 & 1 & 0 & 0 & 0 & 1 & 0 & -20 \\
\hline
9 & 5 & 0 & 0 & 0 & 0 & 1 & 0
\end{array}\right]
\quad
\left[\begin{array}{ccccccc|c}
0 & 13/2 & 1 & 0 & -1/2 & 0 & 0 & 130 \\
0 & 0 & 0 & 1 & 1 & 0 & 0 & 0 \\
1 & 3/4 & 0 & 0 & -1/4 & 0 & 0 & 20 \\
0 & 5/2 & 0 & 0 & -1/2 & 1 & 0 & 20 \\
\hline
0 & -7/4 & 0 & 0 & 9/4 & 0 & 1 & -180
\end{array}\right]$$

$$\left[\begin{array}{ccccccc|c}
0 & 0 & 1 & 0 & 4/5 & 13/5 & 0 & 78 \\
0 & 0 & 0 & 1 & 1 & 0 & 0 & 0 \\
1 & 0 & 0 & 0 & -1/10 & -3/10 & 0 & 14 \\
0 & 1 & 0 & 0 & -1/5 & 2/5 & 0 & 8 \\
\hline
0 & 0 & 0 & 0 & 19/10 & 7/10 & 1 & -166
\end{array}\right]$$
Minimum z is 166 at $x_1 = 14$, $x_2 = 8$

29.

$$\left[\begin{array}{cccccccc|c}
-10 & -12 & -5 & 1 & 0 & 0 & 0 & 0 & -100 \\
5 & 7 & 5 & 0 & 1 & 0 & 0 & 0 & 75 \\
-10 & -2 & -10 & 0 & 0 & 1 & 0 & 0 & -120 \\
10 & 2 & 10 & 0 & 0 & 0 & 1 & 0 & 120 \\
\hline
8 & 10 & 2 & 0 & 0 & 0 & 0 & 1 & 0
\end{array}\right]$$

$$\left[\begin{array}{cccccccc|c}
-5 & -11 & 0 & 1 & 0 & -1/2 & 0 & 0 & -40 \\
0 & 6 & 0 & 0 & 1 & 1/2 & 0 & 0 & 55 \\
1 & 1/5 & 1 & 0 & 0 & -1/10 & 0 & 0 & 12 \\
0 & 0 & 0 & 0 & 0 & 1 & 1 & 0 & 0 \\
\hline
6 & 48/5 & 0 & 0 & 0 & 1/5 & 0 & 1 & -24
\end{array}\right]$$

$$\begin{bmatrix} 1 & 11/5 & 0 & -1/5 & 0 & 1/10 & 0 & 0 & | & 8 \\ 0 & 6 & 0 & 0 & 1 & 1/2 & 0 & 0 & | & 15 \\ 0 & -2 & 1 & 1/5 & 0 & -1/5 & 0 & 0 & | & 4 \\ 0 & 0 & 0 & 0 & 0 & 1 & 1 & 0 & | & 0 \\ 0 & -18/5 & 0 & 6/5 & 0 & -2/5 & 0 & 1 & | & -72 \end{bmatrix}$$

$$\begin{bmatrix} 1 & 0 & 0 & -1/5 & -11/30 & -1/12 & 0 & 0 & | & 5/2 \\ 0 & 1 & 0 & 0 & 1/6 & 1/12 & 0 & 0 & | & 5/2 \\ 0 & 0 & 1 & 1/5 & 1/3 & -1/30 & 0 & 0 & | & 9 \\ 0 & 0 & 0 & 0 & 0 & 1 & 1 & 0 & | & 0 \\ 0 & 0 & 0 & 6/5 & 3/5 & -1/10 & 0 & 1 & | & -63 \end{bmatrix}$$

$$\begin{bmatrix} 1 & 0 & 0 & -1/5 & -11/30 & 0 & 1/12 & 0 & | & 5/2 \\ 0 & 1 & 0 & 0 & 1/6 & 0 & -1/12 & 0 & | & 5/2 \\ 0 & 0 & 1 & 1/5 & 1/3 & 0 & 1/30 & 0 & | & 9 \\ 0 & 0 & 0 & 0 & 0 & 1 & 1 & 0 & | & 0 \\ 0 & 0 & 0 & 6/5 & 3/5 & 0 & 1/10 & 1 & | & -63 \end{bmatrix}$$

Minimum $z = 63$ at $x_1 = 5/2,\ x_2 = 5/2,\ x_3 = 9$

31. Let x = number of Custom, y = number of Executive.

Minimize $C = 70x + 80y$ subject to

$x + y \geq 100$ (Number bought)

$90x + 120y \geq 10800$ (Gross sales)

$4x + 5y \leq 800$ (Storage space)

$x \geq 0, y \geq 0$

$$\begin{bmatrix} -1 & -1 & 1 & 0 & 0 & 0 & | & -100 \\ -90 & -120 & 0 & 1 & 0 & 0 & | & -10800 \\ 4 & 5 & 0 & 0 & 0 & 0 & | & 800 \\ 70 & 80 & 0 & 0 & 0 & 1 & | & 0 \end{bmatrix}$$

33. Let x = number of A, y = number of B, z = number of C.

Maximize $N = 6x + 9y + 6z$ subject to

$x + y + z \geq 6600$ (Size of order)

$20x + 25y + 15z \leq 133000$ (Total cost)

$x + 3y + 2z \leq 13600$ (Storage space)

$8x + 10y + 15z \leq 73000$ (Total weight)

$x \geq 0, y \geq 0, z \geq 0$

$$\begin{bmatrix} -1 & -1 & -1 & 1 & 0 & 0 & 0 & 0 & | & -6600 \\ 20 & 25 & 15 & 0 & 1 & 0 & 0 & 0 & | & 133000 \\ 1 & 3 & 2 & 0 & 0 & 1 & 0 & 0 & | & 13600 \\ 8 & 10 & 15 & 0 & 0 & 0 & 1 & 0 & | & 73000 \\ -6 & -9 & -6 & 0 & 0 & 0 & 0 & 1 & | & 0 \end{bmatrix}$$

35.

$$\begin{bmatrix} -1 & -1 & -1 & 1 & 0 & 0 & 0 & 0 & 0 & | & -6800 \\ 20 & 25 & 15 & 0 & 1 & 0 & 0 & 0 & 0 & | & 133000 \\ 1 & 3 & 2 & 0 & 0 & 1 & 0 & 0 & 0 & | & 13600 \\ 8 & 10 & 15 & 0 & 0 & 0 & 1 & 0 & 0 & | & 80000 \\ 0 & -1 & 0 & 0 & 0 & 0 & 0 & 1 & 0 & | & -2000 \\ -6 & -9 & -6 & 0 & 0 & 0 & 0 & 0 & 1 & | & 0 \end{bmatrix}$$

Section 4.4 Mixed Constraints

37. Let $x_1 =$ minutes jogging, $x_2 =$ minutes playing handball, $x_3 =$ minutes swimming.

(a) Minimize $z = x_1 + x_2 + x_3$ subject to

$$13x_1 + 11x_2 + 7x_3 \geq 660$$
$$x_1 - x_3 = 0$$
$$x_2 \geq 2x_1$$
$$x_1 \geq 0, x_2 \geq 0, x_3 \geq 0$$

$$\begin{bmatrix} -13 & -11 & -7 & 1 & 0 & 0 & 0 & 0 & | & -660 \\ 1 & 0 & -1 & 0 & 1 & 0 & 0 & 0 & | & 0 \\ -1 & 0 & 1 & 0 & 0 & 1 & 0 & 0 & | & 0 \\ 2 & -1 & 0 & 0 & 0 & 0 & 1 & 0 & | & 0 \\ \hline 1 & 1 & 1 & 0 & 0 & 0 & 0 & 1 & | & 0 \end{bmatrix}$$

(b) Maximize $z = 13x_1 + 11x_2 + 7x_3$ subject to

$$x_1 + x_2 + x_3 \leq 90$$
$$x_1 - x_3 = 0$$
$$x_2 \geq 2x_1$$
$$x_1 \geq 0, x_2 \geq 0, x_3 \geq 0$$

$$\begin{bmatrix} 1 & 1 & 1 & 1 & 0 & 0 & 0 & 0 & | & 90 \\ 1 & 0 & -1 & 0 & 1 & 0 & 0 & 0 & | & 0 \\ -1 & 0 & 1 & 0 & 0 & 1 & 0 & 0 & | & 0 \\ 2 & -1 & 0 & 0 & 0 & 0 & 1 & 0 & | & 0 \\ \hline -13 & -11 & -7 & 0 & 0 & 0 & 0 & 1 & | & 0 \end{bmatrix}$$

39. Let $x_1 =$ number of pounds of Lite, $x_2 =$ number of pounds of Trim, $x_3 =$ number of pounds of Regular, $x_4 =$ number of pounds of Health Fare

Maximize profit

$$z = 0.25x_1 + 0.25x_2 + 0.27x_3 + 0.32x_4 \text{ subject to}$$
$$0.75x_1 + 0.50x_2 + 0.80x_3 + 0.15x_4 \leq 2400 \text{ (Wheat)}$$
$$0.20x_1 + 0.25x_2 + 0.20x_3 + 0.50x_4 \leq 1400 \text{ (Oats)}$$
$$0.05x_1 + 0.20x_2 + 0.25x_4 \leq 700 \text{ (Raisins)}$$
$$0.05x_2 \quad 0.10x_4 \leq 250 \text{ (Nuts, maximum)}$$
$$0.05x_2 + 0.10x_4 \geq 200 \text{ (Nuts, minimum)}$$
$$x_1 \geq 0, x_2 \geq 0, x_2 \geq 0, x_4 \geq 0$$

Initial Tableau

$$\begin{bmatrix} 0.75 & 0.50 & 0.80 & 0.15 & 1 & 0 & 0 & 0 & 0 & 0 & 2400 \\ 0.20 & 0.25 & 0.20 & 0.50 & 0 & 1 & 0 & 0 & 0 & 0 & 1400 \\ 0.05 & 0.20 & 0 & 0.25 & 0 & 0 & 1 & 0 & 0 & 0 & 700 \\ 0 & 0.05 & 0 & 0.10 & 0 & 0 & 0 & 1 & 0 & 0 & 250 \\ 0 & -0.05 & 0 & -0.10 & 0 & 0 & 0 & 0 & 1 & 0 & -200 \\ \hline -0.25 & -0.25 & -0.27 & -0.32 & 0 & 0 & 0 & 0 & 0 & 1 & 0 \end{bmatrix}$$

41. Let x_1 = number units of food A, x_2 = number units of food B, x_3 = number units of food C

Minimize cost

$z = 1.40x_1 + 1.65x_2 + 1.95x_3$ subject to

$15x_1 + 10x_2 + 23x_3 \geq 80$ (protein)

$20x_1 + 30x_2 + 11x_3 \geq 95$ (carbohydrates)

$500x_1 + 400x_2 + 200x_3 \geq 1200$ (calories)

$8x_1 + 3x_2 + 6x_3 \leq 35$ (fat)

$x_1 \geq 0, x_2 \geq 0, x_3 \geq 0$

Initial Tableau

$$\begin{bmatrix} -15 & -10 & -23 & 1 & 0 & 0 & 0 & 0 & -80 \\ -20 & -30 & -11 & 0 & 1 & 0 & 0 & 0 & -95 \\ -500 & -400 & -200 & 0 & 0 & 1 & 0 & 0 & -1200 \\ 8 & 3 & 6 & 0 & 0 & 0 & 1 & 0 & 35 \\ \hline 1.40 & 1.65 & 1.95 & 0 & 0 & 0 & 0 & 1 & 0 \end{bmatrix}$$

43. Let x = number of Custom, y = number of Executive.

Minimize $C = 70x + 80y$ subject to

$x + y \geq 100$

$90x + 120y \geq 10800$

$4x + 5y \leq 800$

$x \geq 0, y \geq 0$

$$\begin{bmatrix} -1 & -1 & 1 & 0 & 0 & 0 & -100 \\ -90 & -120 & 0 & 1 & 0 & 0 & -10800 \\ 4 & 5 & 0 & 0 & 1 & 0 & 800 \\ \hline 70 & 80 & 0 & 0 & 0 & 1 & 0 \end{bmatrix} \quad \begin{bmatrix} -1/4 & 0 & 1 & -1/120 & 0 & 0 & -10 \\ 3/4 & 1 & 0 & -1/120 & 0 & 0 & 90 \\ 1/4 & 0 & 0 & 1/24 & 1 & 0 & 350 \\ \hline 10 & 0 & 0 & 2/3 & 0 & 1 & -7200 \end{bmatrix}$$

$$\begin{bmatrix} 1 & 0 & -4 & 1/30 & 0 & 0 & 40 \\ 0 & 1 & 3 & -1/30 & 0 & 0 & 60 \\ 0 & 0 & 1 & 1/30 & 1 & 0 & 340 \\ \hline 0 & 0 & 40 & 1/3 & 0 & 1 & -7600 \end{bmatrix}$$

The manager should order 40 Custom and 60 Executive for a minimum cost of $7600.

45.   Let x = number of A, y = number of B, z = number of C.
Maximize z = 6x + 9y + 6z   subject to

$$x + y + z \geq 6600$$
$$20x + 25y + 15z \leq 133000$$
$$x + 3y + 2z \leq 13600$$
$$8x + 10y + 15z \leq 73000$$
$$x \geq 0, y \geq 0, z \geq 0$$

$$\begin{bmatrix}
-1 & -1 & -1 & 1 & 0 & 0 & 0 & 0 & -6600 \\
20 & 25 & 15 & 0 & 1 & 0 & 0 & 0 & 133000 \\
1 & 3 & 2 & 0 & 0 & 1 & 0 & 0 & 13600 \\
8 & 10 & 15 & 0 & 0 & 0 & 1 & 0 & 73000 \\
\hline
-6 & -9 & -6 & 0 & 0 & 0 & 0 & 1 & 0
\end{bmatrix}$$

$$\begin{bmatrix}
1 & 1 & 1 & -1 & 0 & 0 & 0 & 0 & 6600 \\
-5 & 0 & -10 & 25 & 1 & 0 & 0 & 0 & -32000 \\
-2 & 0 & -1 & 3 & 0 & 1 & 0 & 0 & -6200 \\
-2 & 0 & 5 & 10 & 0 & 0 & 1 & 0 & 7000 \\
\hline
3 & 0 & 3 & -9 & 0 & 0 & 0 & 1 & 59400
\end{bmatrix}$$

$$\begin{bmatrix}
0 & 1 & 1/2 & 1/2 & 0 & 1/2 & 0 & 0 & 3500 \\
0 & 0 & -15/2 & 35/2 & 1 & -5/2 & 0 & 0 & -16500 \\
1 & 0 & 1/2 & -3/2 & 0 & -1/2 & 0 & 0 & 3100 \\
0 & 0 & 6 & 7 & 0 & -1 & 1 & 0 & 13200 \\
\hline
0 & 0 & 3/2 & -9/2 & 0 & 3/2 & 0 & 1 & 50100
\end{bmatrix}$$

$$\begin{bmatrix}
0 & 1 & 0 & 5/3 & 1/15 & 1/3 & 0 & 0 & 2400 \\
0 & 0 & 1 & -7/3 & -2/15 & 1/3 & 0 & 0 & 2200 \\
1 & 0 & 0 & -1/3 & 1/15 & -2/3 & 0 & 0 & 2000 \\
0 & 0 & 0 & 21 & 4/5 & -3 & 1 & 0 & 0 \\
\hline
0 & 0 & 0 & -1 & 1/5 & 1 & 0 & 1 & 46800
\end{bmatrix}$$

$$\begin{bmatrix}
0 & 1 & 0 & 0 & 1/315 & 4/7 & -5/63 & 0 & 2400 \\
0 & 0 & 1 & 0 & -2/45 & 0 & 1/9 & 0 & 2200 \\
1 & 0 & 0 & 0 & 5/63 & -5/7 & 1/63 & 0 & 2000 \\
0 & 0 & 0 & 1 & 4/105 & -1/7 & 1/21 & 0 & 0 \\
\hline
0 & 0 & 0 & 0 & 5/21 & 6/7 & 1/21 & 1 & 46800
\end{bmatrix}$$

Maximum profit is \$46,800 when 2000 of A, 2400 of B, and 2200 of C are ordered.

47.

$$\begin{bmatrix}
-1 & -1 & -1 & 1 & 0 & 0 & 0 & 0 & 0 & -6800 \\
20 & 25 & 15 & 0 & 1 & 0 & 0 & 0 & 0 & 133000 \\
1 & 3 & 2 & 0 & 0 & 1 & 0 & 0 & 0 & 13600 \\
8 & 10 & 15 & 0 & 0 & 0 & 1 & 0 & 0 & 80000 \\
0 & -1 & 0 & 0 & 0 & 0 & 0 & 1 & 0 & -2000 \\
\hline
-6 & -9 & -6 & 0 & 0 & 0 & 0 & 0 & 1 & 0
\end{bmatrix}$$

$$\begin{bmatrix}
1 & 1 & 1 & -1 & 0 & 0 & 0 & 0 & 0 & 6800 \\
-5 & 0 & -10 & 25 & 1 & 0 & 0 & 0 & 0 & -37000 \\
-2 & 0 & -1 & 3 & 0 & 1 & 0 & 0 & 0 & -6800 \\
-2 & 0 & 5 & 10 & 0 & 0 & 1 & 0 & 0 & 12000 \\
1 & 0 & 1 & -1 & 0 & 0 & 0 & 1 & 0 & 4800 \\
\hline
3 & 0 & 3 & -9 & 0 & 0 & 0 & 0 & 1 & 61200
\end{bmatrix}$$

$$\begin{bmatrix}
0 & 1 & 1/2 & 1/2 & 0 & 1/2 & 0 & 0 & 0 & 3400 \\
0 & 0 & -15/2 & 35/2 & 1 & -5/2 & 0 & 0 & 0 & -20000 \\
1 & 0 & 1/2 & -3/2 & 0 & -1/2 & 0 & 0 & 0 & 3400 \\
0 & 0 & 6 & 7 & 0 & -1 & 1 & 0 & 0 & 18800 \\
0 & 0 & 1/2 & 1/2 & 0 & 1/2 & 0 & 1 & 0 & 1400 \\
\hline
0 & 0 & 3/2 & -9/2 & 0 & 3/2 & 0 & 0 & 1 & 51000
\end{bmatrix}$$

$$\begin{bmatrix}
0 & 1 & 0 & 5/3 & 1/15 & 1/3 & 0 & 0 & 0 & 6200/3 \\
0 & 0 & 1 & -7/3 & -2/15 & 1/3 & 0 & 0 & 0 & 8000/3 \\
1 & 0 & 0 & -1/3 & 1/15 & -2/3 & 0 & 0 & 0 & 6200/3 \\
0 & 0 & 0 & 21 & 4/5 & -3 & 1 & 0 & 0 & 2800 \\
0 & 0 & 0 & 5/3 & 1/15 & 1/3 & 0 & 1 & 0 & 200/3 \\
\hline
0 & 0 & 0 & -1 & 1/5 & 1 & 0 & 0 & 1 & 47000
\end{bmatrix}$$

$$\begin{bmatrix}
0 & 1 & 0 & 0 & 0 & 0 & 0 & -1 & 0 & 2000 \\
0 & 0 & 1 & 0 & -1/25 & 4/5 & 0 & 7/5 & 0 & 2760 \\
1 & 0 & 0 & 0 & 2/25 & -3/5 & 0 & 1/5 & 0 & 2080 \\
0 & 0 & 0 & 0 & -1/25 & -36/5 & 1 & -63/5 & 0 & 1960 \\
0 & 0 & 0 & 1 & 1/25 & 1/5 & 0 & 3/5 & 0 & 40 \\
\hline
0 & 0 & 0 & 0 & 6/25 & 6/5 & 0 & 3/5 & 1 & 47040
\end{bmatrix}$$

Maximum profit is \$47,040 when ordering 2080 of A, 2000 of B, and 2760 of C.

49.   Let x_1 = minutes jogging, x_2 = minutes playing handball,

x_3 = minutes swimming.

    (a)   Minimize $z = x_1 + x_2 + x_3$ subject

$$13x_1 + 11x_2 + 7x_3 \geq 660$$

$$x_1 - x_3 = 0$$

$$x_2 \geq 2x_1$$

$$x_1 \geq 0, x_2 \geq 0, x_3 \geq 0$$

$$\begin{bmatrix}
-13 & -11 & -7 & 1 & 0 & 0 & 0 & 0 & -660 \\
1 & 0 & -1 & 0 & 1 & 0 & 0 & 0 & 0 \\
-1 & 0 & 1 & 0 & 0 & 1 & 0 & 0 & 0 \\
2 & -1 & 0 & 0 & 0 & 0 & 1 & 0 & 0 \\
\hline
1 & 1 & 1 & 0 & 0 & 0 & 0 & 1 & 0
\end{bmatrix}$$

$$\begin{bmatrix}
13/11 & 1 & 7/11 & -1/11 & 0 & 0 & 0 & 0 & 60 \\
1 & 0 & -1 & 0 & 1 & 0 & 0 & 0 & 0 \\
-1 & 0 & 1 & 0 & 0 & 1 & 0 & 0 & 0 \\
35/11 & 0 & 7/11 & -1/11 & 0 & 0 & 1 & 0 & 60 \\
\hline
-2/11 & 0 & 4/11 & 1/11 & 0 & 0 & 0 & 1 & -60
\end{bmatrix}$$

$$\begin{bmatrix}
0 & 1 & 20/11 & -1/11 & -13/11 & 0 & 0 & 0 & 60 \\
1 & 0 & -1 & 0 & 1 & 0 & 0 & 0 & 10 \\
0 & 0 & 0 & 0 & 1 & 1 & 0 & 0 & 0 \\
0 & 0 & 42/11 & -1/11 & -35/11 & 0 & 1 & 0 & 60 \\
\hline
0 & 0 & 2/11 & 1/11 & 2/11 & 0 & 0 & 1 & -60
\end{bmatrix}$$

0 minutes jogging, 60 minutes playing handball, 0 minutes swimming.

(b) Maximize $z = 13x_1 + 11x_2 + 7x_3$ subject to

$$x_1 + x_2 + x_3 \le 90$$
$$x_1 - x_3 = 0$$
$$x_2 \ge 2x_1$$
$$x_1 \ge 0, x_2 \ge 0, x_3 \ge 0$$

$$\begin{bmatrix}
1 & 1 & 1 & 1 & 0 & 0 & 0 & 0 & 90 \\
1 & 0 & -1 & 0 & 1 & 0 & 0 & 0 & 0 \\
-1 & 0 & 1 & 0 & 0 & 1 & 0 & 0 & 0 \\
2 & -1 & 0 & 0 & 0 & 0 & 1 & 0 & 0 \\
\hline
-13 & -11 & -7 & 0 & 0 & 0 & 0 & 1 & 0
\end{bmatrix}$$

$$\begin{bmatrix}
0 & 1 & 2 & 1 & -1 & 0 & 0 & 0 & 90 \\
1 & 0 & -1 & 0 & 1 & 0 & 0 & 0 & 0 \\
0 & 0 & 0 & 0 & 1 & 1 & 0 & 0 & 0 \\
0 & -1 & 2 & 0 & -2 & 0 & 1 & 0 & 0 \\
\hline
0 & -11 & -20 & 0 & 13 & 0 & 0 & 1 & 0
\end{bmatrix}$$

$$\begin{bmatrix}
0 & 2 & 0 & 1 & 1 & 0 & -1 & 0 & 90 \\
1 & -1/2 & 0 & 0 & 0 & 0 & 1/2 & 0 & 0 \\
0 & 0 & 0 & 0 & 1 & 1 & 0 & 0 & 0 \\
0 & -1/2 & 1 & 0 & -1 & 0 & 1/2 & 0 & 0 \\
\hline
0 & -21 & 0 & 0 & -7 & 0 & 10 & 1 & 0
\end{bmatrix}$$

$$\begin{bmatrix}
0 & 1 & 0 & 1/2 & 1/2 & 0 & -1/2 & 0 & 45 \\
1 & 0 & 0 & 1/4 & 1/4 & 0 & 1/4 & 0 & 45/2 \\
0 & 0 & 0 & 0 & 1 & 1 & 0 & 0 & 0 \\
0 & 0 & 1 & 1/4 & -3/4 & 0 & 1/4 & 0 & 45/2 \\
\hline
0 & 0 & 0 & 21/2 & 7/2 & 0 & -1/2 & 1 & 945
\end{bmatrix}$$

$$\begin{bmatrix}
2 & 1 & 0 & 1 & 1 & 0 & 0 & 0 & 90 \\
4 & 0 & 0 & 1 & 1 & 0 & 1 & 0 & 90 \\
0 & 0 & 0 & 0 & 1 & 1 & 0 & 0 & 0 \\
-1 & 0 & 1 & 0 & -1 & 0 & 0 & 0 & 0 \\
\hline
2 & 0 & 0 & 11 & 4 & 0 & 0 & 1 & 990
\end{bmatrix}$$

No time jogging or swimming, all 90 minutes playing handball

51.   Initial Tableau

$$\begin{bmatrix} -6 & 10 & 1 & 0 & 0 & 0 & | & -60 \\ 1 & 6 & 0 & 1 & 0 & 0 & | & 72 \\ 10 & 3 & 0 & 0 & 1 & 0 & | & 150 \\ \hline -8 & -10 & 0 & 0 & 0 & 1 & | & 0 \end{bmatrix}$$

Corner $(0, 0)$, $z = 0$
  Not a feasible solution
Tableau 1

$$\begin{bmatrix} 3/5 & 1 & -1/10 & 0 & 0 & 0 & | & 6 \\ -13/5 & 0 & 3/5 & 1 & 0 & 0 & | & 36 \\ 41/5 & 0 & 3/10 & 0 & 1 & 0 & | & 132 \\ \hline -2 & 0 & -1 & 0 & 0 & 1 & | & 60 \end{bmatrix}$$

Corner $(0, 6)$, $z = 60$
  Feasible solution.

Tableau 2

$$\begin{bmatrix} 1 & 5/3 & -1/6 & 0 & 0 & 0 & | & 10 \\ 0 & 13/3 & 1/6 & 1 & 0 & 0 & | & 62 \\ 0 & -41/3 & 5/3 & 0 & 1 & 0 & | & 50 \\ \hline 0 & 10/3 & -4/3 & 0 & 0 & 1 & | & 80 \end{bmatrix}$$

Corner $(10, 0)$, $z = 80$
  Feasible solution

Tableau 3

$$\begin{bmatrix} 1 & 3/10 & 0 & 0 & 1/10 & 0 & | & 15 \\ 0 & 57/10 & 0 & 1 & -1/10 & 0 & | & 57 \\ 0 & -41/5 & 1 & 0 & 3/5 & 0 & | & 30 \\ \hline 0 & -38/5 & 0 & 0 & 4/5 & 1 & | & 120 \end{bmatrix}$$

Corner $(15, 0)$, $z = 120$
  Feasible solution

Tableau 4

$$\begin{bmatrix} 1 & 0 & 0 & -1/19 & 2/19 & 0 & | & 12 \\ 0 & 1 & 0 & 10/57 & -1/57 & 0 & | & 10 \\ 0 & 0 & 1 & 82/57 & 26/57 & 0 & | & 112 \\ \hline 0 & 0 & 0 & 4/3 & 2/3 & 1 & | & 196 \end{bmatrix}$$

Corner $(12, 10)$, $z = 196$
Optimal feasible solution

Corner	z	Feasible?
$(0, 0)$	0	no
$(0, 6)$	60	yes
$(10, 0)$	80	yes
$(15, 0)$	120	yes
$(12, 10)$	196	yes

The simplex method starts at $(0,0)$, which is not feasible, then moves to $(0,6)$ which is feasible. From there on each pivot moves to an adjacent corner that is in the feasible region.

53.   **(a)**

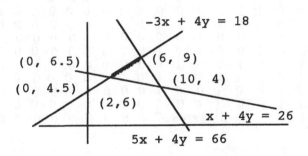

Section 4.4 Mixed Constraints

(b) The feasible region is the portion on the line
$-3x + 4y = 18$ that lies in the region bounded by $5x + 4y \leq 66$, by
$x + 4y \geq 26$. This is simply the line segment joining $(2, 6)$ and $(6, 9)$.

(c) The feasible region is the segment of the line
$a_1x + b_1y = k_1$ where it intersects the feasible region defined by
$a_2x + b_2y \leq k_2$
$a_3x + b_3y \geq k_3$ and
$x \geq 0, y \geq 0$

55. Let the variables be the following:
x_1, x_2, and x_3 = quantity shipped from Cleveland to Chicago,
Dallas, and Atlanta, respectively.
x_4, x_5, and x_6 = quantity shipped from St. Louis to Chicago, Dallas,
and Atlanta, respectively.
x_7, x_8, and x_9 = quantity shipped from Pittsburgh to Chicago,
Dallas, and Atlanta, respectively.

(a) The constraints are
$x_1 + x_2 + x_3 \leq 4200$ (Quantity shipped from Cleveland does not
exceed production)
$x_4 + x_5 + x_6 \leq 4800$ (Quantity shipped from St. Louis)
$x_7 + x_8 + x_9 \leq 3700$ (Quantity shipped from Pittsburgh)
$x_1 + x_4 + x_7 = 5400$ (Quantity shipped to Chicago)
$x_2 + x_5 + x_8 = 3800$ (Quantity shipped to Dallas)
$x_3 + x_6 + x_9 = 2700$ (Quantity shipped to Atlanta)
with all variables nonnegative.
Since we are to minimize shipping costs, the objective function is
Minimize $z = 35x_1 + 64x_2 + 60x_3 + 37x_4 + 59x_5 + 51x_6 + 49x_7 + 68x_8 + 57x_9$

(b) The initial simplex tableau is

$$
\begin{array}{ccccccccccccccccccccc|c}
x_1 & x_2 & x_3 & x_4 & x_5 & x_6 & x_7 & x_8 & x_9 & s_1 & s_2 & s_3 & s_4 & s_5 & s_6 & s_7 & s_8 & s_9 & z & \\
1 & 1 & 1 & 0 & 0 & 0 & 0 & 0 & 0 & 1 & 0 & 0 & 0 & 0 & 0 & 0 & 0 & 0 & 0 & 4200 \\
0 & 0 & 0 & 1 & 1 & 1 & 0 & 0 & 0 & 0 & 1 & 0 & 0 & 0 & 0 & 0 & 0 & 0 & 0 & 4800 \\
0 & 0 & 0 & 0 & 0 & 0 & 1 & 1 & 1 & 0 & 0 & 1 & 0 & 0 & 0 & 0 & 0 & 0 & 0 & 3700 \\
1 & 0 & 0 & 1 & 0 & 0 & 1 & 0 & 0 & 0 & 0 & 0 & 1 & 0 & 0 & 0 & 0 & 0 & 0 & 5400 \\
-1 & 0 & 0 & -1 & 0 & 0 & -1 & 0 & 0 & 0 & 0 & 0 & 0 & 1 & 0 & 0 & 0 & 0 & 0 & -5400 \\
0 & 1 & 0 & 0 & 1 & 0 & 0 & 1 & 0 & 0 & 0 & 0 & 0 & 0 & 1 & 0 & 0 & 0 & 0 & 3800 \\
0 & -1 & 0 & 0 & -1 & 0 & 0 & -1 & 0 & 0 & 0 & 0 & 0 & 0 & 0 & 1 & 0 & 0 & 0 & -3800 \\
0 & 0 & 1 & 0 & 0 & 1 & 0 & 0 & 1 & 0 & 0 & 0 & 0 & 0 & 0 & 0 & 1 & 0 & 0 & 2700 \\
0 & 0 & -1 & 0 & 0 & -1 & 0 & 0 & -1 & 0 & 0 & 0 & 0 & 0 & 0 & 0 & 0 & 1 & 0 & -2700 \\
35 & 64 & 60 & 37 & 59 & 51 & 49 & 68 & 57 & 0 & 0 & 0 & 0 & 0 & 0 & 0 & 0 & 0 & 1 & 0 \\
\end{array}
$$

57.   Initial tableau   Final tableau

$$\begin{bmatrix} 1 & 4 & 3 & 1 & 0 & 0 & 0 & 134 \\ -2 & -10 & -5 & 0 & 1 & 0 & 0 & -280 \\ 5 & 1 & 3 & 0 & 0 & 1 & 0 & 100 \\ -7 & -7 & -3 & 0 & 0 & 0 & 1 & 0 \end{bmatrix} \quad \begin{bmatrix} 0 & 0 & 2.26 & 2.53 & 1 & -0.11 & 0 & 48 \\ 0 & 1 & 0.63 & 0.26 & 0 & -0.05 & 0 & 30 \\ 1 & 0 & 0.47 & -0.05 & 0 & 0.21 & 0 & 14 \\ 0 & 0 & 4.74 & 1.47 & 0 & 1.11 & 1 & 308 \end{bmatrix}$$

Maximum z = 308 at the corner (14, 30, 0).

59.   Initial tableau

$$\begin{bmatrix} 4 & 3 & 2 & 5 & 1 & 0 & 0 & 0 & 0 & 113 \\ 1 & 2 & 6 & 4 & 0 & 1 & 0 & 0 & 0 & 92 \\ 8 & 4 & 4 & 10 & 0 & 0 & 1 & 0 & 0 & 212 \\ 2 & 3 & 4 & 5 & 0 & 0 & 0 & 1 & 0 & 107 \\ -16 & -14 & -22 & -28 & 0 & 0 & 0 & 0 & 1 & 0 \end{bmatrix}$$

Final tableau

$$\begin{bmatrix} 0 & 1 & 0 & 0 & 1 & 0 & -0.5 & 0 & 0 & 7 \\ 1 & 0 & 0 & 0 & 0.82 & 0.45 & -0.09 & -1 & 0 & 8 \\ 0 & 0 & 0 & 1 & -1.18 & -0.55 & -0.41 & 1 & 0 & 10 \\ 0 & 0 & 1 & 0 & 0.32 & 0.45 & -0.09 & -0.5 & 0 & 5 \\ 0 & 0 & 0 & 0 & 1 & 2 & 1 & 1 & 1 & 616 \end{bmatrix}$$

The maximum value of z is 616 and it occurs at (8, 7, 5, 10).

Section 4.5

1.   The -20 in the last column with nonnegative entries in the rest of the first row indicates no feasible solutions.

3.   There is an unbounded feasible region, so no solution, because all entries in column 4 are negative.

5.

$$\begin{bmatrix} 3 & 5 & 1 & 0 & 0 & 0 & 60 \\ 1 & 1 & 0 & 1 & 0 & 0 & 14 \\ 2 & 1 & 0 & 0 & 1 & 0 & 24 \\ -15 & -15 & 0 & 0 & 0 & 1 & 0 \end{bmatrix} \quad \begin{bmatrix} 3/5 & 1 & 1/5 & 0 & 0 & 0 & 12 \\ 2/5 & 0 & -1/5 & 1 & 0 & 0 & 2 \\ 7/5 & 0 & -1/5 & 0 & 1 & 0 & 12 \\ -6 & 0 & 3 & 0 & 0 & 1 & 180 \end{bmatrix}$$

$$\begin{bmatrix} 0 & 1 & 1/2 & -3/2 & 0 & 0 & 9 \\ 1 & 0 & -1/2 & 5/2 & 0 & 0 & 5 \\ 0 & 0 & 1/2 & -7/2 & 1 & 0 & 5 \\ 0 & 0 & 0 & 15 & 0 & 1 & 210 \end{bmatrix}$$

This gives maximum of 210 at (5, 9).

$$\begin{bmatrix} 0 & 1 & 0 & 2 & -1 & 0 & 4 \\ 1 & 0 & 0 & -1 & 1 & 0 & 10 \\ 0 & 0 & 1 & -7 & 2 & 0 & 10 \\ 0 & 0 & 0 & 15 & 0 & 1 & 210 \end{bmatrix}$$

This gives maximimum of 210 also at (10, 4).

Maximum $z = 210$ at (5, 9), (10, 4), and on the line segment between.

7.

$$\begin{bmatrix} 2 & 1 & 2 & 1 & 0 & 0 & 0 & 20 \\ 1 & 2 & 2 & 0 & 1 & 0 & 0 & 20 \\ 1 & 1 & 4 & 0 & 0 & 1 & 0 & 20 \\ -3 & -3 & -4 & 0 & 0 & 0 & 1 & 0 \end{bmatrix}$$

$$\begin{bmatrix} 3/2 & 1/2 & 0 & 1 & 0 & -1/2 & 0 & 0 & 10 \\ 1/2 & 3/2 & 0 & 0 & 1 & -1/2 & 0 & 10 \\ 1/4 & 1/4 & 1 & 0 & 0 & 1/4 & 0 & 5 \\ -2 & -2 & 0 & 0 & 0 & 1 & 1 & 20 \end{bmatrix}$$

$$\begin{bmatrix} 1 & 1/3 & 0 & 2/3 & 0 & -1/3 & 0 & 20/3 \\ 0 & 4/3 & 0 & -1/3 & 1 & -1/3 & 0 & 20/3 \\ 0 & 1/6 & 1 & -1/6 & 0 & 1/3 & 0 & 10/3 \\ 0 & -4/3 & 0 & 4/3 & 0 & 1/3 & 1 & 100/3 \end{bmatrix}$$

$$\begin{bmatrix} 1 & 0 & 0 & 3/4 & -1/4 & -1/4 & 0 & 5 \\ 0 & 1 & 0 & -1/4 & 3/4 & -1/4 & 0 & 5 \\ 0 & 0 & 1 & -1/8 & -1/8 & 3/8 & 0 & 5/2 \\ 0 & 0 & 0 & 1 & 1 & 0 & 1 & 40 \end{bmatrix}$$

Maximum = 40 at (5, 5, 2.5)

$$\begin{bmatrix} 1 & 0 & 2/3 & 2/3 & -1/3 & 0 & 0 & 20/3 \\ 0 & 1 & 2/3 & -1/3 & 2/3 & 0 & 0 & 20/3 \\ 0 & 0 & 8/3 & -1/3 & -1/3 & 1 & 0 & 20/3 \\ 0 & 0 & 0 & 1 & 1 & 0 & 1 & 40 \end{bmatrix}$$

Maximum = 40 also at (20/3, 20/3, 0)

Maximum $z = 40$ at (5, 5, 2.5), (20/3, 20/3, 0), and on the line segment between.

9.

$$\begin{bmatrix} 2 & -5 & 1 & 0 & 0 & 10 \\ 2 & -1 & 0 & 1 & 0 & -2 \\ -8 & -3 & 0 & 0 & 1 & 0 \end{bmatrix} \quad \begin{bmatrix} -8 & 0 & 1 & -5 & 0 & 20 \\ -2 & 1 & 0 & -1 & 0 & 2 \\ -14 & 0 & 0 & -3 & 1 & 6 \end{bmatrix}$$

All entries in column 1 are negative, unbounded feasible region

11.

$$\begin{bmatrix} 1 & -3 & 2 & 1 & 0 & 0 & 50 \\ -2 & 4 & 5 & 0 & 1 & 0 & 40 \\ -8 & -6 & -2 & 0 & 0 & 1 & 0 \end{bmatrix} \quad \begin{bmatrix} 1 & -3 & 2 & 1 & 0 & 0 & 50 \\ 0 & -2 & 9 & 2 & 1 & 0 & 140 \\ 0 & -30 & 14 & 8 & 0 & 1 & 400 \end{bmatrix}$$

All entries in column 2 are negative, unbounded feasible region

13.

$$\begin{bmatrix} 1 & -1 & 1 & 0 & -13 \\ 2 & 9 & 0 & 1 & 72 \\ -12 & -20 & 0 & 0 & 0 \end{bmatrix} \quad \begin{bmatrix} -1 & 1 & -1 & 0 & 13 \\ 11 & 0 & 9 & 1 & -45 \\ -32 & 0 & -20 & 0 & 260 \end{bmatrix}$$

Pivoting leads to a row with a negative entry in row 2 of the last column and all other entries in the row are nonnegative. Thus, there is no feasible solution.

15.

$$\left[\begin{array}{ccccccc|c} 6 & 4 & 3 & 1 & 0 & 0 & 0 & 60 \\ 3 & 6 & 4 & 0 & 1 & 0 & 0 & 48 \\ 1 & 1 & -2 & 0 & 0 & 1 & 0 & -60 \\ \hline -20 & -30 & -15 & 0 & 0 & 0 & 1 & 0 \end{array}\right]$$

$$\left[\begin{array}{ccccccc|c} 15/2 & 11/2 & 0 & 1 & 0 & 3/2 & 0 & -30 \\ 5 & 8 & 0 & 0 & 1 & 2 & 0 & -72 \\ -1/2 & -1/2 & 1 & 0 & 0 & -1/2 & 0 & 30 \\ \hline -55/2 & -75/2 & 0 & 0 & 0 & -15/2 & 1 & 450 \end{array}\right]$$

Pivoting leads to a row with a negative entry the last column of row 2 and all other entries in the row are nonnegative. Thus, there is no feasible solution.

17.

$$\left[\begin{array}{ccccc|c} -4 & 1 & 1 & 0 & 0 & 2 \\ 2 & -1 & 0 & 1 & 0 & 1 \\ \hline -1 & -4 & 0 & 0 & 1 & 0 \end{array}\right] \quad \left[\begin{array}{ccccc|c} -4 & 1 & 1 & 0 & 0 & 2 \\ -2 & 0 & 1 & 1 & 0 & 3 \\ \hline -17 & 0 & 4 & 0 & 1 & 8 \end{array}\right]$$

All entries in column 1 are negative, unbounded feasible region

19.

$$\left[\begin{array}{ccccccc|c} 9 & -4 & -6 & 1 & 0 & 0 & 0 & -36 \\ 4 & 5 & 8 & 0 & 1 & 0 & 0 & 40 \\ 6 & -2 & 1 & 0 & 0 & 1 & 0 & 18 \\ \hline -18 & -15 & -8 & 0 & 0 & 0 & 1 & 0 \end{array}\right] \quad \left[\begin{array}{ccccccc|c} -3/2 & 2/3 & 1 & -1/6 & 0 & 0 & 0 & 6 \\ 16 & -1/3 & 0 & 4/3 & 1 & 0 & 0 & -8 \\ 15/2 & -8/3 & 0 & 1/6 & 0 & 1 & 0 & 12 \\ \hline -30 & -29/3 & 0 & -4/3 & 0 & 0 & 1 & 48 \end{array}\right]$$

$$\left[\begin{array}{ccccccc|c} 61/2 & 0 & 1 & 5/2 & 2 & 0 & 0 & -10 \\ -48 & 1 & 0 & -4 & -3 & 0 & 0 & 24 \\ -241/2 & 0 & 0 & -21/2 & -8 & 1 & 0 & 76 \\ \hline -494 & 0 & 0 & -40 & -29 & 0 & 1 & 280 \end{array}\right]$$

The last entry in row 1 is negative and the rest of the row is nonnegative, no feasible solution

21.

$$\left[\begin{array}{ccccc|c} 5 & 3 & 1 & 0 & 0 & 30 \\ -2 & -1 & 0 & 1 & 0 & -20 \\ \hline -15 & -9 & 0 & 0 & 1 & 0 \end{array}\right] \quad \left[\begin{array}{ccccc|c} 0 & 1/2 & 1 & 5/2 & 0 & -20 \\ 1 & 1/2 & 0 & -1/2 & 0 & 10 \\ \hline 0 & -3/2 & 0 & -15/2 & 1 & 150 \end{array}\right]$$

The last entry in row 1 is negative and the rest of the row is nonnegative, no feasible solution

23.

$$\left[\begin{array}{ccccc|c} 10 & 15 & 1 & 0 & 0 & 150 \\ -6 & -3 & 0 & 1 & 0 & -180 \\ \hline -4 & -12 & 0 & 0 & 1 & 0 \end{array}\right] \quad \left[\begin{array}{ccccc|c} 0 & 10 & 1 & 5/3 & 0 & -150 \\ 1 & 1/2 & 0 & -1/6 & 0 & 30 \\ \hline 0 & -10 & 0 & -2/3 & 1 & 120 \end{array}\right]$$

The last entry in row 1 is negative and the rest of the row is nonnegative, no feasible solution

25.

$$\begin{bmatrix} -2 & -3 & -4 & 1 & 0 & 0 & 0 & | & -60 \\ -2 & -3 & 6 & 0 & 1 & 0 & 0 & | & -30 \\ 0 & 6 & -5 & 0 & 0 & 1 & 0 & | & 30 \\ \hline 2 & 3 & 1 & 0 & 0 & 0 & 1 & | & 0 \end{bmatrix} \quad \begin{bmatrix} 1/2 & 3/4 & 1 & -1/4 & 0 & 0 & 0 & | & 5 \\ -5 & -15/2 & 0 & 3/2 & 1 & 0 & 0 & | & -120 \\ 5/2 & 39/4 & 0 & -5/4 & 0 & 1 & 0 & | & 105 \\ \hline 3/2 & 9/4 & 0 & 1/4 & 0 & 0 & 1 & | & -15 \end{bmatrix}$$

$$\begin{bmatrix} 0 & 0 & 1 & -1/10 & 1/10 & 0 & 0 & | & 3 \\ 2/3 & 1 & 0 & -1/5 & -2/15 & 0 & 0 & | & 16 \\ -4 & 0 & 0 & 7/10 & 13/10 & 1 & 0 & | & -51 \\ \hline 0 & 0 & 0 & 7/10 & 3/10 & 0 & 1 & | & -51 \end{bmatrix}$$

$$\begin{bmatrix} 0 & 0 & 1 & -1/10 & 1/10 & 0 & 0 & | & 3 \\ 0 & 1 & 0 & -1/12 & 1/12 & 1/6 & 0 & | & 15/2 \\ 1 & 0 & 0 & -7/40 & -13/40 & -1/4 & 0 & | & 51/4 \\ \hline 0 & 0 & 0 & 7/10 & 3/10 & 0 & 1 & | & -51 \end{bmatrix}$$

Column 6 indicates multiple solutions.

$$\begin{bmatrix} 0 & 0 & 1 & -1/10 & 1/10 & 0 & 0 & | & 3 \\ 0 & 6 & 0 & -1/2 & 1/2 & 1 & 0 & | & 45 \\ 1 & 3/2 & 0 & -3/10 & -1/5 & 0 & 0 & | & 24 \\ \hline 0 & 0 & 0 & 7/10 & 3/10 & 0 & 1 & | & -51 \end{bmatrix}$$

Multiple solutions, minimum $z = 51$ at $(51/4, 15/2, 3)$, $(24, 0, 3)$, and points on the line segment between.

27.

$$\begin{bmatrix} 3 & -10 & 6 & 1 & 0 & 0 & 0 & | & 60 \\ -15 & 4 & 6 & 0 & 1 & 0 & 0 & | & 60 \\ 4 & 5 & -20 & 0 & 0 & 1 & 0 & | & 100 \\ \hline -8 & -5 & -8 & 0 & 0 & 0 & 1 & | & 0 \end{bmatrix} \quad \begin{bmatrix} 1/2 & -5/3 & 1 & 1/6 & 0 & 0 & 0 & | & 10 \\ -18 & 14 & 0 & -1 & 1 & 0 & 0 & | & 0 \\ 14 & -85/3 & 0 & 10/3 & 0 & 1 & 0 & | & 300 \\ \hline -4 & -55/3 & 0 & 4/3 & 0 & 0 & 1 & | & 80 \end{bmatrix}$$

$$\begin{bmatrix} -23/14 & 0 & 1 & 1/21 & 5/42 & 0 & 0 & | & 10 \\ -9/7 & 1 & 0 & -1/14 & 1/14 & 0 & 0 & | & 0 \\ -157/7 & 0 & 0 & 55/42 & 85/42 & 1 & 0 & | & 300 \\ \hline -193/7 & 0 & 0 & 1/42 & 55/42 & 0 & 1 & | & 80 \end{bmatrix}$$

All entries in column 1 are negative, unbounded feasible region

29. Let x = number of Petite, y = number of Deluxe.
Maximize $z = 5x + 6y$ subject to
$$x + 2y \le 3950$$
$$4x + 3y \le 9575$$
$$y \ge 2000$$
$$x \ge 0, y \ge 0$$

$$\begin{bmatrix} 1 & 2 & 1 & 0 & 0 & 0 & | & 3950 \\ 4 & 3 & 0 & 1 & 0 & 0 & | & 9575 \\ 0 & -1 & 0 & 0 & 1 & 0 & | & -2000 \\ \hline -5 & -6 & 0 & 0 & 0 & 1 & | & 0 \end{bmatrix} \quad \begin{bmatrix} 1 & 0 & 1 & 0 & 2 & 0 & | & -50 \\ 4 & 0 & 0 & 1 & 3 & 0 & | & 3575 \\ 0 & 1 & 0 & 0 & -1 & 0 & | & 2000 \\ \hline -5 & 0 & 0 & 0 & -6 & 1 & | & 12000 \end{bmatrix}$$

The last entry in row 1 is negative and the rest of the row is nonnegative, no feasible solution

31. Let x = amount invested in stocks and y = amount invested in bonds.
Maximize expected return z = 0.10x + 0.07y subject to

$+ y \geq 10,000$ (amount invested)

$x \geq 5000 + y$ (5000 more in stocks than bonds)

$x \geq 0, y \geq 0$

The simplex tableau is

$$\left[\begin{array}{ccccc|c} -1 & -1 & 1 & 0 & 0 & -10,000 \\ -1 & 1 & 0 & 1 & 0 & -5,000 \\ \hline -0.10 & -0.07 & 10 & -0 & 1 & 0 \end{array}\right] \quad \left[\begin{array}{ccccc|c} 1 & 1 & -1 & 0 & 0 & 10,000 \\ 0 & 2 & -1 & 1 & 0 & 5,000 \\ \hline 0 & 0.03 & -0.10 & 0 & 1 & 1,000 \end{array}\right]$$

Column 3 contains all negative entries which indicates an unbounded feasible region so there is no maximum expected return.

33. Let x_1 = number units of food A, x_2 = number units of food B,

x_3 = number units of food C.

Minimize cost z = $1.40x_1 + x_2 + 1.50x_3$ subject to

$15x_1 + 10x_2 + 23x_3 \geq 80$ (protein)

$20x_1 + 30x_2 + 11x_3 \geq 95$ (carbohydrates)

$200x_1 + 160x_2 + 160x_3 \geq 1400$ (calories)

$10x_1 + 5x_2 + 6x_3 \leq 40$ (fat)

$x_1 \geq 0, x_2 \geq 0, x_3 \geq 0$

$$\left[\begin{array}{cccccccc|c} -15 & -10 & -23 & 1 & 0 & 0 & 0 & 0 & -80 \\ -20 & -30 & -11 & 0 & 1 & 0 & 0 & 0 & -95 \\ -200 & -160 & -160 & 0 & 0 & 1 & 0 & 0 & -1400 \\ 10 & 5 & 6 & 0 & 0 & 0 & 1 & 0 & 40 \\ \hline 1.4 & 1.0 & 1.5 & 0 & 0 & 0 & 0 & 1 & 0 \end{array}\right]$$

$$\left[\begin{array}{cccccccc|c} 0 & 2 & -11 & 1 & 0 & -0.08 & 0 & 0 & 25 \\ 0 & -14 & 5 & 0 & 1 & -0.10 & 0 & 0 & 45 \\ 1 & 0.8 & 0.8 & 0 & 0 & -0.01 & 0 & 0 & 7 \\ 0 & -3 & -2 & 0 & 0 & 0.05 & 1 & 0 & -30 \\ \hline 0 & -0.12 & 0.38 & 0 & 0 & 0.01 & 0 & 1 & -9.8 \end{array}\right]$$

$$\left[\begin{array}{cccccccc|c} 0 & 0 & -12.33 & 1 & 0 & -0.04 & 0.67 & 0 & 5 \\ 0 & 0 & 14.33 & 0 & 1 & -0.33 & -4.67 & 0 & 185 \\ 1 & 0 & 0.27 & 0 & 0 & -0.01 & 0.27 & 0 & -1 \\ 0 & 1 & 0.67 & 0 & 0 & -0.02 & -0.33 & 0 & 10 \\ \hline 0 & 0 & 0.46 & 0 & 0 & 0.01 & -0.04 & 1 & -8.6 \end{array}\right]$$

$$\left[\begin{array}{cccccccc|c} 5 & 0 & -11 & 1 & 0 & 0 & 2 & 0 & 0 \\ 40 & 0 & 25 & 0 & 1 & 0 & 6 & 0 & 145 \\ 120 & 0 & 32 & 0 & 0 & 1 & 32 & 0 & -120 \\ 2 & 1 & 1.2 & 0 & 0 & 0 & 0.2 & 0 & 8 \\ \hline -0.6 & 0 & 0.3 & 0 & 0 & 0 & -0.2 & 1 & -8 \end{array}\right]$$

Notice row 3. The last entry is negative and all other entries are nonnegative. This indicates no feasible solution.

Section 4.6

1. **(a)** For $(0, 0)$, $s_1 = 17$. For $(2, 2)$ $s_1 = 3$. For $(5, 10)$, $s_1 = -33$. For $(2, 1)$, $s_1 = 6$. For $(2, 3)$, $s_1 = 0$.
   (b) $(0, 0)$, $(2, 2)$, $(2, 1)$, and $(2, 3)$ are in the feasible region.
   (c) $(2, 3)$ is on the line.

3. Let $x_2 = 0$, $s_2 = 0$ in the second constraint to obtain $5x_1 = 30$, $x_1 = 6$. Thus, the corner point is $(6, 0)$.

5. For $(0, 0, 0)$, $s_1 = 40$. For $(1, 2, 3)$, $s_1 = 26$.
   For $(0, 10, 0)$, $s_1 = 0$. For $(4, 2, 7)$, $s_1 = 13$.

7.

Point	s_1	s_2	s_3	Is point on boundary?	Is point in feasible region?
$(5, 10)$	15	32	18	No	Yes
$(8, 10)$	12	14	-3	No	No
$(5, 13)$	0	29	0	Yes	Yes
$(11, 13)$	-6	-7	-42	No	No
$(10, 12)$	0	0	-29	No	No
$(15, 11)$	0	-29	-58	No	No

9. **(i)** **(a)** $x_1 = 0$, $x_2 = 0$, $s_1 = 900$, $s_2 = 2800$, $z = 0$
   (b) Intersection of $x_1 = 0$ and $x_2 = 0$
   (ii) **(a)** $x_1 = 180$, $x_2 = 0$, $s_1 = 0$, $s_2 = 1360$, $z = 540$
   (b) Intersection of $5x_1 + 2x_2 = 900$ and $x_2 = 0$
   (iii) **(a)** $x_1 = 100$, $x_2 = 200$, $s_1 = 0$, $s_2 = 0$, $z = 700$
   (b) Intersection of $5x_1 + 2x_2 = 900$ and $8x_1 + 10x_2 = 2800$

11. **(a)** For $(3, 5)$ to be on the boundary line $s_1 = 0$ and $6(3) + 4(5) = 24$. Since the latter is false, $(3, 5)$ is not on the boundary line.
    (b) For $(3, 5)$ to be in the feasible region, $6(3) + 4(5) + s_1 = 24$ for some nonnegative value of s_1. However, the equation is true only when $s_1 = -14$ so $(3, 5)$ cannot be in the feasible region.

13.   **(a)**   Both $x_2 \le 36/7$ and $x_2 \le 32/5$ must hold, so x_2 must be the smaller, $36/7$.

   (b) $x_1 \le 6$ and $x_1 \le 16$, so x_1 is the smaller, 6.

15.   Initial tableau

$$\begin{bmatrix} -7 & 10 & 1 & 0 & 0 & 0 & 0 & | & 50 \\ 3 & 5 & 0 & 1 & 0 & 0 & 0 & | & 90 \\ 4 & 5 & 0 & 0 & 1 & 0 & 0 & | & 105 \\ 1 & 0 & 0 & 0 & 0 & 1 & 0 & | & 20 \\ \hline -17 & -20 & 0 & 0 & 0 & 0 & 1 & | & 0 \end{bmatrix}$$

Corner $(0, 0)$, $z = 0$, slack $(50, 90, 105, 20)$

Tableau 1

$$\begin{bmatrix} -0.7 & 1 & 0.1 & 0 & 0 & 0 & 0 & | & 5 \\ 6.5 & 0 & -0.5 & 1 & 0 & 0 & 0 & | & 65 \\ 7.5 & 0 & -0.5 & 0 & 1 & 0 & 0 & | & 80 \\ 1.0 & 0 & 0 & 0 & 0 & 1 & 0 & | & 20 \\ \hline -31 & 0 & 2 & 0 & 0 & 0 & 1 & | & 100 \end{bmatrix}$$

Corner $(0, 5)$, $z = 100$,)
slack $(0, 65, 80, 20$

Tableau 2

$$\begin{bmatrix} 0 & 1 & 0.046 & 0.108 & 0 & 0 & 0 & | & 12 \\ 1 & 0 & -0.077 & 0.154 & 0 & 0 & 0 & | & 10 \\ 0 & 0 & 0.077 & -1.154 & 1 & 0 & 0 & | & 5 \\ 0 & 0 & 0.077 & 0.154 & 0 & 1 & 0 & | & 10 \\ \hline 0 & 0 & -0.385 & 4.769 & 0 & 0 & 1 & | & 410 \end{bmatrix}$$

Corner $(10, 12)$, $z = 410$,
slack $(0, 0, 5, 10)$

Tableau 3

$$\begin{bmatrix} 0 & 1 & 0 & 0.80 & -0.6 & 0 & 0 & | & 9 \\ 1 & 0 & 0 & -1.00 & 1.0 & 0 & 0 & | & 15 \\ 0 & 0 & 1 & -15.00 & 13.0 & 0 & 0 & | & 65 \\ 0 & 0 & 0 & 1.00 & -1.0 & 1 & 0 & | & 5 \\ \hline 0 & 0 & 0 & -1.00 & 5.0 & 0 & 1 & | & 435 \end{bmatrix}$$

Corner $(15, 9)$, $z = 435$,
slack $(65, 0, 0, 5)$

Tableau 4

$$\begin{bmatrix} 0 & 1 & 0 & 0 & 0.2 & -0.8 & 0 & | & 5 \\ 1 & 0 & 0 & 0 & 0 & 1.0 & 0 & | & 20 \\ 0 & 0 & 1 & 0 & -2.0 & 15.0 & 0 & | & 140 \\ 0 & 0 & 0 & 1 & -1.0 & 1.0 & 0 & | & 5 \\ \hline 0 & 0 & 0 & 0 & 4.0 & 1.0 & 1 & | & 440 \end{bmatrix}$$

Optimal solution reached
Corner $(20, 5)$, $z = 440$,
slack $(140, 5, 0, 0)$

Corner (x,y)	Value of z	Increase in z	Slack (s_1, s_2, s_3, s_4)
$(0, 0)$	0		$(50, 90, 105, 20)$
$(0, 5)$	100	100	$(0, 65, 80, 20)$
$(10, 12)$	410	310	$(0, 0, 5, 10)$
$(15, 9)$	435	25	$(65, 0, 0, 5)$
$(20, 5)$	440	5	$(140, 5, 0, 0)$

17.   Initial tableau

$$\begin{bmatrix} 1 & 6 & 3 & 1 & 0 & 0 & 0 & | & 36 \\ 3 & 6 & 6 & 0 & 1 & 0 & 0 & | & 45 \\ 5 & 6 & 1 & 0 & 0 & 1 & 0 & | & 46 \\ \hline -10 & -24 & -13 & 0 & 0 & 0 & 1 & | & 0 \end{bmatrix}$$

Corner $(0, 0)$, $z = 0$ slack $(36, 45, 46)$

Tableau 1

$$\begin{bmatrix} 0.167 & 1 & 0.5 & 0.167 & 0 & 0 & 6 \\ 2.000 & 0 & 3.0 & -1.000 & 1 & 0 & 9 \\ 4.000 & 0 & -2.0 & -1.000 & 0 & 1 & 10 \\ \hline -6.000 & 0 & -1.0 & 4.000 & 0 & 0 & 144 \end{bmatrix}$$

Corner $(0, 6, 0)$, $z = 144$, slack $(0, 9, 10)$

Tableau 2

$$\begin{bmatrix} 0 & 1 & 0.583 & 0.208 & 0 & -0.042 & 0 & 5.583 \\ 0 & 0 & 4.000 & -0.500 & 1 & -0.500 & 0 & 4.000 \\ 1 & 0 & -0.500 & -0.250 & 0 & 0.250 & 0 & 2.500 \\ \hline 0 & 0 & -4.000 & 2.500 & 0 & 1.500 & 1 & 159 \end{bmatrix}$$

Corner $(2.5, 5.583, 0)$, $z = 159$, slack $(0, 4, 0)$

Tableau 3

$$\begin{bmatrix} 0 & 1 & 0 & 0.281 & -0.146 & 0.031 & 0 & 5 \\ 0 & 0 & 1 & -0.125 & 0.250 & -0.125 & 0 & 1 \\ 1 & 0 & 0 & -0.313 & 0.125 & 0.188 & 0 & 3 \\ \hline 0 & 0 & 0 & 2.000 & 1.000 & 1.000 & 1 & 163 \end{bmatrix}$$

Corner $(3, 5, 1)$, $z = 163$, slack $(0, 0, 0)$
Optimal solution

Corner (x_1, x_2, x_3)	Value of z	Increase in z	Slack (s_1, s_2, s_3)
$(0, 0, 0)$	0		$(36, 45, 46)$
$(0, 6, 0)$	144	144	$(0, 9, 10)$
$(2.5, 5.583, 0)$	159	15	$(0, 4, 0)$
$(3, 5, 1)$	163	4	$(0, 0, 0)$

19.   Initial Tableau

$$\begin{bmatrix} 2 & 1 & 2 & 1 & 0 & 0 & 0 & 330 \\ 1 & 2 & 2 & 0 & 1 & 0 & 0 & 330 \\ -2 & -2 & 1 & 0 & 0 & 1 & 0 & 132 \\ \hline -1 & -2 & -3 & 0 & 0 & 0 & 1 & 0 \end{bmatrix}$$

Corner $(0, 0, 0)$, $z = 0$, slack $(330, 330, 132)$

Tableau 1

$$\begin{bmatrix} 6 & 5 & 0 & 1 & 0 & -2 & 0 & 66 \\ 5 & 6 & 0 & 0 & 1 & -2 & 0 & 66 \\ -2 & -2 & 1 & 0 & 0 & 1 & 0 & 132 \\ \hline -7 & -8 & 0 & 0 & 0 & 3 & 1 & 396 \end{bmatrix}$$

Corner $(0, 0, 132)$, $z = 396$,
slack $(66, 66, 0)$

Tableau 2

$$\begin{bmatrix} 1.833 & 0 & 0 & 1 & -0.833 & -0.333 & 0 & 11 \\ 0.833 & 1 & 0 & 0 & 0.167 & -0.333 & 0 & 11 \\ -0.333 & 0 & 1 & 0 & 0.333 & 0.333 & 0 & 154 \\ \hline -0.333 & 0 & 0 & 0 & 1.333 & 0.333 & 1 & 484 \end{bmatrix}$$

Corner $(0, 11, 154)$, $z = 484$,
slack $(11, 0, 0)$

Tableau 3

$$\begin{bmatrix} 1 & 0 & 0 & 0.545 & -0.455 & -0.182 & 0 & 6 \\ 0 & 1 & 0 & -0.455 & 0.545 & -0.182 & 0 & 6 \\ 0 & 0 & 1 & 0.182 & 0.182 & 0.273 & 0 & 156 \\ 0 & 0 & 0 & 0.182 & 1.182 & 0.273 & 1 & 486 \end{bmatrix}$$

Corner $(6, 6, 156)$, $z = 486$, slack $(0, 0, 0)$

Corner (x_1, x_2, x_3)	Value of z	Increase in z	Slack (s_1, s_2, s_3)
$(0, 0, 0)$	0		$(330, 330, 132)$
$(0, 0, 132)$	396	396	$(66, 66, 0)$
$(0, 11, 154)$	484	88	$(11, 0, 0)$
$(6, 6, 156)$	486	2	$(0, 0, 0)$

21.   There is no change in z because the the process remains at that corner point but changes from one boundary line to another at that point..

Section 4.7

1.   The slope of the objective function must be between the slopes of the two lines for the maximum to occur at their intersection, that is, between $-\dfrac{3}{2} = -1.5$ and $-\dfrac{5}{2} = -2.5$.

(a)   Since the slope of the objective function is $-\dfrac{7}{7} = -1$ and -1 is not between -1.5 and -2.5, the objective function is not maximum at the intersection.

(b)   The slope of the objective function is $-\dfrac{8}{4} = -2$ which is between -1.5 and -2.5 so the objective function attains its maximum value at the intersection.

(c)   The slope of the objective function is $-\dfrac{34}{20} = -1.7$ which is between -1.5 and -2.5 so the objective function attains its maximum value at the intersection.

3.   The slope of $Ax + By$, $m = -\dfrac{A}{B}$, must lie between the slopes of $10x + 15y = 255$, $m = -\dfrac{10}{15}$, and $18x + 5y = 261$, $m = -\dfrac{18}{5}$, that is, $-\dfrac{18}{5} \le -\dfrac{A}{B} \le -\dfrac{10}{15}$. A slope of -2 serves as one example.

Section 4.7 Sensitivity Analysis

5. The slope of the objective function must be closer to zero than the slope of $10x + 7y = 280$, $m = -\dfrac{10}{7} = -1.429$.

(a) The slope of $5x + 5y$ is -1 and $-1.429 < -1 < 0$ so z reaches a maximum at $(0, 28)$.

(b) The slope of $9x + 20y$ is $-\dfrac{9}{20} = -0.45$ and $-1.429 < -0.45 < 0$ so z reaches a maximum at $(0, 28)$.

(c) The slope of $8x + 3y$ is $-\dfrac{8}{3} = -2.67$ and $-2.67 < -1.429 < 0$ so z does not reach a maximum at $(0, 28)$.

7. The slope of the objective function, $-\dfrac{A}{B}$, must be farther from zero than the slope of $5x + 3y = 150$, that is, $-\dfrac{A}{B} \le -\dfrac{5}{3} \le 0$ or $\dfrac{5}{3} \le \dfrac{A}{B}$ or $A \ge \dfrac{5}{3}B$.

(a) $-\dfrac{4}{7}$ is not less than $-\dfrac{5}{3}$ so the objective function does not reach maximum at $(30, 0)$.

(b) $-\dfrac{5}{2} \le -\dfrac{5}{3}$ so the objective function reaches maximum at $(30, 0)$.

(c) $-2 \le -\dfrac{5}{3}$ so the objective function reaches maximum at $(30, 0)$.

9. For $z = Ax + By$ to attain a maximum at $(21, 0)$ it must be steeper than the line $6x + 7y = 126$, that is, the slope $m = -\dfrac{A}{B}$, must be farther from zero than $-\dfrac{6}{7}$ so $-\dfrac{A}{B} \le -\dfrac{6}{7} \le 0$. A slope of -3 is one example.

11. For the objective function $z = Ax + 20y$ the slope is $-\dfrac{A}{20}$ and it must be between the slopes $-\dfrac{9}{8} = -1.125$ and $-\dfrac{4}{10} = -0.4$ for the optimal point to stay the same. Thus, $-1.125 \le -\dfrac{A}{20} \le -0.4$ or $1.125 \ge \dfrac{A}{20} \ge 0.4$

Multiply each term by 20 gives $22.5 \ge A \ge 8$
The coefficient of x must lie between 8 and 22.5.

13. The objective function $z = 6x + By$ has slope $-\dfrac{6}{B}$ which must lie between $-\dfrac{4}{5}$ and $-\dfrac{8}{4}$ so $-\dfrac{8}{4} \le -\dfrac{6}{B} \le -\dfrac{4}{5}$ which we can write as $\dfrac{4}{5} \le \dfrac{6}{B} \le \dfrac{8}{4} = 2$

Taking reciprocals and reversing inequality signs we have $\dfrac{5}{4} \ge \dfrac{B}{6} \ge \dfrac{1}{2}$

Section 4.7 Sensitivity Analysis

Multiply each term by 6 to eliminate fractions $7.5 \geq B \geq 3$ so B lies between 3 and 7.5.

15. The new feasible region is

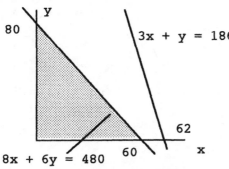

8x + 6y = 480

with the corners (0, 0), (0, 80), (60, 0).

Check for the value of z.

Corner	$z = 52x + 25y$
(0, 0)	0
(0, 80)	2000
(60, 0)	3120

Maximum z = 3120 at (60, 0).

17. **(a)** Solve the system
$$4x + 5y = 140$$
$$6x + 2y = 144 + D$$

$$8x + 10y = 280$$
$$\underline{30x + 10y = 720 + 5D}$$
$$22x \quad\;\; = 440 + 5D$$
$$x = 20 + \frac{5}{22}D$$

Substitute x into the first equation to obtain y.
$$y = 12 - \frac{4}{22}D$$

The point of intersection is $(20 + \frac{5}{22}D, 12 - \frac{4}{22}D)$

(b) $z = 10x + 4y = 10(20 + \frac{5}{22}D) + 4(12 - \frac{4}{22}D) = 248 + \frac{34}{22}D$

(c) Each unit change in D creates a change of $\frac{34}{22} = 1.55$ units in z.

(d) Since neither x nor y can be negative, $x = 20 + \frac{5}{22}D \geq 0$ and

$y = 12 - \frac{4}{22}D \geq 0$.

$20 + \frac{5}{22}D \geq 0$ gives $D \geq -88$

$12 - \frac{4}{22}D \geq 0$ gives $66 \geq D$

D is restricted to the interval $-88 \leq D \leq 66$

143

Chapter 4 Review

1.
$$6x_1 + 4x_2 + 3x_3 + s_1 = 220$$
$$x_1 + 5x_2 + x_3 + s_2 = 162$$
$$7x_1 + 2x_2 + 5x_3 + s_3 = 139$$

3.
$$6x_1 + 5x_2 + 3x_3 + 3x_4 + s_1 = 89$$
$$7x_1 + 4x_2 + 6x_3 + 2x_4 + s_2 = 72$$

5.
$$10x_1 + 12x_2 + 8x_3 + s_1 = 24$$
$$7x_1 + 13x_2 + 5x_3 + s_2 = 35$$
$$-20x_1 - 36x_2 - 19x_3 + z = 0$$

7.
$$3x_1 + 7x_2 + s_1 = 14$$
$$9x_1 + 5x_2 + s_2 = 18$$
$$x_1 - x_2 + s_3 = 21$$
$$-9x_1 - 2x_2 + z = 0$$

9.
$$x_1 + x_2 + x_3 + s_1 = 15$$
$$2x_1 + 4x_2 + x_3 + s_2 = 44$$
$$-6x_1 - 8x_2 - 4x_3 + z = 0$$

11. **(a)**

$$\begin{bmatrix} 5 & 3 & 2 & 1 & 0 & 0 & 0 & | & 600 \\ 4 & 6 & 1 & 0 & 1 & 0 & 0 & | & 900 \\ 1 & 2 & 3 & 0 & 0 & 1 & 0 & | & 800 \\ -5 & -8 & -4 & 0 & 0 & 0 & 1 & | & 0 \end{bmatrix} \begin{matrix} 600/3 = 200 \\ 900/6 = 150\,\text{Pivot row} \\ 800/2 = 400 \\ \; \end{matrix}$$

Column 2 is the pivot column.
The 6 in row 2, column 2 is the pivot element.

(b)

$$\begin{bmatrix} 1 & 4 & 0 & 3 & 0 & -2 & 0 & | & 60 \\ 0 & 6 & 1 & 5 & 0 & 4 & 0 & | & 60 \\ 0 & -3 & 0 & 1 & 1 & 2 & 0 & | & 60 \\ 0 & -1 & 0 & -2 & 0 & 3 & 1 & | & 48 \end{bmatrix} \begin{matrix} 60/3 = 20 \\ 60/5 = 12\,\text{Pivot row} \\ 60/1 = 60 \\ \; \end{matrix}$$

Column 4 is pivot column.
The 5 in row 2, column 4 is the pivot element.

13. **(a)** $x_1 = 0, \; x_2 = 80, \; s_1 = 0, \; s_2 = 42, \; z = 98$
(b) $x_1 = 73, \; x_2 = 42, \; x_3 = 15, \; s_1 = 0, \; s_2 = 0, \; s_3 = 0, \; z = 138$

15. **(a)**

$$\begin{bmatrix} 11 & 5 & 3 & 1 & 0 & 0 & 0 & | & 142 \\ -3 & -4 & -7 & 0 & 1 & 0 & 0 & | & -95 \\ 2 & 15 & 1 & 0 & 0 & 1 & 0 & | & 124 \\ -3 & -5 & -4 & 0 & 0 & 0 & 1 & | & 0 \end{bmatrix}$$

(b)

$$\begin{bmatrix} 7 & 4 & 1 & 0 & 0 & | & 28 \\ -1 & -3 & 0 & 1 & 0 & | & -6 \\ 14 & 22 & 0 & 0 & 1 & | & 0 \end{bmatrix}$$

17. **(a)**

$$\begin{bmatrix} -15 & -8 & 1 & 0 & 0 & 0 & 0 & | & -120 \\ 10 & 12 & 0 & 1 & 0 & 0 & 0 & | & 120 \\ 15 & 5 & 0 & 0 & 1 & 0 & 0 & | & 75 \\ -15 & -5 & 0 & 0 & 0 & 0 & 1 & | & -75 \\ \hline -5 & -12 & 0 & 0 & 0 & 0 & 1 & | & 0 \end{bmatrix}$$

(b)

$$\begin{bmatrix} 14 & 9 & 1 & 0 & 0 & 0 & 0 & | & 126 \\ -10 & -11 & 0 & 1 & 0 & 0 & 0 & | & -110 \\ -5 & 1 & 0 & 0 & 1 & 0 & 0 & | & 9 \\ 5 & -1 & 0 & 0 & 0 & 1 & 0 & | & -9 \\ \hline 3 & 2 & 0 & 0 & 0 & 0 & 1 & | & 0 \end{bmatrix}$$

19.

$$\begin{bmatrix} 2 & 4 & 2 & 1 & 0 & 0 & 0 & | & 34 \\ 3 & 6 & 4 & 0 & 1 & 0 & 0 & | & 57 \\ 2 & 5 & 1 & 0 & 0 & 1 & 0 & | & 30 \\ \hline -3 & -5 & -2 & 0 & 0 & 0 & 1 & | & 0 \end{bmatrix} \quad \begin{bmatrix} 2/5 & 0 & 6/5 & 1 & 0 & -4/5 & 0 & | & 10 \\ 3/5 & 0 & 14/5 & 0 & 1 & -6/5 & 0 & | & 21 \\ 2/5 & 1 & 1/5 & 0 & 0 & 1/5 & 0 & | & 6 \\ \hline -1 & 0 & -1 & 0 & 0 & 1 & 1 & | & 30 \end{bmatrix}$$

$$\begin{bmatrix} 0 & -1 & 1 & 1 & 0 & -1 & 0 & | & 4 \\ 0 & -3/2 & 5/2 & 0 & 1 & -3/2 & 0 & | & 12 \\ 1 & 5/2 & 1/2 & 0 & 0 & 1/2 & 0 & | & 15 \\ \hline 0 & 5/2 & -1/2 & 0 & 0 & 3/2 & 1 & | & 45 \end{bmatrix} \quad \begin{bmatrix} 0 & -1 & 1 & 1 & 0 & -1 & 0 & | & 4 \\ 0 & 1 & 0 & -5/2 & 1 & 1 & 0 & | & 2 \\ 1 & 3 & 0 & -1/2 & 0 & 1 & 0 & | & 13 \\ \hline 0 & 2 & 0 & 1/2 & 0 & 1 & 1 & | & 47 \end{bmatrix}$$

$$\begin{bmatrix} 0 & -1 & 1 & 1 & 0 & -1 & 0 & | & 4 \\ 0 & 1 & 0 & -5/2 & 1 & 1 & 0 & | & 2 \\ 1 & 3 & 0 & -1/2 & 0 & 1 & 0 & | & 13 \\ \hline 0 & 2 & 0 & 1/2 & 0 & 1 & 1 & | & 47 \end{bmatrix}$$

Maximum $z = 47$ at $(13, 0, 4)$

21.

$$\begin{bmatrix} -4 & 1 & 1 & 0 & 0 & | & 3 \\ 1 & -2 & 0 & 1 & 0 & | & 12 \\ \hline -10 & -15 & 0 & 0 & 1 & | & 0 \end{bmatrix} \quad \begin{bmatrix} -4 & 1 & 1 & 0 & 0 & | & 3 \\ -7 & 0 & 2 & 1 & 0 & | & 18 \\ \hline -70 & 0 & 15 & 0 & 1 & | & 45 \end{bmatrix}$$

Unbounded feasible region, no maximum.

23.

$$\begin{bmatrix} 4 & 1 & 1 & 1 & 0 & 0 & | & 372 \\ 1 & 8 & 6 & 0 & 1 & 0 & | & 1116 \\ \hline -1 & -3 & -1 & 0 & 0 & 1 & | & 0 \end{bmatrix} \quad \begin{bmatrix} 31/8 & 0 & 1/4 & 1 & -1/8 & 0 & | & 465/2 \\ 1/8 & 1 & 3/4 & 0 & 1/8 & 0 & | & 279/2 \\ \hline -5/8 & 0 & 5/4 & 0 & 3/8 & 1 & | & 837/2 \end{bmatrix}$$

$$\begin{bmatrix} 1 & 0 & 2/31 & 8/31 & -1/31 & 0 & | & 60 \\ 0 & 1 & 23/31 & -1/31 & 4/31 & 0 & | & 132 \\ \hline 0 & 0 & 40/31 & 5/31 & 11/31 & 1 & | & 456 \end{bmatrix}$$

Maximum $z = 456$ at $(60, 132, 0)$

25. **(a)** x_3 **(b)** x_2

27.
$$\begin{bmatrix} 3 & 4 & 5 \\ 1 & 0 & 7 \\ -2 & 6 & 8 \end{bmatrix} \qquad \begin{bmatrix} 4 & -5 \\ 3 & 0 \\ 2 & 12 \\ 1 & 9 \end{bmatrix}$$

29.
$$\left[\begin{array}{cccccc|c} -2 & -1 & 0 & 1 & 0 & 0 & -6 \\ 0 & -1 & -2 & 0 & 1 & 0 & -8 \\ \hline 6 & 8 & 16 & 0 & 0 & 1 & 0 \end{array}\right] \qquad \left[\begin{array}{cccccc|c} 1 & 1/2 & 0 & -1/2 & 0 & 0 & 3 \\ 0 & -1 & -2 & 0 & 1 & 0 & -8 \\ \hline 0 & 5 & 16 & 3 & 0 & 1 & -18 \end{array}\right]$$

$$\left[\begin{array}{cccccc|c} 2 & 1 & 0 & -1 & 0 & 0 & 6 \\ -1 & 0 & 1 & 1/2 & -1/2 & 0 & 1 \\ \hline 6 & 0 & 0 & 0 & 8 & 1 & -64 \end{array}\right] \qquad \left[\begin{array}{cccccc|c} 0 & 1 & 2 & 0 & -1 & 0 & 8 \\ -2 & 0 & 2 & 1 & -1 & 0 & 2 \\ \hline 6 & 0 & 0 & 0 & 8 & 1 & -64 \end{array}\right]$$

Multiple solutions, minimum $z = 64$ at $(0, 6, 1)$ and $(0, 8, 0)$.

31.
$$\left[\begin{array}{ccccccc|c} 2 & 5 & 4 & 1 & 0 & 0 & 0 & 40 \\ 40 & 45 & 30 & 0 & 1 & 0 & 0 & 430 \\ -6 & -3 & -4 & 0 & 0 & 1 & 0 & -48 \\ \hline -6 & -11 & -8 & 0 & 0 & 0 & 1 & 0 \end{array}\right] \qquad \left[\begin{array}{ccccccc|c} 0 & 4 & 8/3 & 1 & 0 & 1/3 & 0 & 24 \\ 0 & 25 & 10/3 & 0 & 1 & 20/3 & 0 & 110 \\ 1 & 1/2 & 2/3 & 0 & 0 & -1/6 & 0 & 8 \\ \hline 0 & -8 & -4 & 0 & 0 & -1 & 1 & 48 \end{array}\right]$$

$$\left[\begin{array}{ccccccc|c} 0 & 0 & 32/15 & 1 & -4/25 & -11/15 & 0 & 32/5 \\ 0 & 1 & 2/15 & 0 & 1/25 & 4/15 & 0 & 22/5 \\ 1 & 0 & 3/5 & 0 & -1/50 & -3/10 & 0 & 29/5 \\ \hline 0 & 0 & -44/15 & 0 & 8/25 & 17/15 & 1 & 416/5 \end{array}\right]$$

$$\left[\begin{array}{ccccccc|c} 0 & 0 & 1 & 15/32 & -3/40 & -11/32 & 0 & 3 \\ 0 & 1 & 0 & -1/16 & 1/20 & 5/16 & 0 & 4 \\ 1 & 0 & 0 & -9/32 & 1/40 & -3/32 & 0 & 4 \\ \hline 0 & 0 & 0 & 11/8 & 1/10 & 1/8 & 1 & 92 \end{array}\right]$$

Maximum $z = 92$ at $(4, 4, 3)$

33.
$$\left[\begin{array}{ccccccc|c} -1 & -1 & -1 & 1 & 0 & 0 & 0 & -6 \\ 2 & 1 & 3 & 0 & 1 & 0 & 0 & 10 \\ 0 & 2 & -1 & 0 & 0 & 1 & 0 & 5 \\ \hline -2 & -5 & -3 & 0 & 0 & 0 & 1 & 0 \end{array}\right] \qquad \left[\begin{array}{ccccccc|c} 1 & 1 & 1 & -1 & 0 & 0 & 0 & 6 \\ 0 & -1 & 1 & 2 & 1 & 0 & 0 & -2 \\ 0 & 2 & -1 & 0 & 0 & 1 & 0 & 5 \\ \hline 0 & -3 & -1 & -2 & 0 & 0 & 1 & 12 \end{array}\right]$$

$$\left[\begin{array}{ccccccc|c} 1 & 0 & 2 & 1 & 1 & 0 & 0 & 4 \\ 0 & 1 & -1 & -2 & -1 & 0 & 0 & 2 \\ 0 & 0 & 1 & 4 & 2 & 1 & 0 & 1 \\ \hline 0 & 0 & -4 & -8 & -3 & 0 & 1 & 18 \end{array}\right]$$

$$\left[\begin{array}{ccccccc|c} 1 & 0 & 7/4 & 0 & 1/2 & -1/4 & 0 & 15/4 \\ 0 & 1 & -1/2 & 0 & 0 & 1/2 & 0 & 5/2 \\ 0 & 0 & 1/4 & 1 & 1/2 & 1/4 & 0 & 1/4 \\ \hline 0 & 0 & -2 & 0 & 1 & 2 & 1 & 20 \end{array}\right] \qquad \left[\begin{array}{ccccccc|c} 1 & 0 & 0 & -7 & -3 & -2 & 0 & 2 \\ 0 & 1 & 0 & 2 & 1 & 1 & 0 & 3 \\ 0 & 0 & 1 & 4 & 2 & 1 & 0 & 1 \\ \hline 0 & 0 & 0 & 8 & 5 & 4 & 1 & 22 \end{array}\right]$$

Maximum $z = 22$ at $(2, 3, 1)$

35.
$$\begin{bmatrix} 1 & -3 & 1 & 0 & 0 & | & 24 \\ -5 & 4 & 0 & 1 & 0 & | & 20 \\ -20 & -32 & 0 & 0 & 1 & | & 0 \end{bmatrix} \quad \begin{bmatrix} -11/4 & 0 & 1 & 3/4 & 0 & | & 39 \\ -5/4 & 1 & 0 & 1/4 & 0 & | & 5 \\ -60 & 0 & 0 & 8 & 1 & | & 160 \end{bmatrix}$$

Unbounded feasible region, no maximum.

37.
$$\begin{bmatrix} -3 & -2 & 1 & 0 & 0 & 0 & | & -24 \\ -5 & -4 & 0 & 1 & 0 & 0 & | & -46 \\ -4 & -9 & 0 & 0 & 1 & 0 & | & -60 \\ 18 & 36 & 0 & 0 & 0 & 1 & | & 0 \end{bmatrix} \quad \begin{bmatrix} 1 & 2/3 & -1/3 & 0 & 0 & 0 & | & 8 \\ 0 & -2/3 & -5/3 & 1 & 0 & 0 & | & -6 \\ 0 & -19/3 & -4/3 & 0 & 1 & 0 & | & -28 \\ 0 & 24 & 6 & 0 & 0 & 1 & | & -144 \end{bmatrix}$$

$$\begin{bmatrix} 1 & 0 & -2 & 1 & 0 & 0 & | & 2 \\ 0 & 1 & 5/2 & -3/2 & 0 & 0 & | & 9 \\ 0 & 0 & 29/2 & -19/2 & 1 & 0 & | & 29 \\ 0 & 0 & -54 & 36 & 0 & 1 & | & -360 \end{bmatrix} \quad \begin{bmatrix} 1 & 0 & 0 & -9/29 & 4/29 & 0 & | & 6 \\ 0 & 1 & 0 & 4/29 & -5/29 & 0 & | & 4 \\ 0 & 0 & 1 & -19/29 & 2/29 & 0 & | & 2 \\ 0 & 0 & 0 & 18/29 & 108/29 & 1 & | & -252 \end{bmatrix}$$

Minimum $z = 252$ at $(6, 4)$

39.
$$\begin{bmatrix} 4 & 1 & 1 & 0 & 0 & 0 & 0 & | & 180 \\ -1 & -3 & 0 & 1 & 0 & 0 & 0 & | & -120 \\ -1 & 3 & 0 & 0 & 1 & 0 & 0 & | & 150 \\ 1 & -3 & 0 & 0 & 0 & 1 & 0 & | & -150 \\ -5 & -15 & 0 & 0 & 0 & 0 & 1 & | & 0 \end{bmatrix} \quad \begin{bmatrix} 11/3 & 0 & 1 & 1/3 & 0 & 0 & 0 & | & 140 \\ 1/3 & 1 & 0 & -1/3 & 0 & 0 & 0 & | & 40 \\ -2 & 0 & 0 & 1 & 1 & 0 & 0 & | & 30 \\ 2 & 0 & 0 & -1 & 0 & 1 & 0 & | & -30 \\ 0 & 0 & 0 & -5 & 0 & 0 & 1 & | & 600 \end{bmatrix}$$

$$\begin{bmatrix} 13/3 & 0 & 1 & 0 & 0 & 1/3 & 0 & | & 130 \\ -1/3 & 1 & 0 & 0 & 0 & -1/3 & 0 & | & 50 \\ 0 & 0 & 0 & 0 & 1 & 1 & 0 & | & 0 \\ -2 & 0 & 0 & 1 & 0 & -1 & 0 & | & 30 \\ -10 & 0 & 0 & 0 & 0 & -5 & 1 & | & 750 \end{bmatrix} \quad \begin{bmatrix} 1 & 0 & 3/13 & 0 & 0 & 1/13 & 0 & | & 30 \\ 0 & 1 & 1/13 & 0 & 0 & -4/13 & 0 & | & 60 \\ 0 & 0 & 0 & 0 & 1 & 1 & 0 & | & 0 \\ 0 & 0 & 6/13 & 1 & 0 & -11/13 & 0 & | & 90 \\ 0 & 0 & 30/13 & 0 & 0 & -55/13 & 1 & | & 1050 \end{bmatrix}$$

$$\begin{bmatrix} 1 & 0 & 3/13 & 0 & -1/13 & 0 & 0 & | & 30 \\ 0 & 1 & 1/13 & 0 & 4/13 & 0 & 0 & | & 60 \\ 0 & 0 & 0 & 0 & 1 & 1 & 0 & | & 0 \\ 0 & 0 & 6/13 & 1 & 11/13 & 0 & 0 & | & 90 \\ 0 & 0 & 30/13 & 0 & 55/13 & 0 & 1 & | & 1050 \end{bmatrix}$$

Maximum $z = 1050$ at $(30, 60)$

41.
$$\begin{bmatrix} 1 & 3 & 1 & 0 & 0 & | & 9 \\ 1 & -1 & 0 & 1 & 0 & | & -2 \\ -2 & -1 & 0 & 0 & 1 & | & 0 \end{bmatrix} \quad \begin{bmatrix} 4 & 0 & 1 & 3 & 0 & | & 3 \\ -1 & -1 & 0 & -1 & 0 & | & 2 \\ -3 & 0 & 0 & -1 & 1 & | & 2 \end{bmatrix}$$

$$\begin{bmatrix} 1 & 0 & 1/4 & 3/4 & 0 & | & 3/4 \\ 0 & 1 & 1/4 & -1/4 & 0 & | & 11/4 \\ 0 & 0 & 3/4 & 5/4 & 1 & | & 17/4 \end{bmatrix}$$

Maximum $z = 17/4$ at $(3/4, 11/4)$

43.   Let x_1 = number of hunting jackets, x_2 = number of all-weather jackets, x_3 = number of ski jackets. Maximize $z = 7.5x_1 + 9x_2 + 11x_3$ subject to

$$3x_1 + 2.5x_2 + 3.5x_3 \le 3200 \text{ (Hours labor)}$$
$$26x_1 + 20x_2 + 22x_3 \le 18000 \text{ (Operating costs)}$$
$$x_1 \ge 0, x_2 \ge 0, x_3 \ge 0$$

$$\begin{bmatrix} 3 & 2.5 & 3.5 & 1 & 0 & 0 & 3200 \\ 26 & 20 & 22 & 0 & 1 & 0 & 18000 \\ -7.5 & -9 & -11 & 0 & 0 & 1 & 0 \end{bmatrix}$$

45.   The feasible region of the problem is

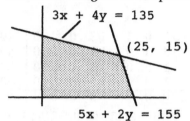

The slope of the objective function, $m = -\dfrac{A}{8}$, must be between the slopes of the constraints.

For $5x + 2y = 155$, $m = -\dfrac{5}{2}$. For $3x + 4y = 135$, $m = -\dfrac{3}{4}$. Thus, $-\dfrac{5}{2} < -\dfrac{A}{8} < -\dfrac{3}{4}$

Multiply through by -8 to obtain $20 > A > 6$

A can be any value 6 through 20.

148

Chapter 5
Mathematics of Finance

Section 5.1

1. $I = (1100)(0.08)(9/12) = \$66$ 3. $I = (600)(0.10)(18/12) = \$90$

5. $I = (745)(0.085)(6/12) = 31.6625 = \$31.66$

7. $I = (300)(0.06)(1) = \$18$
    $A = P + I = 300 + 18 = \$318$

9. $I = (500)(0.06)(3) = \$90$ 11. $I = (700)(0.015)(5) = \$52.50$

13. $I = (950)(0.0175)(7) = 116.375 = \$116.38$

15. $P = I/rt = \dfrac{21.45}{(0.065)(0.5)} = \$660$ 17. $P = I/rt = \dfrac{115.50}{(0.07)(1.5)} = \$1100$

19. $r = I/Pt = \dfrac{49.4}{(1140)(8/12)} = 0.065 = 6.5\%$

21. $r = I/Pt = \dfrac{616.25}{(5800)(1.25)} = 0.085 = 8.5\%$

23. $A = (2700)[1 + (0.04)(1.5)] = \$2862$

25. $A = (6500)[1 + (0.036)(1.75)] = \$6909.50$

27. $P = \dfrac{1800}{1 + (0.06)(1.5)} = 1651.376 = \$1651.38$

29. $D = (1850)(0.045)(1) = \$83.25,$ $PR = M - D = \$1766.75$

31. $D = (485)(0.038)(1.5) = \$27.65,$ $PR = \$457.35$

Section 5.1 Simple Interest

33. $I = 3500(0.046)(2.5) = 402.5$ Total interest paid was $402.50.

35.   One year's simple interest is $I = 17500(0.0685)(1) = \$1198.75$

37.   **(a)**   For one month $I = 85000(0.051)(\frac{1}{12}) = \$361.25$

        (b)   Total interest payment
               $I = 85000(0.051)(5) = 21,675$ Total interest = $21,675

39. $I = Prt$ fit $= I/Pr = 144/(800)(0.06) = 3$ years

41. $t = I/Pr = \dfrac{18}{(800)(0.06)} = 0.375$ years $= 4.5$ months

43. $P = I/rt = \dfrac{48.75}{(0.075)(1)} = \$650$

45. $I = (20,000)(0.09)(1/12) = \$150$ toward interest,
$179.95 - 150 = \$29.95$ toward principal

47. $P = 860, A = 900$
$A = P(1 + rt)$
$900 = 860(1 + r(0.5))$
$40 = 430r$
$r = 0.0930 = 9.3\%$

49. $A = 1000, P = 950, t = 0.5, I = 50$
$50 = 950(r)(0.5)$ $r = \dfrac{50}{475} = 0.10526 = 10.53\%$

51.   Simple interest: $I = (3000)(0.103)(4/12) = \$103$
Simple discount: $M = \dfrac{PR}{1 - dt} = \dfrac{3000}{1-(0.101)(4/12)} = 3104.5188$
$D = \$3104.52 - \$3000.00 = \$104.52$. Simple interest at 10.3% results in the lower fee.

53. $I = (450)(0.11)(1)$ billion $= \$49.5$ billion

55. $I = (500,000)(0.065)(5) = \$162,500$

57.   **(a)**   Semiannual interest $= 625000(0.055)(0.5) = \$17,187.50$
        (b)   Total interest $= 625000(0.055)(8) = \$275,000.$

59. $PR = M(1 - dt) = (1 \text{ million})[1 - (0.07)(90/360)] = \$982,500$

61. $982,000 = 1,000,000(1 - d\left(\dfrac{90}{360}\right))$

$-18000 = -250,000d$

$d = \dfrac{18,000}{250,000} = 0.072 = 7.2\%$

63. $PR = M(1 - dt) = (2 \text{ million})[1 - (0.05125)(30/360)] = \$1,991,458$

65. $\text{Quarterly interest} = 350,000(0.042)(\dfrac{1}{4}) = \$3675$

$\text{Total interest} = 24(3675) = \$88,200$

Section 5.2

1. **(a)** $i = 0.066/2 = 0.033 = 3.3\%$
　　(b) $i = 0.066/4 = 0.0165 = 1.65\%$
　　(c) $i = 0.066/12 = 0.0055 = 0.55\%$

3. **(a)** $A = 4500(1.075)^6 = \$6,944.86$
　　(b) $I = A - P = \$6,944.86 - \$4500 = \$2,444.86$

5. **(a)** $A = 31,000(1.075)^8 = \$55,287.81$
　　(b) $I = \$55,287.81 - \$31,000 = \$24,287.81$

7. **(a)** $A = 5000(1.07)^4 = \$6,553.98$
　　(b) $I = \$6,553.98 - \$5000 = \$1,553.98$

9. First quarter: $A = 800(1.015) = \$812$
　　Second quarter: $A = 800(1.015)^2 = \$824.18$
　　Third quarter: $A = 800(1.015)^3 = \$836.54$
　　Fourth quarter: $A = 800(1.015)^4 = \$849.09$

11. $A = 1800(1.045)^2 = \$1,965.65$

13. **(a)** $A = (12,000)(1.10)^3 = \$15,972$
　　(b) $A = (12,000)(1.05)^6 = \$16,081.15$
　　(c) $A = (12,000)(1.025)^{12} = \$16,138.67$

15. $A = (10{,}000)(1.015)^{20} = \$13{,}468.55$ **17.** $A = 460(1.0045)^6 = \$472.56$

19. $A = 640(1.02)^6 = \$720.74$ **21.** $A = 232.75(1.01)^4 = \$242.20$

23. $P = \dfrac{25000}{(1.03)^{30}} = \$10{,}299.67$

25. $A = 3600$, $i = 0.005$ per month, $n = 42$ months.

$$P = \frac{A}{(1+i)^n} = \frac{3600}{(1.005)^{42}} = \$2919.63$$

27. $x = (1.0135)^4 - 1 = 0.0551 = 5.51\%$

29. Let x = nominal quarterly rate, then

$0.05302 = (1 + x)^4 - 1$

$1.05302 = (1 + x)^4$

$(1.05302)^{1/4} = 1 + x$

$1.012999 = 1 + x$

$x = 0.012999$

The annual nominal rate $= 4(0.012999) = 0.051996$ which we round to 0.052.
The annual nominal rate is 5.2%,

31. Let x = nominal monthly rate, then

$0.07422 = (1 + x)^{12} - 1$

$1.07422 = (1 + x)^{12}$

$(1.07422)^{1/12} = 1 + x$

$1.005984 = 1 + x$

$x = 0.005984$

The annual nominal rate $= 12(0.005984) = 0.07180$ which we round to 0.072.
The annual nominal rate is 7.2%,

33. $2 = (1 + i)^{28}$ The 28th root of 2, $2^{1/28}$, is 1.02506 so we can say 2.5% is the quarterly rate which gives 10% annual rate.

35. $A = 260(1.016)^5 = \$281.48$

37. Ken had $1000(1.013)^{20} = \$1294.76$
Barb had $1000(1.004275)^{60} = 1291.69$
Ken had \$3.07 more.

Section 5.2 Compound Interest

39.   **(a)** $I = 5400(0.042)(25) = 5670$
The value after 25 years is $5400 + \$5670 = \$11,070$

(b)   A $\$5400(1.0105)^{100} = \$15,347.07$
The value at compound interest is $4277.07 greater than the value from simple interest.

41.   The effective rate for 6.8% is $(1.034)^2 - 1 = 0.069156 = 6.9156\%$.
The effective rate for 6.6% is $(1.0165)^4 - 1 = 0.067651 = 6.7651\%$, so 6.8% compounded semiannually is better.

43.   The effective rate of 6.4% compounded annually is 6.4%.
The effective rate of 6.2% compounded quarterly is $(1.0155)^4 - 1 = 0.06345 = 6.345\%$, so 6.4% is better.

45.   The effective rate of 5.0% compounded quarterly is $(1.0125)^4 - 1 = 0.05094 = 5.094\%$, which is not as good as 5.1% compounded annually.

47. $A = 15,000(1.014)^{28} = \$22,138.80$

49. $1485.95 = 1000(1 + r)^{20}$, so $(1 + r)^{20} = 1.48595$.
$1 + r = 1.48595^{1/20}$ which is approximately 1.02,
$i = 0.02$, so the annual rate is 8%.

51. $633.39 = 500(1 + i)^8$, so $1.26678 = (1 + i)^8$
$(1.26678)^{1/8} = 1 + i$
$1.030001 = 1 + i$
$i = 0.030001$ which we round to $= i = 0.03$, 6% per year.

53.   For $6000 invested for 10 years, $A = 6,000(1.025)^{40} = \$16,110.38$
The $6000 investment is somewhat better.

55. $A = 18,000, P = 8,000,$
$n = 28$ semiannual periods
$18,000 = 8000(1 + i)^{28}$
$2.25 = (1 + i)^{28}$
$2.25^{1/28} = 1 + i$
$1.0294 = 1 + i$
$i = 0.0294$ per half year
$i = 2(0.0294) = 0.0588$ per year.
The annual interest rate is 5.88%

57. $240,000 = P(1.02)^{20}$
$P = \dfrac{240,000}{(1.02)^{20}} = \$161,513.12$

59.   **(a)** $20,000 is the future value and $10,250 is the present value,
i = annual rate, so
$20,000 = 10250(1 + i)^{10}$
$1.9512 = (1 + i)^{10}$
$\sqrt[10]{1.9512} = 1 + i$
$1.0691 = 1 + i$
$i = 6.91\%$

   (b) $10000 = P(1.062)^{6}$
$P = \dfrac{10,000}{1.062^{6}} = \dfrac{10,000}{1.43465} = 6970.32$
She should pay $6970.32.

61.   The number of years elapsed was $2002 - 1845 = 157$.
$A = 500(1.07)^{157} = 20,522,161.50$
The amount required is over $20.5 million.

63. $i = r/12$
$(1 + i)^{12} - 1 = 0.0585$
$(1 + i)^{12} = 1.0585$
$1 + i = (1.0585)^{1/12} = 1.004749$
$i = 0.004749$
$r = 12(= 0.004749) = 0.056988$
The annual rate is 5.7%

65.

Interest rate	Years to double	Actual value of $1
3%	24	2.03
5%	14.4	2.02
6%	12	2.01
7&	10.3	2.01
8%	9	2.00
9%	8	1.99
10%	7.2	1.99

154

67.

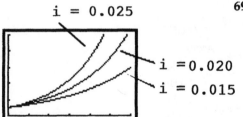

$i = 0.025$

$i = 0.020$

$i = 0.015$

69.

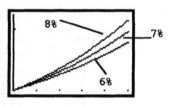

8%

7%

6%

71. For 5% interest

Years	Interest
50	52,750
100	604,906
150	6,936,696
200	79,545,872

For 4.5%, the amount after 200 years is $30,620,757.
For 15.5%, the amount after 200 years is $205,707,326.

73.

0.09

0.07

x represents the number of times interest is compounded annually. As x increases, and compounding occurs more frequently, the effective rate increases but tends to level off.

75. **(a),(b)**

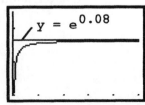

$y = e^{0.08}$

(c) As x increases and the frequency of compounding increases, the amount of $1 increases with a limiting value of $e^{0.08}$.

77. **(a)** $162.89
(b) 6,728,970
(c)

Kenya

U.S.

Spain

79.

Year	P 1000	r 0.052
	Lauren	Anna
1	1052.00	1052
2	1106.70	1104
3	1164.25	1156
4	1224.79	1208
5	1288.48	1260
6	1355.48	1312
7	1425.97	1364
8	1500.12	1416
9	1578.13	1468
10	1660.19	1520

Using your TI-83

1.

Year	Amount
0	200.00
1	210.20
2	220.92
3	232.19
4	244.03
5	256.47
6	269.55
7	283.30
8	297.75
9	312.94
10	328.89

3. Find the intersection of
$Y1 = 5000(1.01375)^{\wedge}x$ and
$Y2 = 12000$. This occurs when
$x = 64.11$. The investment
reaches \$12,000 in the 65th
quarter, about 16 years.

Using EXCEL

1.

Year	Amount
0	500.00
1	530.00
2	561.80
3	595.51
4	631.24
5	669.11
6	709.26

3.

Year	6%	5.50%
0	2000.00	2000.00
1	2120.00	2110.00
2	2247.20	2226.05
3	2382.03	2348.48
4	2524.95	2477.65
5	2676.45	2613.92

5.   Use **Goal Seek.** Enter 0 (for x) in A2 and =2500*(1.013)^A2 in B2, and seek the value of A2 so that B2 is 4000.
A2 is 36.38 when B2 is 4000 so the investment reaches $4000 in the 37th quarter, just over three years.

Section 5.3

1.
Year Deposited	Value at End of 4 Years
1	$600(1.05)^3 = \$\ 694.58$
2	$600(1.05)^2 = \$\ 661.50$
3	$600(1.05)^1 = \$\ 630.00$
4	$600(1.05)^0 = \$\ \underline{600.00}$
	$2,586.08$ Final Value

3. $n = 15, i = 0.07, R = 16,000$

$$A = 16,000 \left[\frac{1.07^{15} - 1}{0.07} \right] = \$402,064.35$$

5. $n = 20, i = 0.04/4 = 0.01, R = 250$

$$A = 250 \left[\frac{1.01^{20} - 1}{0.01} \right] = \$5504.75$$

7. $n = 20, i = 0.01, R = 200$

$$A = 200 \left[\frac{1.01^{20} - 1}{0.01} \right] = \$4403.80$$

9. $n = 20, i = 0.029, R = 4000$

$$A = 4000 \left[\frac{1.029^{20} - 1}{0.029} \right]$$
$$= \$106,394.86$$

11. $n = 16, i = 0.068/4 = 0.017, R = 750$

$$A = 750 \left[\frac{1.017^{16} - 1}{0.017} \right] = \$13,658.37$$

13. $n = 4, i = 0.065, A = 2500$

$$2500 = R \left[\frac{1.065^4 - 1}{0.065} \right] \qquad R = \$567.26$$

15. $n = 20, i = 0.03, A = 14,500$

$$14,500 = R \left[\frac{1.03^{20} - 1}{0.03} \right] \qquad R = \$539.63$$

17. $n = 12, i = 0.015, A = 10,000$

$$10,000 = R\left[\frac{1.015^{12} - 1}{0.015}\right] \qquad R = \$766.80$$

19. $n = 16, i = 0.018, A = 75,000$

$$75,000 = R\left[\frac{1.018^{16} - 1}{0.018}\right] \quad R = \$4086.63$$

21. $n = 18, i = 0.00525, A = 15,000$

$$15,000 = R\left[\frac{1.00525^{18} - 1}{0.00525}\right] \qquad R = \$796.76$$

23. $n = 24, i = 0.004, R = 100$

$$A = 100\left[\frac{1.004^{24} - 1}{0.004}\right] = \$2513.71$$

25. $n = 10, i = 0.04, R = 400$

$$A = 400\left[\frac{1.04^{10} - 1}{0.04}\right] = \$4,802.44$$

27. $n = 12, i = 0.015, A = 4000$

$$4000 = R\left[\frac{1.015^{12} - 1}{0.015}\right] \qquad R = \$306.72$$

29. $n = 9, i = 0.07, R = 5000$

$$A = 5,000\left[\frac{1.07^{9} - 1}{0.07}\right] = \$59,889.94$$

31. $n = 6, i = 0.07, A = 150,000$

$$150,000 = R\left[\frac{1.07^{6} - 1}{0.07}\right] \qquad R = \$20,969.37$$

33. $n = 32, i = 0.01575, A = 750,000$

$$750,000 = R\left[\frac{1.01575^{32} - 1}{0.01575}\right] \qquad R = \$18,205.56$$

Section 5.3 Annuities and Sinking Funds

35. $n = 5, i = 0.06, R = 4{,}000,$

The amount accumulated after 5 years $= 4000 \left[\dfrac{1.06^5 - 1}{0.06} \right] = 22{,}548.37.$

This is less than the required amount.

37. **(a)** $R = 1000, n = 10, i = 0.08$

$$A = 1000 \left[\dfrac{1.08^{10} - 1}{0.08} \right] = \$14{,}486.56$$

(b) $0.08(14486.56) = \$1158.92$ The annual interest will make the $1000 payment.

39. $n = 30, i = 0.016, R = 100 \qquad A = 100 \left[\dfrac{(1.016)^{30} - 1}{0.016} \right] = \$3812.16$

41. $n = 60, i = 0.0075, A = 15{,}000$

$$15{,}000 = R \left[\dfrac{(1.0075)^{60} - 1}{0.0075} \right] \qquad R = \$198.88$$

43. $n = 6, i = 0.081, A = 100{,}000$

$$100{,}000 = R \left[\dfrac{(1.081)^6 - 1}{0.081} \right] \qquad R = \$13{,}597.20$$

45. In each case $i = \dfrac{0.06}{12} = 0.005$

(a) For Evelyn's fund, $R = 100$ and $n = 120$

$$A = 100 \left[\dfrac{1.005^{120} - 1}{0.005} \right] = 16{,}387.93$$

The value of Evelyn's fund at retirement was $16,387.93.
For Esther's fund $R = 50$ and $n = 240$

$$A = 50 \left[\dfrac{1.005^{240} - 1}{0.005} \right] = 23{,}102.04$$

The value of Esther's fund at retirement was $23,102.04.
For Lois's fund $R = 25$ and $n = 360$

$$A = 25 \left[\dfrac{1.005^{360} - 1}{0.005} \right] = 25{,}112.88$$

The value of Lois's fund at retirement was $25,112.88.

(b) Start your savings program early for the best return.

Section 5.3 Annuities and Sinking Funds

47. The monthly payments are 45(3.40) = $153, i = 0.051/12 = 0.00425, and n = 12(35) = 420.

$$A = 153\left[\frac{1.00425^{420} - 1}{0.00425}\right] = 153(1161.67) = 177,735.51$$

At age 65 the annuity will be worth $177,735.

49. i = 0.00575, R = 200

(a)	n = 300, A = $159,473	**(b)**	n = 312, A = $173,309
(c)	n = 324, A = $188,130	**(d)**	n = 336, A = $204,007
(e)	n = 348, A = $221,015	**(f)**	n = 360, A = $239,234

51. R = 2000, i = 0.06

(a)	n = 40, A = $309,524	**(b)**	n = 35, A = $222,870
(c)	n = 25, A = $109,729	**(d)**	n = 15, A = $46,552

53.

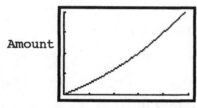

Number of periods

This shows how an annuity grows as the number of time periods increases.

55. Window: X:[0, 0.01], Y:[7000, 9000]

The monthly interest rate is 0.006238, 0.6238% per month.

57. **(a)** Window: X: [0, 100], Y: [0. 20000]

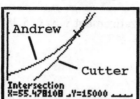

Cutter's investment catches up at x = 55.48 quarters when y = $15,000.

(b)

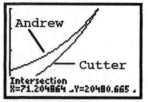

Cutter's investment catches up at x = 71.21 quarters when y = $20,480

(c)   Window: X:[0, 125], Y:[0, 50000]

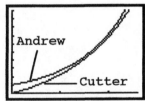

Cutter's investment never catches up, Andrew's is increasing faster.

59. $R = 50$, $n = 60$, $i = 0.045/12 = 0.00375$

$$A = 50 \left[\frac{(1.00375)^{61} - 1}{0.00375} \right] = 3419.87$$

Using Your TI-83

1.

Month	Amount
1	200.00
2	401.50
3	604.51
4	809.05

3.

Year	Amount
1	1000.00
2	2072.00
3	3221.18
4	4453.11

Using Excel

1. $R = 1500$, $i = 0.057$, $n = 10$.

Month	Amount
0	0.00
1	1,500.00
2	3,085.50
3	4,761.37
4	6,532.77
5	8,405.14
6	10,384.23
7	12,476.13
8	14,687.27
9	17,024.45
10	19,494.84

3. $R = 75$, $i = 0.0045$, $n = 12$

Month	Amount
0	0.00
1	75.000
2	150.34
3	226.01
4	302.03
5	378.39
6	455.09
7	532.14
8	609.54
9	687.28
10	765.37
11	843.82
12	922.61

Section 5.4

1. $n = 8, i = 0.07, R = 4000$

$$P = 4000\left[\frac{1-(1.07)^{-8}}{0.07}\right] = \$23,885.19$$

The present value is $16,800.

3. $R = 508.80, i = 0.00475, \text{ and } n = 36$

$$P = 508.8\left[\frac{1-(1.00475)^{-36}}{0.00475}\right] = 16,799.73$$

5. $R = 240, i = 0.0055, n = 120.$

$$P = 240\left[\frac{1 - (1.0055)^{-120}}{0.0055}\right] = 21,042.03$$

The present value is $21,042.03.

7. $n = 22, i = 0.017, R = 300$

$$P = 300\left[\frac{1-(1.017)^{-22}}{0.017}\right] = \$5468.07$$

9. $n = 12, i = 0.014, R = 218.66$

$$P = 218.66\left[\frac{1-(1.014)^{-12}}{0.014}\right] = \$2,399.96 \text{ rounded to } \$2,400$$

11. $R = 381.04, i = 0.0045, n = 24$

$$P = 381.04\left[\frac{1-(1.0045)^{-24}}{0.0045}\right] = 8,650.02$$

The amount borrowed was $8650.

13. $n = 5, i = 0.069, P = 9500$

$$9500 = R\left[\frac{1-(1.069)^{-5}}{0.069}\right] = R\left[\frac{1.069^5 - 1}{0.069(1.069)^5}\right]$$

$$R = \frac{9500(0.069)}{1-1.069^{-5}} = 2310.76$$

Annual payments are $2310.76.

Section 5.4 Present Value and Amortization

15. $n = 60, i = 0.018, P = 7500$

$$7500 = R\left[\frac{1-(1.018)^{-60}}{0.018}\right] \qquad R = \$205.44$$

The quarterly payments are $205.44.

17. $n = 96, i = 0.006, P = 32,000$

$$32000 = R\left[\frac{1-(1.006)^{-96}}{0.006}\right] \qquad R = 439.47$$

The monthly payments are $439.47.

19. $n = 180, i = 0.00475, P = 96,000$

$$96000 = R\left[\frac{1-(1.00475)^{-180}}{0.00475}\right] \qquad R = 794.63$$

Monthly payments are $794.63.

21. $i = 0.069/12 = 0.00575, n = 48, P = 12,750$

$$12750 = R\left[\frac{1-(1.00575)^{-48}}{0.00575}\right] \qquad R = 304.72$$

The monthly payments are $304.72.

23. $i = 0.066/12 = 0.0055, n = 360, P = 68,000$

$$68000 = R\left[\frac{1-(1.0055)^{-360}}{0.0055}\right] \qquad R = 434.29$$

The monthly payments are $434.29.

25.   **(a)** $360(550.32) = \$198,115.20$
(b) $198,115.20 - 75,000 = \$123,115.20$

27.   **(a)** $I = 68,000(0.069)(1/12) = \$391$ interest
(b)   Total paid $= (25)(12)(476.28) = \$142,884$
$476.28 - 391 = \$85.28$ to principal

29. $n = 20, i = 0.0145, P = 75,000, R = 2484.21$

$$\text{Bal} = 75,000(1.0145)^{20} - 2,484.21\left[\frac{1.0145^{20}-1}{0.0145}\right]$$

$$= 100,023.56 - 57,162.09 = \$42,861.47$$

Section 5.4 Present Value and Amortization

31. $n = 24, i = 0.108/12 = 0.009, R = 400.03$

$$P = 400.03 \left[\frac{1 - 1.009^{-24}}{0.009} \right] = 8600.0145 \qquad \text{She borrowed \$8600}$$

33. $n = 16, i = 0.025, P = 7500$

$$7500 = R \left[\frac{1 - 1.025^{-16}}{0.025} \right] \qquad R = \$574.49$$

35. $n = 20, i = 0.05, R = 10,000$

$$P = 10,000 \left[\frac{1 - 1.05^{-20}}{0.05} \right] = \$124,622.10$$

37. $n = 24, i = 0.0.525, P = 12,000$

$$12,000 = R \left[\frac{1 - 1.00525^{-24}}{0.00525} \right] \qquad R = \$533.47$$

39. **(i)** Find amount needed at age 18 to pay \$15,000 per year for 4 years.
$n = 4, i = 0.08, R = 15,000$

$$P = 15,000 \left[\frac{1 - 1.08^{-4}}{0.08} \right] = \$49,681.90 \text{ needed at 18}$$

(ii) Find periodic payments of an annuity that will accumulate to \$49,681.90 in 18 years.
$n = 18, i = 0.08, P = 49,681.90$

$$49,681.90 = R \left[\frac{1.08^{18} - 1}{0.08} \right] \qquad R = \$1,326.61$$

41. $n = 60, i = 0.005, R = 200$

$$P = 200 \left[\frac{(1.005)^{60} - 1}{0.005(1.005)^{60}} \right] = \$10,345.11$$

43. $n = 60, i = 0.00525, P = 9700$

$$9700 = R \left[\frac{(1.00525)^{60} - 1}{0.00525(1.00525)^{60}} \right] \qquad R = \$188.88$$

Section 5.4 Present Value and Amortization

45. For this exercise, $R = 2083.33$, $i = \dfrac{0.06}{12} = 0.005$, $n = 480$

The present value is

$$P = 2083.33\left[\frac{1 - 1.005^{-480}}{0.005}\right] = 378{,}640.1947$$

The insurance company should provide $378,640.20

47. **(a)** $P = 75{,}000$, $i = 0.027/12 = 0.00225$, $n = 24$

$A = 75{,}000(1.00225)^{24} = 79{,}156.54$

(b) Using the present value formula

$$79{,}156.54 = R\left[\frac{1 - (1.00225)^{-96}}{0.00225}\right] \qquad R = 917.73$$

Her monthly payments amount to $917.73.

49. $n = 300$, $i = 0.0075$, $P = 80{,}000$

$$\text{Monthly payments} = \frac{80000(0.0075)}{1 - 1.0075^{-300}} = \$671.36$$

Total payments over two years $= 24(671.36) = \$16{,}112.64$

Balance after two years $= 80000(1.0075)^{24} - 671.36\left[\dfrac{1.0075^{24} - 1}{0.0075}\right]$

$$= 95{,}713.08 - 17{,}581.89 \ = \$78{,}131.19$$

Costs of owning the house

Down payment	$10,000.00
Payment of balance	78,131.19
Monthly payments	16,112.64
Real estate agent	5,400.00
Closing Costs	1,350.00
	110,993.83
Income from sale of house	90,000.00
Net cost	20,993.83

Average monthly cost $= 20{,}993.83/24 = \$874.74$.

55. $A = 75{,}000$, $i = 0.057/12 = 0.00475$

First, find the monthly payments from

$$75{,}000 = R\left[\frac{1 - (1.00475)^{-360}}{0.00475}\right] \qquad R = \$435.30$$

(a) The first month's interest on $75,000 is $356.25 so $79.05 is repaid the first month.

(b)   We need to find the balance after 180 months. It is

$$Bal = 75{,}000(1.00475)^{180} - 435.30\left[\frac{1.00475^{180}-1}{0.00475}\right]$$

$$= 175{,}996.46 - 123{,}407.05 = 52{,}589.41$$

The monthly interest on \$52,589.41 is \$249.80 so the principal repaid is \$435.30 - \$249.80 = \$185.50.

57. $n = 360, i = 0.00875$

(a)   Monthly payments; $R = \dfrac{125000(0.00875)}{1-1.00875^{-360}} = 1143.42$

Balance after 10 years:

$$Bal = 125000(1.00875)^{120} - 1143.42\left[\frac{1.00875^{120}-1}{0.00875}\right]$$

$$= 355{,}578.70 - 241{,}049.87 = \$114{,}528.83$$

(b) $n = 240, i = 0.00675, A = 114{,}528.83$

Monthly payments: $R = \dfrac{114528.83(0.00675)}{1-1.00675^{-240}} = 965.11$

(c)   Cost of 240 payments at 1143.42 = \$274,420.80
Cost of 240 payments at 965.11 = 231,626.40
The 20 year total is reduced by
274,420.80 - 231,626.40 = \$42,794.40

63.   Graph $y = 500\left[\dfrac{(1+x)^5-1}{x}\right]$ and $y = 2886.87$ and find their intersection.

Window: X: [0, 0.1], Y: [2000, 3000]

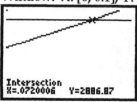

They intersect at $x = 0.072$. The annual rate is 7.2%.

65.   Graph the balance function and $y = 10{,}000$.
Window: X: [0, 0.025], Y: [5000, 12000]

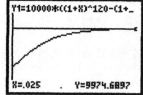

As the interest rate increases, the balance increases to about \$10,000. When $x = 0.025$ per month, the balance after 10 years is \$9974.69 so less than \$26 has been repaid.

69. **(a)** $R = 50{,}000$, $I = 3\%$, $r = 7\%$, $n = 15$

$$P = 50000\left(\frac{1.03}{0.04}\right)\left[1 - \left(\frac{1.03}{1.07}\right)^{15}\right] = 560{,}476$$

(b) $R = 50{,}000$, $i = 7\%$, $n = 15$ for the present value of an annuity

$$P = 50000\left[\frac{1 - (1.07)^{-15}}{0.07}\right] = 455{,}396$$

(c) $R = 35{,}000$, $I = 5\%$, $r = 8\%$, $n = 10$

$$P = 35000\left(\frac{1.05}{0.03}\right)\left[1 - \left(\frac{1.05}{1.08}\right)^{10}\right] = 300{,}746$$

(d) $R = 35{,}000$, $I = 10\%$, $r = 12\%$, $n = 10$

$$P = 35000\left(\frac{1.10}{0.02}\right)\left[1 - \left(\frac{1.10}{1.12}\right)^{10}\right] = 317{,}402$$

(e) $R = 35{,}000$, $I = 1\%$, $r = 4\%$, $n = 10$

$$P = 35000\left(\frac{1.01}{0.03}\right)\left[1 - \left(\frac{1.01}{1.04}\right)^{10}\right] = 299{,}010$$

(f) $R = 35{,}000$, $I = 10\%$, $r = 13\%$, $n = 10$

$$P = 35{,}000\left(\frac{1.10}{0.03}\right)\left[1 - \left(\frac{1.10}{1.13}\right)^{10}\right] = 302{,}756$$

(g) Part (e) with the lowest inflation and interest rates requires the smallest initial fund amount and part (f) with high inflation and interest rates requires the largest initial amount so it appears that low inflation and interest rates are preferable.

(h) $R = 35{,}000$, $I = 0$, $r = 3\%$, $n = 10$

$$P = 35000\left(\frac{1}{0.03}\right)\left[1 - \left(\frac{1}{1.03}\right)^{10}\right] = 298{,}557$$

This requires the smallest present value.

71.

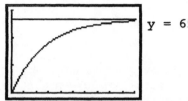

$y = 65$

This represents the present value of $1 periodic payment for x time periods at 1.5% per period. As x approaches 250, the value of y, the present value, tends to level off and approaches 65. Actually, if x continues beyond 250, the value of y will approach, but never exceed, 66.67. This implies that an investment of $66.67 at 1.5% per payment period will provide $1 per period forever.

73. $n = 360$, $P = 100{,}000$, monthly rates are $\dfrac{0.06}{12} = 0.005$, $\dfrac{0.081}{12} = 0.00675$,

$\dfrac{0.102}{12} = 0.0085$.

(a) For 6%, $R = \dfrac{100000(0.005)}{1 - 1.005^{-360}} = 599.55$

For 8.1%, $R = \dfrac{100000(0.00675)}{1 - 1.00675^{-360}} = 740.75$

For 10.2%, $R = \dfrac{100000(0.0085)}{1 - 1.0085^{-360}} = 892.39$

(b) Graph $y = 100000(1.005)^X - 599.55(1.005^X - 1)/0.005$

$y = 100000(1.00675)^X - 740.75(1.00675^X - 1)/0.00675$

$y = 100000(1.0085)^X - 892.39(1.0085^X - 1)/0.0085$

(c) Months required to reduce the balance to 75%, 50%, and 25% of the original loan.

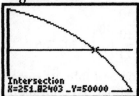

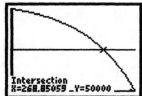

6% interest, 50% balance   8.1% interest, 50% balance

Balance	Interest Rate		
	6%	8.1%	10.2%
75%	163 mo.	189 mo.	212 mo.
50%	252 mo.	269 mo.	283 mo.
25%	313 mo.	321 mo.	328 mo.

(d) As the interest rate increases, it takes longer to reduce the balance to the 75%, 50%, and 25% levels.

75. $P = 7600, R = 60,$ and $i = \dfrac{0.075}{12} = 0.00625$ so $7600 = 60\left[\dfrac{1-1.00625^{-n}}{0.00625}\right]$

Find the intersection of the graphs $y = 60\left[\dfrac{1-1.00625^{-x}}{0.00625}\right]$ and $y = 7600$

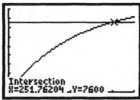

Intersection
X=251.76204 Y=7600

This occurs at $x = 251.76$ It will take about 252 months or 21 years to repay the loan.

77. $P = 1426, R = 29,$ and $i = \dfrac{0.195}{12} = 0.01625$ so $1426 = 29\left[\dfrac{1-1.01625^{-n}}{0.01625}\right]$

Find the intersection of the graphs $y = 29\left[\dfrac{1-1.01625^{-x}}{0.01625}\right]$ and $y = 1426$

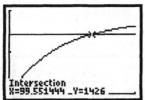

Intersection
X=99.551444 Y=1426

This occurs at $x = 99.6$ months.
It will take over eight years to pay off the credit card bill.

79. $P = 2500, R = 150, n = 18$

$2500 = 150\left[\dfrac{1-(1+i)^{18}}{i}\right]$

Find the intersection of $y = 150\left[\dfrac{1-(1+x)^{-18}}{x}\right]$ and $y = 2500$

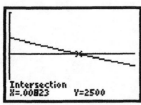

Intersection
X=.00823   Y=2500

This occurs at $x = 0.00823$ so the monthly interest rate is about 0.823% and the annual interest rate (APR) is about 9.88%.

81. $P = 1800$, $R = 161.25$, and $n = 12$ so $1800 = 161.25\left[\dfrac{1-(1+i)^{-12}}{i}\right]$

Find the intersection of $y = 161.25\left[\dfrac{1-(1+x)^{-12}}{x}\right]$ and $y = 1800$

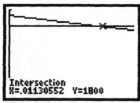

This occurs at $x = 0.0113$ or 1.13% per month which gives an annual rate of 13.56%.

83. Find where $y = 115\left[\dfrac{1-(1+x)^{-18}}{x}\right]$

and $y = 1800$ intersect.

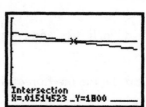

$x = 0.01515$ per month $= 18.18\%$ annual

85. Find where $y = 112\left[\dfrac{1-(1+x)^{-15}}{x}\right]$

and $y = 1600$ intersect.

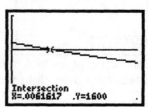

$x = 0.00616$ per month $= 7.39\%$ annual

TI-83 Exercises

1. Monthly payment $= 174.90$

Month	Interest	Principal Repaid	Balance
1	15.00	159.90	1840.10
2	13.80	161.10	1679.00
3	12.59	162.31	1516.69
4	11.38	163.52	1353.17

3. Monthly payment = 802.52

Month	Interest	Repaid	Balance
1	552.50	250.02	84,749.98
2	550.87	251.65	84,498.33
3	549.24	253.28	84,245.05
4	547.59	254.93	83,990.12

Chapter 5 Review

1. $I = (500)(0.09)(2) = \$90$

3. $P = \dfrac{A}{1 + rt} = \dfrac{1190.40}{1 + (0.08)(3)} = \$960$ (Rounded to nearest dollar)

5. $I = (3000)(0.09)(5) = \$1,350$

7. $D = (8500)(0.09)(2) = \$1530,\ PR = 8500 - 1530 = \$6970$

9. $A = 5000(1.07)^3 = 6125.22$ so interest = $\$1,125.22$

11. After ten years the value is $A = 1000(1.017)^{40} = 1962.63$ so it will not double in value.

13. $x = (1.03)^2 - 1 = 0.0609 = 6.09\%$

15. $x = (1.006)^{12} - 1 = 0.07442 = 7.442\%$

17. Effective rate of 5.6% = $(1.014)^4 - 1 = 0.0572 = 5.72\%$.
Effective rate of 5.7% = $(1.057)^1 - 1 = 0.057\ 5.7\%$, so 5.6% compounded quarterly is better.

19. **(a)** $A = (8,000)(1.06)^4 = \$10,099.82$
    (b) $I = 10099.82 - 8000 = \$2099.82$

21. **(a)** $A = 5,000(1.0145)^{24} = \$7063.49$
    (b) $I = 7063.49 - 5000 = \$2063.49$

23. $A = 15,000(1.055)^2 = \$16,695.38$

25. $50,000 = P(1.02)^{20} = P(1.485947),$
     $P = \$33,648.57$

27. $P = 2000, i = 0.0165, n = 24$
     $A = 2000(1.0165)^{24} = 2962.15.$
     The investment will not reach $3500 in six years.

29. $A = 1,000 \left[\dfrac{1.06^5 - 1}{0.06} \right] = \$5,637.09$

31. $n = 10, i = 0.032$
     $A = 600 \left[\dfrac{1.032^{10} - 1}{0.032} \right] = \$6942.02$

33. $n = 24, i = 0.16$
     $A = 250 \left[\dfrac{1.016^{24} - 1}{0.016} \right] = \$7245.15$

35. $n = 10, i = 0.048$
     $2,000,000 = R \left[\dfrac{1.048^{10} - 1}{0.048} \right]$
     $R = \$160,500$ (Rounded to nearest dollar)

37. $P = 5000/(1.06)^5 = \$3,736.29$

39. $n = 20, i = 0.0155$
     $6000 = P(1.0155)^{20}$
     $P = \$4411.16$

41. $n = 20, i = 0.02$
     $P = 50,000/(1.02)^{20} = \$33,648.57$

43. Maximum total payment $= 297,000, i = 0.02, P = 200,000, n = 20$
     $A = 200,000(1.02)^{20} = 297,289.48$
     After five years the total payment is $297,289.48 so the term should be less than five years.

45. $n = 12, i = 0.016,$
     $3000 = R \left[\dfrac{1 - 1.016^{-12}}{0.016} \right] \quad R = \$276.76$

47. $n = 5, i = 0.074,$
     $4800 = R \left[\dfrac{1 - 1.074^{-5}}{0.074} \right]$
     $R = \$1183.24$

49. $n = 20, i = 0.04$
     $7.8 \text{ million} = R \left[\dfrac{1 - 1.04^{-20}}{0.04} \right]$
     $R = \$573,937.65$

51. $n = 8, i = 0.09$

$$98,000 = R\left[\frac{1 - 1.09^{-8}}{0.09}\right]$$

$R = \$17,706.09$

53. Amount at end of first five years
$A = 1000(1.0155)^{20} = \$1360.19$
Amount at end of second five years
$A = \$1360.19(1.016)^{20} = \$1868.42$

55. $n = 40, i = 0.0145, P = 1700$
$A = 1700(1.0145)^{40} = \$3023.65$

57. $n = 160, i = 0.016$
$500,000 = R(1.016)^{160}$
$R = \$39,443.62$

59. $n = 72, i = 0.005$

$$100,000 = R\frac{(1.005)^{72} - 1}{0.005}$$

$R = \$1157.29$

Chapter 6
Sets and Counting

Section 6.1

1. **(a)** True   **(b)** False   **(c)** False
    (d) False   **(e)** True   **(f)** True
    (g) False   **(h)** True   **(i)** False

3. $B = \{M, I, S, P\}$ 5. $C = \{16, 18, 20, 22, ...\}$

7. Not equal   9. Equal   11. Not equal

13. $A \subseteq B$ 15. $A \not\subseteq B$ 17. $A \not\subseteq B$

19. $A \not\subseteq B$

21. **(a)** $\varnothing$, $\{-1\}$, $\{2\}$, $\{4\}$, $\{-1, 2\}$, $\{-1, 4\}$, $\{2, 4\}$, $\{-1, 2, 4\}$
     (b) $\varnothing$, $\{4\}$
     (c) $\varnothing$, $\{-3\}$, $\{5\}$, $\{6\}$, $\{8\}$, $\{-3, 5\}$, $\{-3, 6\}$, $\{-3, 8\}$, $\{5, 6\}$, $\{5, 8\}$, $\{6, 8\}$,
              $\{-3, 5, 6\}$, $\{-3, 5, 8\}$, $\{-3, 6, 8\}$, $\{6, 5, 8\}$, $\{-3, 5, 6, 8\}$

23. $\varnothing$ 25. Not empty   27. $\varnothing$

29. $\varnothing$ 31. $\varnothing$ 33. $\{1, 2, 4, 6, 7\}$

35. $\{a, b, c, d, x, y, z\}$ 37. $\{9, 12\}$ 39. $A' = \{17, 18, 19\}$

41. $A' = \{11, 12, 13\}$

43. **(a)** $A' = \{-1, 0, 12\}$ **(b)** $B' = \{1, 12, 13\}$
     (c) $(A \cup B)' = \{12\}$ **(d)** $(A \cap B)' = \{-1, 0, 1, 12, 13\}$

45. $A \cap B = \{1, 2, 3\}$ **47.** $A \cup B = \{1, 2, 3, 6, 9\}$

49. $A \cap B \cap C = \{2, 3\}$ **51.** $A \cap \varnothing = \varnothing$

53. $(A \cap C) \cup B = \{1, 2, 3, 6, 9\}$ **55.** $\{13, 22, 33, ...\}$

57. $\{1, 2, 3, 4\}$

59.   **(a)** $\subset$ **(b)** $\subset$ **(c)** $=$
       (d) $\not\subset$ **(e)** $\subseteq$

61. $A \cap B = \{x \mid x \text{ is an integer that is a multiple of } 35\}$

63. $A \cup B = \{5, 6, 10, 12, 15, 18, 20, 24, 25, 30\}$ **65.**   Disjoint

67. $A \cap B$ is the set of students at Miami Bay Univ. who are taking bothfinite math and American history.

69.   **(a)** $A \cap B$ is the set of students at Winfield College who had a 4.0 GPA for both the Fall, 1997 and Spring, 1998 semester.
       (b)   A is not necessarily a subset of B because some students may have a 4.0 GPA in the fall semester and not in the spring semester.
       (c) $A \cup B$ is the set of students who had a 4.0 GPA for the fall semester, or the spring semester, or both.

Section 6.2

1.   **(a)** $n(A) = 10$ **(b)** $n(B) = 5$
       (c) $n(A \cap B) = 4$ **(d)** $n(A \cup B) = 10 + 5 - 4 = 11$

3. $n(A \cup B) = 120 + 100 - 40 = 180$

5. $30 = 15 + 22 - n(A \cap B)$ $n(A \cap B) = 7$

7. $28 = 14 + n(B) - 5$ **9.** $100 = 60 + 75 - n(A \cap B)$
      $n(B) = 28 - 14 + 5 = 19$ $n(A \cap B) = 60 + 75 - 100 = 35$

11. (a) 42   (b) 60   (c) 84
     (d) 57   (e) 39   (f) 81
     (g) 15   (h) 15   (i) 81

13. (a) 35   (b) 9   (c) 6
     (d) 38   (e) 38   (f) 12

15. $n(A \cup B) = 27; n(A \cup C) = 26.$
Since $n(A \cup B \cup C) = 33,$ the
number in C is only $33 - 27 = 6.$
The number in B is only $33 - 26 = 7$

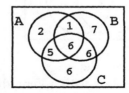

17. $460 = 240 + n(B) - 55, n(B) = 275$ so $n(B') = 500 - 275 = 225$

19. Represent the information with a Venn diagram.

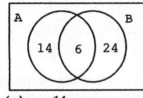

(a) 14   (b) 24   (c) $14 + 24 = 38$
(d) $14 + 6 + 24 = 44$

21. Represent the information with a Venn diagram.

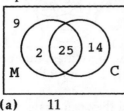

(a) 11   (b) 14   (c) 9   (d) 16

23. (a) 21   (b) 10   (c) 8
     (d) 56   (e) 29   (f) 62

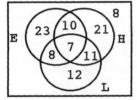

25. Represent the information with a Venn diagram
(a) 28 (b) 9 (c) 20 (d) 7

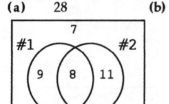

27. Represent the information with a Venn diagram

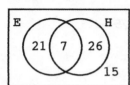

$21 + 7 + 26 + 15 = 69$
69 students were present.

29. It must be the case that $n(A \cup B) \geq n(A)$, but it is not.

31. Represent the information with a Venn diagram

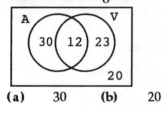

(a) 30 (b) 20

33. Represent the information with a Venn diagram

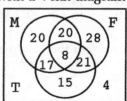

Yes, since these numbers total only 133, not 135.

35. The information reported a total of 56 people and they can be represented in the Venn diagram like this.

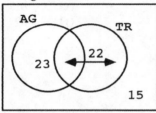

$n(TR \cup AG) = 22 + 23 = 45$

Total number of people = $45 + 15 = 60$ which is not consistent with the reported 56 total.

Section 6.3 Basic Counting Principles

37. **(a)** $15 = n(A \cup B) = n(A) + n(B) - n(A \cap B) = 15 - n(A \cap B),\ n(A \cap B) = 0$

(b) $A \cap B$ = empty set.

39. When $A \subseteq B$ the Venn diagram is

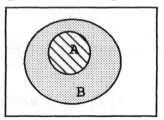

$A \cup B$ is the area occupied by B.

41. $n(A \cup B \cup C) = 12 + 10 + 14 - 7 - 5 - 3 + 2 = 23$

43. $n(A \cup B \cup C) = 425 + 680 + 855 - 275 - 505 - 375 + 240 = 1045$

Section 6.3

1.

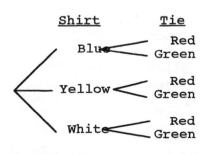

3.

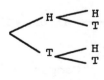

5. $3 \times 4 = 12$ **7.** $5 \times 9 = 45$ **9.** $6 \times 7 = 42$

11. $4 \times 7 = 28$

13. **(a)** $13 \times 13 = 169$ **(b)** $13 \times 13 \times 13 \times 13 = 28,561$

15. The first book can be selected in 22 ways. The second book can be selected from the 21 remaining books. $22 \times 21 = 462$ ways to select the two books.

17. $6 \times 5 \times 3 = 90$

19. Since Reshma buys just one item, there are $13 + 9 = 22$ choices.

21. $4^6 = 4{,}096$

23. Eight choices for the first digit and ten choices for the other 6. $8 \times 10^6 = 8{,}000{,}000$

25. **(a)** $6^4 = 1296$ **27.** $26 \times 10 \times 10 \times 10 = 26{,}000$
(b) $6 \times 5 \times 4 \times 3 = 360$

29. The number of games is the number of ways a home team and a visiting team can be chosen, $8 \times 7 = 56$ ways.

31. $6 \times 3 = 18$

33.

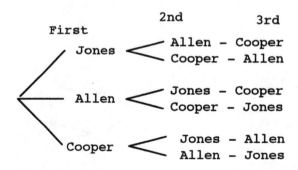

35. $32 \times 18 = 576$

37. They have 5 choices for a museum and 4 choices for an art exhibit. They can select one of each in $5 \times 4 = 20$ ways.

39. **(a)** $7 \times 6 \times 5 = 210$ **(b)** $1 \times 6 \times 5 = 30$
(c) $2 \times 6 \times 5 = 60$ **(d)** $6 \times 2 \times 5 = 60$
(e) Vowel in first position in $2 \times 5 \times 4$ ways, vowel in second positon in $5 \times 2 \times 4$ ways, and a vowel in the third position in $5 \times 4 \times 2$ ways. Total = $2 \times 5 \times 4 + 5 \times 2 \times 4 + 5 \times 4 \times 2 = 3 \times 40 = 120$

Section 6.3 Basic Counting Principles

41. $8\times1\times6\times1\times4\times1\times2\times1 = 384$

43. Men must sit in the first and seventh seat so $4\times3\times3\times2\times2\times1\times1 = 144$

45. **(a)** $6\times5\times4\times3 = 360$ **(b)** $6\times5\times5\times5 = 750$

47. Number of arrangements with art first + number of ways with music first =
$5\times4\times3\times2\times1\times3\times2\times1 + 3\times2\times1\times5\times4\times3\times2\times1 = 2\times720 = 1440$

49. $2^7 = 128$

51.

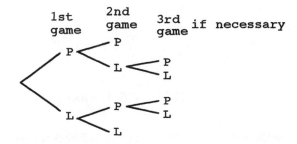

53. There are $26 \times 26 \times 26 = 17{,}576$ different sets of three initials so it is not possible to index all slides with three initials. If four initials are used, there are $26^4 = 456{,}976$ possibilities which is more than adequate for the 22,500 slides.

55. **(a)** Since there are 10 choices for each digit, there are $10^9 = 1{,}000{,}000{,}000$ possible social security numbers.

     (b) There are 26 choices for each of the six characters so there are $26^6 = 308{,}915{,}776$ possible social security numbers, less than the current system.

     (c) If all ten digits and all 26 letters are used, there are $36^6 = 2{,}176{,}782{,}336$ possible social security numbers, over twice the possibilities of the current system.
If the 26 letters of the alphabet and the 8 digits 2-9 are used, there are $34^6 = 1{,}544{,}804{,}416$ possibilities. This system will also provide a greater number of possibilities than the current system.

57. We need to find the number of six-character, seven-character, and eight character passwords and add the results.

Six-character	$26^6 = 308{,}915{,}776$
Seven-character	$26^7 = 8{,}031{,}810{,}176$
Eight character	$26^8 = 2.088270646 \times 10^{11}$
Total	2.1716×10^{11} which is about 217 billion

Section 6.4

1. $3! = 6$

3. $5! = 120$

5. $5!3! = 120 \times 6 = 720$

7. $\dfrac{7!}{3!} = 840$

9. $\dfrac{12!}{7!} = 95{,}040$

11. $P(6, 4) = 6 \times 5 \times 4 \times 3 = 360$

13. $P(100, 3) = 100 \times 99 \times 98 = 970{,}200$

15. $P(7, 4) = 7 \times 6 \times 5 \times 4 = 840$

17. $P(6, 3) = 6 \times 5 \times 4 = 120$

19. $P(7, 5) = 7 \times 6 \times 5 \times 4 \times 3 = 2{,}520$

21. $P(8, 3) = 8 \times 7 \times 6 = 336$

23. (a) $P(7, 7) = 7! = 5{,}040$
    (b) $P(7, 3) = 7 \times 6 \times 5 = 210$

25. $P(3, 3) = 3! = 6$

`27. $P(22, 3) = 22 \times 21 \times 20 = 9{,}240$

29. (a) $P(4, 4) = 4! = 24$
    (b) $P(7, 4) = 7 \times 6 \times 5 \times 4 = 840$

31. $7 \times 6 \times 5 = 210$

33. $P(5, 5) = 5! = 120$

35. $P(4, 4) = 4! = 24$

37. (a) $P(4, 3) = 4 \times 3 \times 2 = 24$
    (b) $4^3 = 64$

39. $P(12, 2) \times P(11, 5) \times P(10, 2) = (12 \times 11)(11 \times 10 \times 9 \times 8 \times 7)(10 \times 9) = (132)(55440)(90)$
    $= 658{,}627{,}200$

Section 6.4 Permutations

41. $\dfrac{8!}{2!} = 20{,}160$

43. $\dfrac{6!}{3!} = 120$

45. **(a)** $\dfrac{8!}{2!\,2!} = 10{,}080$

(b) $\dfrac{11!}{4!\,4!\,2!} = 34{,}650$

(c) $\dfrac{10!}{2!\,2!\,3!} = 151{,}200$

(d) $\dfrac{10!}{3!\,2!\,2!} = 151{,}200$

47. **(a)** $\dfrac{7!}{4!\,3!} = 35$

(b) $\dfrac{7!}{5!\,2!} = 21$

49. **(a)** The series ends when team A wins 4 games so the total number of games played may range from 4 to 7. However, assume that A wins 4 and B wins 3 games (B "wins" the unplayed games). The number of sequences of games possible if A wins the series is the number of sequences of 4 wins by A and 3 wins by B, $\dfrac{7!}{4!\,3!} = 35$

(b) 35

(c) $35 + 35 = 70$

51. $P(10, 4) = 10 \times 9 \times 8 \times 7 = 5{,}040$

53. $P(28, 3) = 28 \times 27 \times 26 = 19{,}656$

55. **(a)** $P(n, 2) = n(n - 1)$

(b) $P(n, 3) = n(n - 1)(n - 2)$

(c) $P(n, 1) = n$

(d) $P(n, 5) = n(n - 1)(n - 2)(n - 3)(n - 4)$

57. $26 \times 9 \times 26 \times 9 \times 26 = 1{,}423{,}656$

59. $P(15, 4) = 15 \times 14 \times 13 \times 12 = 32{,}760$

61. **(a)** Any path will require 4 moves to the right and 3 down in some order. The question is the number of sequences of 4 R's and 3 D's, $\dfrac{7!}{3!\,4!} = 35$

(b) Any path will require 5 moves to the right and 3 down in some order. The question is the number of sequences of 5 R's and 3 D's, $\dfrac{8!}{3!\,5!} = 56$

65. **(a)**

n	n!	(n − 1)!	n(n − 1)!
2	2	1	$2 \times 1 = 2$
3	6	2	$3 \times 2 = 6$
4	24	6	$4 \times 6 = 24$
5	120	24	$5 \times 24 = 120$

(b) $n! = n(n-1)\,(n-2)\,(n-3)\ldots 3{\times}2{\times}1$

$(n-1)! = (n-1)\,(n-2)\,(n-3)\ldots 3{\times}2{\times}1$

$n(n-1)! = n(n-1)\,(n-2)\,(n-3)\ldots 3{\times}2{\times}1 = n!$

69. **(a)** The number of possible order of finish was $P(19, 3) = 5{,}814$.

(b) The number of possible order of finish were $P(12, 3) = 1{,}320$.

(c) The number of possible order of finish were $P(13, 3) = 1{,}716$.

TI-83 Exercises
1. 95,040

3. 201,600

Excel Exercises

1. 120

3. 1.15585 E+29

Section 6.5

1. $C(6, 2) = \dfrac{6\times 5}{2} = 15$

3. $C(13, 3) = \dfrac{13\times 12\times 11}{3\times 2} = 286$

5. $C(9, 5) = \dfrac{9\times 8\times 7\times 6\times 5}{5\times 4\times 3\times 2} = 126$

7. $C(4, 4) = \dfrac{4!}{4!} = 1$

9. $\{a, b\}, \{a, c\}, \{a, d\}, \{b, c\}, \{b, d\}, \{c, d\}$

11. $\{a, b, c, d\}, \{a, b, c, e\}, \{a, b, d, e\}, \{a, c, d, e\}, \{b, c, d, e\}$

13. $(x + y)^5 = C(5, 0)x^5 + C(5, 1)x^4y + C(5, 2)x^3y^2 + C(5, 3)x^2y^3 + C(5, 4)x y^4 + C(5, 5)y^5$
$= x^5 + 5x^4y + 10x^3y^2 + 10x^2y^3 + 5xy^4 + y^5$

15. $(x + y)^{10} = C(10, 0)x^{10} + C(10, 1)x^9y + C(10, 2)x^8y^2 + C(10, 3)x^7y^3 + C(10, 4)x^6y^4 + C(10, 5)x^5y^5 + C(10, 6)x^4y^6 + C(10, 7)x^3y^7 + C(10, 8)x^2y^8 + C(10, 9)xy^9 + C(10, 10)y^{10}$
$= x^{10} + 10x^9y + 45x^8y^2 + 120x^7y^3 + 210x^6y^4 + 252x^5y^5 + 210x^4y^6 + 120x^3y^7 + 45x^2y^8 + 10xy^9 + y^{10}$

17. The number of subsets is $2^5 = 32$

19. $C(6, 2) = \dfrac{6 \times 5}{2 \times 1} = 15$

21. $C(15, 3) = \dfrac{15 \times 14 \times 13}{3 \times 2} = 455$

23. **(a)** $C(7, 4) = \dfrac{7 \times 6 \times 5 \times 4}{4 \times 3 \times 2} = 35$

     (b) $P(7, 4) = 7 \times 6 \times 5 \times 4 = 840$

25. $C(5, 2) \times C(6, 3) \times C(8, 2) = (10)(20)(28) = 5600$

27. $C(6, 3) \times C(10, 4) = (20)(210) = 4200$

29. The selection is from one of two disjoint sets.

31. The selection is from one of two disjoint sets.
$C(9, 3) + C(11, 3) = 84 + 165 = 249$ $C(7, 3) + C(6, 3) = 35 + 20 = 55$

33. This asks for the number of subsets with one or more elements. That number is the number of all subsets minus the number of subsets with no elements, that is, $2^6 - C(6, 0) = 2^6 - 1 = 64 - 1 = 63$ ways.

35. This asks for the number of subsets with two through six elements selected from a set of seven elements. This can be done in:
Number of all subsets minus the number of subsets with zero elements and the number with one element and the number with seven elements, that is
$2^7 - C(7, 0) - C(7, 1) - C(7,7) = 128 - 1 - 7 - 1 = 119$ ways.

37. The committee may be composed of: 3 men and 2 women, 4 men and 1 woman, or 5 men.
$C(5, 3)C(6, 2) + C(5, 4) C(6, 1) + C(5, 5) = 10 \times 15 + 5 \times 6 + 1 = 181$

39. The selection may be composed of: 2 American and 2 English, 3 American and 1 English, or 4 American.
$C(8, 2)C(6, 2) + C(8, 3)C(6, 1) + C(8, 4) = 28 \times 15 + 56 \times 6 + 70 = 826$

41. $P(40, 2)C(10, 3) = 1560 \times 120 = 187{,}200$

43. $P(8, 2)C(12, 4) = 56 \times 495 = 27{,}720$

Section 6.5 Combinations

45. $C(28, 3) = 3276$ **47.** $C(20, 5) = 15,504$

49.   **(a)**   {Alice}, {Bianca}, {Cal}, {Dewayne}
    (b)   {Alice, Bianca}, {Alice, Cal}, {Alice, Dewayne}, {Bianca, Cal}, {Bianca, Dewayne}, {Cal, Dewayne}
    (c)   {Alice, Bianca, Cal}, {Alice, Bianca, Dewayne}, {Bianca, Cal, Dewayne}, {Alice, Cal, Dewayne},
    (d)   {Alice, Bianca, Cal, Dewayne}
    (e) $\varnothing$ **(f)**   16

51.   **(a)** $2 \times 4 \times 3 = 24$
    (b) $2 \times C(4, 2)\, C(3, 2) \times 4 = 2 \times 6 \times 3 \times 4 = 144$

53.   **(a)** $C(10, 6)C(8, 5) = (210)(56) = 11,760$
    (b)   The selection can be 6 multiple choice and 4 true-false, 7 multiple choice and 3 true-false, or 8 multiple choice and 2 true-false.
$C(8, 6)C(10, 4) + C(8, 7)C(10, 3) + C(8, 8)C(10, 2)$
$= 5880 + 960 + 45 = 6885$

55.   **(a)**   It will take $C(32, 3) = 4960$ days.
    (b)   It will take $32/3 = 10.66$ days which must be rounded up to 11 days with two students being invited the last day.

59.   **(a)**   20 at $x = 3$ **(b)**   35 at $x = 3$ and 4
    (c)   70 at $x = 4$ **(d)**   126 at $x = 4$ and 5
    (e)   252 at $x = 5$
    (f) $x = 25$ makes $C(50, x)$ largest. $x = 100$ makes $C(200, x)$ largest.
    (g) $x = 27$ and 28 makes $C(55, x)$ largest. $x = 150$ and 151 makes $C(301, x)$ largest.

63.   **(a)**   Since the order of selection is irrelevant, the six numbers can be chosen in $C(54, 6) = 25,827,165$ different ways.
    (b)   Measure a small stack of pennies and you will find 17 pennies makes a stack one inch tall. Thus, a stack of 25,827,165 pennies makes a stack $\dfrac{25,827,165}{17} = 1,519,245$ inches tall.

$$\text{Height in feet} = \frac{1,519,245}{12} = 126,603.75 \text{ feet.}$$

$$\text{Height in miles} = \frac{134,516}{5280} = 23.978 \text{ miles.}$$

65. There are 26 cards that are not red (all black) so there are C(26, 5) = 65,780 such hands.

69. The number is $C(52, 13) = 6.35 \times 10^{11}$.

71. **(a)** The number is P(45, 6) = 5,864,443,200
     (b) The number is C(45, 6) = 8,145,060

TI-83 Exercises

1. 3003   **3.** 2,882,880

5. $\dfrac{33,649}{462} = 72.833$

Excel Exercises

1. 20   **3.** 3.1.80535 E+13

Section 6.6

1. This is a selection from one of two sets` so it is an addition rule problem with the number of possible selections = 21 + 23 = 44.

3. A combination problem with C(8, 3) = 56 ways the three can be selected.

5. This is a multiplication rule with one item selected from each of 3 sets. The number of ways the selection can be made is $56 \times 8 \times 15 = 6720$ ways.

7. This is a two level problem with the first level being a multiplication rule problem since selections are made from each of two sets. The selection from each set is a permutation problem since three different positions are being filled in each class.
        Number of selections of the student council
        = P(110, 3) P(90, 3) =
        = $1,294,920 \times 704,880$ which is about 9.128×10^{11}.

9. C(11, 2) + 23 = 78   **11.** C(10, 3) = 120

13. $5 \times 8 \times 3 = 120$ **15.** $8! = 40{,}320$

17. $n(A \cup B) = 28 + 21 - 4 = 45$

19. $C(15, 3) \; C(12, 4) \; 455 \times 495 = 225{,}225$

21.   **(a)**   There are 8 possible selections for the first digit and 10 for each of the other six so the total is 8,000,000.

   (b)   Since the digits are all different there are
   $8 \times 9 \times 8 \times 7 \times 6 \times 5 \times 4 = 483{,}840$ numbers with all digits different.

   (c)   The number with at least one digit repeated is the number in (a) minus the number in (b)
   $8{,}000{,}000 - 483{,}840 = 7{,}516{,}160$

23.   The selection is to be made from one of the three disjoint sets of books so we determine the number of ways the selection can be made from each set and add. For each set the number is
   Handy Man: $C(7, 2) = 21$ choices.
   Gardening: $C(10, 3) = 120$ choices.
   Interior Decorating: $C(5, 1) = 5$
The number ways the free gift can be selected is $21 + 120 + 5 = 146$.

25.   The number of ways the selection can be made without regard to gender is $P(11, 3)$ and the number of ways the selection can be made with all men is $P(6, 3)$ so the number of ways the selection can be made with no more than two men is $P(11, 3) - P(6, 3) = 990 - 120 = 870$.

27.   Since selections are made from each of the four clubs, the Multiplication Rule applies. The number of possible selections is $C(22, 2) \; C(19, 2) \; C(25, 2)$
   $C(14, 2) = 231 \times 171 \times 300 \times 91 = 1{,}078{,}377{,}300$ ways.

29.   The selection will be made from just one of the clubs so the Addition Rule applies. The number of possible selections is
   $C(22, 2) + C(19, 2) + C(25, 2) + C(14, 2) = 231 + 171 + 300 + 91 = 793$ ways.

31.   The number of committees possible regardless of gender is $C(13, 5)$ and the number of committees possible with no women is $C(6, 5)$ so the number of committees possible with at least one woman is
   $C(13, 5) - C(6, 5) = 1287 - 6 = 1281$.

Section 6.7

1. $\dfrac{12!}{(3!)4} = 369{,}600$

3. $\dfrac{7!}{3!\,4!} = 35$

5. $\dfrac{9!}{2!\,3!\,4!} = 1260$

7. $\dfrac{6!}{2!\,4!} = 15$

9. $\dfrac{9!}{(3!)^3} = 1680$

11. $\dfrac{14!}{3!\,5!\,6!} = 168{,}168$

13. $\dfrac{18!}{(6!)^3} = 17{,}153{,}136$

15. $\dfrac{15!}{6!\,4!\,5!} = 630{,}630$

17. $\dfrac{10!}{4!\,4!\,2!} = 3{,}150$

19. $\dfrac{10!}{(5!)^2} = 252$

21. $\dfrac{18!}{(6!)^3} = 17{,}153{,}136$

23. $\dfrac{9!}{(3!)^4} = 280$

25. $\dfrac{12!}{3!\,(4!)^3} = 5775$

27. $\dfrac{15!}{2!\,(5!)^2\,3!\,2!} = 3{,}783{,}780$

29. $\dfrac{16!}{2\,(6!)(6!)(4!)} = 840{,}840$

31. $\dfrac{22!}{3!\,(2!)^3\,4!\,(4!)^4}$

33. $\dfrac{9!}{3!\,4!\,2!} = 1{,}260$

35. $\dfrac{50!}{10!\,(5!)^{10}}$

Chapter 6 Review

1.

(a)	True	**(b)**	False	**(c)**	False
(d)	False	**(e)**	True	**(f)**	False
(g)	True	**(h)**	False	**(i)**	True
(j)	False	**(k)**	True	**(l)**	False
(m)	True	**(n)**	False	**(o)**	False

3.

(a) Equal   **(b)** Not equal

(c) Equal, both are empty

Chapter 6 Review

5. $58 = 32 + 40 - n(A \cap B)$
$n(A \cap B) = 32 + 40 - 58 = 14$

7.

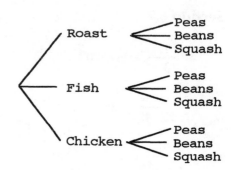

9.   **(a)** $10^4 \cdot 26^2 = 6,760,000$

(b) $10 \times 9 \times 8 \times 7 \times 26 \times 25 = 3,276,000$

11. $C(15, 5) = 3,003$

13. $(40)(27)(85)(34) = 3,121,200$

15. $4 \cdot 5! \times 3 = 1,440$

17.   **(a)** $P(11, 3) = 990$
(b) $11^3 = 1,331$

19. $P(10, 2)C(8, 4) = (90)(70) = 6300$

21.   This totals 58, not 60

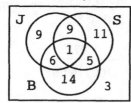

23.   **(a)**   20   **(b)**   6
(c)   23

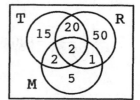

25. $5 \times 8 \times 7 = 280$

27. $C(5, 2) = 10$

29.   **(a)** $(20)^3 = 8,000$

(b) $20 \times 19 \times 18 = 6,840$

31. $C(15, 4)C(20, 4)C(25, 3)C(11, 1)$

33. $P(9, 5) = 15,120$

35. $C(22, 5) = 26,334$

37.   **(a)** $P(8, 4) = 8 \times 7 \times 6 \times 5 = 1,680$

(b) $C(9, 5) = \dfrac{9 \times 8 \times 7 \times 6 \times 5}{6 \times 5 \times 4 \times 3} = 126$

(c) $P(7, 7) = 7! = 5,040$

(d) $C(5, 5) = 1$

    (e) $4! = 24$ **(f)** $\dfrac{7!}{3!\,4!} = 35$

    (g) $\dfrac{8!}{4!} = 1{,}680$

39. $P(20, 2)P(15, 1)P(25, 1) = 380 \times 15 \times 25 = 142{,}500$

41. $83 = 46 + 51 - n(A \cap B)$ **43.** $C(10, 4) = 210$
        $n(A \cap B) = 46 + 51 - 83 = 14$

45. $\varnothing$, {red}, {white}, {blue}, {red, white}, {red, blue}, {white, blue},
        {red, white, blue}

47. $P(10, 2) \cdot C(8, 3) = (90)(56) = 5{,}040$ **49.** $C(6, 4) + C(8, 4) = 15 + 70 = 85$

Chapter 7
Probability

Section 7.1

1. **(a)** {True, False}
   (b) {vowel, consonant} or the letters {a,b,c, ... ,x,y,z}
   (c) {1, 2, 3, 4, 5, 6}
   (d) {HHH, HHT, HTH, THH, TTH, THT, HTT, TTT}
   (e) {Mon, Tue, Wed, Thurs, Fri, Sat, Sun}
   (f) {Grand Canyon wins, Bosque wins}, or {Bosque wins, Bosque loses}
   (g) {pass, fail}, {A, B, C, D, F}
   (h) {Susan & Leah, Susan & Dana, Susan & Julie, Leah & Dana, Leah & Julie, Dana and Julie}
   (i) {Mon, Tue, Wed, Thur, Fri, Sat, Sun}, or {1, 2, 3, 4,...,31}
   (j) {male, female}, {passing student, failing student}

3. **(a)** Male of normal weight, female of normal weight.
   (b) Female underweight, female of normal weight, female overweight.
   (c) Male underweight, female underweight, male overweight, female overweight.

5. **(a)** Valid   **(b)** Valid
   (c) Invalid $P(T) + P(D) + P(H) = 0.95$
   (d) Invalid $P(2) + P(4) + P(6) + P(8) + P(10) = 1.15$
   (e) Valid   **(f)** Invalid $P(Utah) = -0.4 < 0$
   (g) Valid   **(h)** Invalid $P(maybe) = 1.1 > 1$
   (i) Invalid $P(True) + P(False) = 0$

7. **(a)** $P(\{A, B\}) = 0.1 + 0.2 = 0.3$ **(b)** $P(\{B, D\}) = 0.2 + 0.4 = 0.6$
   (c) $P(\{A, C, D\}) = 0.1 + 0.3 + 0.4 = 0.8$
   (d) $P(\{A, B, C, D\}) = 1$

9. **(a)** $0.37 + 0.28 = 0.65$ **(b)** $0.19 + 0.11 + 0.05 = 0.35$

11. **(a)** $0.05 + 0.10 + 0.06 = 0.21$ **(b)** $0.10 + 0.15 + 0.18 = 0.43$
    (c) $0.05 + 0.10 + 0.06 + 0.11 + 0.18 + 0.14 = 0.64$
    (d) $0.12 + 0.15 = 0.27$

Section 7.1 Introduction to Probability

13.   **(a)**   The table of relative frequencies is obtained by dividing each entry by 81.

	Ms. Busby	Mr. Butler	Mrs. Hutchison
Calculator	0.160	0.136	0.210
No Calculator	0.185	0.198	0.111

     (b)   P(Calculator) = 0.160 + 0.136 + 0.210 = 0.506
     (c)   P(Ms. Busby's student and no calculator) = 0.185
     (d)   P(Mr. Butler) = 0.136 + 0.198 = 0.334

15.   P(A) + P(B) + P(C) + P(D)
     = 2×P(D) + 3×Π(D) + 4×P(D) + P(D) = 1
         10×P(D) = 1 so P(D) = 0.1
         P(A) = 0.20, P(B) = 0.30, P(C) = 0.40, P(D) = 0.10

17.   P(A) + P(B) + P(C) = 1
     P(A) + P(B)   = 0.75
         P(B) + P(C) = 0.45
     Subtract second equation from the first
         P(A) + P(B) + P(C) = 1
         P(A) + P(B)   = 0.75
                P(C) = 0.25
     From third equation P(B) + 0.25 = 0.45 gives P(B) = 0.20, so P(A) = 0.55.

19.   P(Mini) = 140/800 = 0.175, P(Burger) = 345/800 = 0.431,
     P(Big B) = 315/800 = 0.394

21.   P(below 40) = 160/1800 = 0.089,
     P(40-49) = 270/1800 = 0.150
     P(50-65) = 1025/1800 = 0.569,
     P(over 65) = 345/1800 = 0.192

23.   **(a)**   P({2, 4}) = 0.16 + 0.30 = 0.46
     (b)   P({1, 3, 5}) = 0.15 + 0.19 + 0.20 = 0.54
     (c)   P(prime) = 0.16 + 0.19 + 0.20 = 0.55

25.   **(a)**   4/52 = 1/13   **(b)**   (4 + 4)/52 = 2/13
     (c)   13/52 = 1/4   **(d)**   26/52 = 1/2

Section 7.2

1. $4/15 = 0.267$

3. n = the total number of sweaters, 36. s = the number of ways to select one small size, 22.
    $$P(\text{small}) = \frac{22}{36} = \frac{11}{18} = 0.611.$$

5. There are $6(6) = 36$ possible ways for two dice to turn up. Of these, a seven occurs in 6 ways so $p = 6/36 = 1/6$

7. $\dfrac{C(10,\ 2)}{C(17,\ 2)} = \dfrac{45}{136} = 0.33$

9. $\dfrac{C(4,\ 2)}{C(52,\ 2)} = \dfrac{6}{1326} = \dfrac{1}{221} = 0.0045$

11. (a) $P(\text{stamped}) = \dfrac{10}{18} = \dfrac{5}{9}$ (b) $P(\text{not stamped}) = \dfrac{8}{18} = \dfrac{4}{9}$

    (c) $P(\text{not stamped or stamped}) = \dfrac{18}{18} = 1$

13. Two tosses can occur in 4 ways. Except for the one case when both are tails, at least one is a head. Thus, $p = 3/4$

15. $\dfrac{C(12,\ 2) \times C(9,\ 1)}{C(21,\ 3)} = \dfrac{594}{1330} = \dfrac{297}{665} = 0.447$

17. $n(E) = 26$, $n(E \cap F) = 9$, $n(E \cup F) = 49$, and $n(S) = 95$
    (a) $P(E) = 26/95 = 0.274$ (b) $P(E \cap F) = 9/95 = 0.095$
    (c) $P(E \cup F) = 49/95 = 0.516$

19. (a) $\dfrac{C(4,\ 4)}{C(15,4)} = \dfrac{1}{1365} = 0.00073$

    (b) $\dfrac{C(4,\ 3) \times C(11,\ 1)}{C(15,\ 4)} = \dfrac{4 \times 11}{1365} = \dfrac{44}{1365} = 0.0322$

    (c) $\dfrac{C(4,\ 2) \times C(11,\ 2)}{C(15,\ 4)} = \dfrac{6 \times 55}{1365} = \dfrac{22}{91} = 0.242$

    (d) $\dfrac{C(11,\ 4)}{C(15,\ 4)} = \dfrac{330}{1365} = \dfrac{22}{91} = 0.242$

21.   (a) $\dfrac{P(10,\ 2)P(6,\ 3)}{P(16,\ 5)} = \dfrac{90 \times 120}{524160} = \dfrac{90}{4368} = \dfrac{15}{728} = 0.0206$

       (b) $\dfrac{10 \times 6 \times 9 \times 5 \times 8}{P(16,\ 5)} = \dfrac{10 \times 6 \times 9 \times 5 \times 8}{524160} = \dfrac{15}{364} = 0.0412$

23.   (a) $\dfrac{P(4,\ 4)}{P(8,\ 4)} = \dfrac{24}{1680} = \dfrac{1}{70} = 0.0143$

       (b) $\dfrac{P(4,\ 2)P(4,\ 2)}{P(8,\ 4)} = \dfrac{12 \times 12}{1680} = \dfrac{3}{35} = 0.0857$

25.   The three pieces can be selected in $8 \times 7 \times 6 = 336$ ways and the vocal, quartet, and piano pieces can be selected in $4 \times 3 \times 1 = 12$ ways. The probability of playing a vocal, quartet, and piano piece in that order is $\dfrac{12}{336} = \dfrac{1}{28} = 0.0357$

27.   There are $C(10,4)$ ways the boxes could be selected. There were $C(6,3)$ ways three boxes for the bedroom and $C(4,1)$ ways one box for the kitchen could be selected. The probability of three bedroom boxes and one kitchen box is $\dfrac{C(6,3) \times C(4,1)}{C(10,4)} = \dfrac{20 \times 4}{210} = \dfrac{8}{21} = 0.381$

29.   Begins with a 5, 6, 7, 8 or 9: $\dfrac{5 \times 8 \times 7}{P(9,\ 3)} = \dfrac{280}{504} = \dfrac{5}{9} = 0.555$

31.   (a) $\dfrac{1}{P(5,\ 5)} = \dfrac{1}{5!} = \dfrac{1}{120}$

       (b) $\dfrac{P(2,\ 2)P(3,\ 3)}{P(5,\ 5)} = \dfrac{2 \times 6}{120} = \dfrac{1}{10}$

33. $\dfrac{C(8,\ 3)\ C(6,\ 1)}{C(14,\ 4)} = \dfrac{56 \times 6}{1001} = \dfrac{48}{143} = 0.336$

35.   (a) $6/28 = 3/14$
       (b) $3/28$
       (c) $10/28 = 5/14$

37.   (a)   P(red and green) $= \dfrac{41 \times 22}{C(63,2)} = 0.462$

       (b)   P(both green) $= \dfrac{C(22,2)}{C(63,2)} = 0.118$

39.   (a) $400/500 = 4/5$ (b) $100/500 = 1/5$

Section 7.2 Equally Likely Events

41. **(a)** $\dfrac{P(8,\ 3)}{P(15,\ 3)} = \dfrac{8 \times 7 \times 6}{15 \times 14 \times 13} = \dfrac{8}{65} = 0.123$

(b) $\dfrac{8 \times 7 \times 7}{P(15,\ 3)} = \dfrac{8 \times 7 \times 7}{15 \times 14 \times 13} = \dfrac{28}{195} = 0.144$

(c) $\dfrac{5 \times 4 \times 3}{P(15,\ 3)} = \dfrac{5 \times 4 \times 3}{15 \times 14 \times 13} = \dfrac{2}{91} = 0.022$

43. $324/570 = 54/95 = 0.568$

45. Five diamonds can be selected in C(13,5) and a selection of any five cards can occur in C(52, 5) ways. Thus, $P(5 \text{ diamonds}) = \dfrac{C(13,\ 5)}{C(52,\ 5)} = \dfrac{1287}{2598960} = 0.000495$

47. The dice can turn up in 6^6 different ways.
If all turn up a different number, then the first die will turn up some number, six possible ways. Then the second die can turn up any of the five remaining numbers, the third any of the remaining four numbers, and so on.
The six different numbers then can occur in $6 \times 5 \times 4 \times 3 \times 2 \times 1 = 720$ ways.
The probability of six different numbers is $p = \dfrac{720}{6^6} = 0.0154$.

49. **(a)** The draw can occur in C(40, 6) = 3,838,380 ways.
(b) The probability of winning is 1/3838380.

51. Let x = number of dimes dated before 1960
$p = x/250 = 0.2$ so $x = 0.2(250) = 50$

53. **(a)** $\dfrac{6 \times 3 \times 1}{10 \times 9 \times 8} = \dfrac{1}{40} = 0.025$ **(b)** $\dfrac{6 \times 5 \times 3}{10 \times 9 \times 8} = \dfrac{1}{8} = 0.125$

(c) $\dfrac{6 \times 5 \times 4}{10 \times 9 \times 8} = \dfrac{1}{6} = 0.167$

55. $P(2 \text{ pennies before } 2000) = \dfrac{C(220,\ 2) \times C(140,\ 3)}{C(360,5)} = 0.220$

The calculated probability indicates 2 pennies dated before 200 should occur about 22% of the time compared to 25% found by the class.

57. **(a)** To win, the participant must select the six numbers drawn. This can be done in C(6, 6) = 1 way.
The six numbers can be drawn in C(54, 6) = 25,827,165 ways
$P(\text{Win first prize}) = \dfrac{1}{25,827,165} = 0.0000000387$

(b) To win second prize, the participant must select 5 of the six numbers drawn and select one of the 48 not drawn. This can be done in $C(6, 5) \times C(48, 1) = 6 \times 48 = 288$ ways.

$$P(\text{second prize}) = \frac{288}{25,827,165} = 0.0000112$$

Section 7.3

1. $E \cup F = \{1, 2, 3, 4, 5, 7, 9\}$ $E \cap F = \{1, 3\}$ $E' = \{2, 4, 6, 8, 10\}$

3. $E \cup F$ = the set of students who are passing English or failing chemistry (or both)
$E \cap F$ = the set of students who are passing English and failing chemistry
E' = the set of students who are failing English

5. $1 - 3/5 = 2/5$ **7.** $P(E) = 1 - 0.7 = 0.3$

9. $1 - 0.3 = 0.7$ **11.**   Mutually exclusive

13.   Not mutually exclusive because a student can be taking both.

15.   Mutually exclusive since there is no February 30.

17.   **(a)** $\dfrac{C(4, 4)}{C(52, 4)} = \dfrac{1}{270725}$ **19.** $\dfrac{14 + 22}{60} = \dfrac{36}{60} = 0.600$

         (b) $\dfrac{1}{270725} + \dfrac{1}{270725} = \dfrac{2}{270725}$

21.   **(a)**   The complement of at least one male is all females.
Find 1 - probability all females.
$1 - \dfrac{C(5,3)}{C(12, 3)} = 1 - \dfrac{10}{220} = 1 - \dfrac{1}{22} = \dfrac{21}{22} = 0.955$

         (b)   The complement of at most 2 males is 3 males.
Find 1 - probability all males.
$1 - \dfrac{C(7,3)}{C(12, 3)} = 1 - \dfrac{35}{220} = 1 - \dfrac{7}{44} = \dfrac{37}{44} = 0.841$

23.   The complement of at least one of them wins is none of them win.
$1 - \dfrac{C(46, 3)}{C(50, 3)} = 1 - \dfrac{15180}{19600} = \dfrac{4420}{19600} = \dfrac{221}{980} = 0.226$ (1 - probability none of 4 wins)

Section 7.3 Compound Events

25. The Venn diagram of this information is

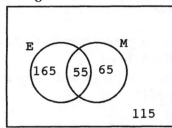

115

(a) $120/400 = 3/10$ **(b)** $220/400 = 11/20 = 0.550$ **(c)** $55/400 = 11/80 = 0.138$

(d) $\dfrac{120 + 220 - 55}{400} = \dfrac{285}{400} = \dfrac{57}{80} = 0.713$

(e) $220 - 55 = 165$ are in enrolled in English but not in mathematics so

$p = \dfrac{165}{400} = 0.413$

(f) 120 are enrolled in mathematics and $115 + 65 = 180$ are not in English, and 65 are both enrolled in mathematics and are not enrolled in English, so n(not in English or in math) $= 180 + 120 - 65 = 235$ so $p = \dfrac{235}{400} = 0.588$

27. **(a)** $4/35$ **(b)** $P(E \cup F) = \dfrac{10 + 14 - 4}{35} = \dfrac{20}{35} = \dfrac{4}{7}$

29. $P(E \cup F) = 0.75 + 0.24 - 0.18 = 0.81$

31. $P(E \cup F) = 0.36 + 0.26 - 0.11 = 0.51$

33. **(a)** $P(E \cup F) = 0.85 + 0.60 - 0.55 = 0.90$
(b) P(neither) = 1 - P(at least one) $= 1 - 0.90 = 0.10$

35. **(a)** $P(E \cup F) = 0.75 + 0.63 - 0.54 = 0.84$
(b) $1 - 0.84 = 0.16$

37. **(a)** $3/6 = 1/2$ **(b)** $2/6 = 1/3$
(c) $3/6 = 1/2$ **(d)** $4/6 = 2/3$

39. Mutually exclusive
41. Mutually exclusive, the coin cannot land on both heads and tails at the same time.

43. **(a)** $5/6$ **(b)** $1/6$
(c) $3/6 = 1/2$ **(d)** $2/6 = 1/3$
(e) $2/6 = 1/3$

45. **(a)** 1/10   **(b)** 1/10
        (c) $5/10 = 1/2$ **(d)** $5/10 = 1/2$
        (e) $5/10 = 1/2$
        (f) There are 5 even numbers, six greater than 4, and 3 even greater than 4. $P(E \cup F) = (5 + 6 - 3)/10 = 8/10 = 4/5$
        (g) There are 5 even numbers, 7 less than 8, and 3 even less than 8. $P(E \cup F) = (5 + 7 - 3)/10 = 9/10$

47. There are 4 aces, 13 spades, and one card that is both an ace and a spade. $P(E \cup F) = (4 + 13 - 1)/52 = 16/52 = 4/13$

49. $P(E \cup F) = 0.89 = 0.76 + 0.62 - P(\text{passing both})$
        $P(\text{passing both}) = 0.49$

51. P(Female or Executive)
        = P(Female) + P(Executives) - P(Females and Executives)
            0.65 = 0.55 + P(Executives) - 0.05, so
            P (Executives) = 0.15
            Since 0.05 are female executives, the remainder
            0.10 = 10%, are male executives.

53. $(3 + 1)/6 = 4/6 = 2/3$

55. **(a)** $C(43, 3)/C(115, 3) = 12341/246905 = 0.050$
        (b) $C(43, 1)C(64, 2)/C(115, 3) = 43(2016)/246905 = 0.351$
        (c) $64(43)(8)/C(115, 3) = 22016/246905 = 0.089$

57. P(At least one has the part) = 0.85 + 0.93 - 0.81 = 0.97

59. **(a)** Probability of at least one 6 = 1 – probability of no sixes $= 1 - \left(\dfrac{5}{6}\right)^4 =$ 0.518
        (b) Probability of a pair of sixes = 1 – probability of no pair of sixes
            $= 1 - \left(\dfrac{35}{36}\right)^{26} = 0.519.$
        They are essentially equally likely.

Section 7.4

1.   **(a)**   3/4   **(b)**   1/2

3.   **(a)**   16/45
   (b)   P(correct | from Section 1) = 7/24
   (c)   P(Section 2 | correct) = 9/16

5.   **(a)**   8 bills are not $1 of which 3 are $5.   p = 3/8
   (b)   10 bills are smaller than $10 of which 7 are $1.   p = 7/10

7.   P(K|F) is the probability of selecting Kate from the group of 10 girls.
   P(F|K) is the probability the selection is a female when Kate is selected.
   P(K|F) = 1/10, P(F|K) = 1

9.   **(a)**   28/60 = 7/15   **(b)**   6/20 = 3/20
   (c)   6/28 = 3/14

11.   P(E|F) = 0.3/0.7 = 3/7   P(F|E) = 0.3/0.6 = 1/2

13.   P(E|F) = 0.24/0.40 = 3/5   P(F|E) = 0.24/0.60 = 2/5

15.   **(a)**   0.20/0.45 = 4/9 = 0.444   **(b)**   0.20/0.75 = 4/15 = 0.267

17.   **(a)**   41/497 = 0.0825   **(b)**   260/497 = 0.523
   (c)   145/497 = 0.292   **(d)**   46/260 = 0.177
   (e)   46/115 = 0.400
   (f)   (23 + 145)/(41 + 341) = 168/382 = 0.440

19.   (4/52)(4/51) = 4/663 = 0.006

21.   **(a)** $(4/52)(4/52) = (1/13)^2 = 1/169$
   (b) $(13/52)(13/52) = (1/4)^2 = 1/16$
   (c) $(26/52)(26/52) = (1/2)^2 = 1/4$

23.   (4/12)(2/11)(6/10) = 2/55   25.   (6/14)(5/13) = 15/91 = 0.165

27.   (3/32)(3/32) = 9/1024 = 0.0088

29.   **(a)**   (4/15)(4/15) = 16/225 = 0.0711   **(b)**   (4/15)(3/15) = 4/75 = 0.0533

Section 7.4 Conditional Probability

31. (a) $(6/15)(4/15)(6/15) = 16/375 = 0.0427$
     (b) $(5/15)(5/15)(5/15) = 1/27 = 0.037$

33. If one die shows a 2, the other die can show five different faces.
If the second die shows a 2, the first one can show five different faces. Thus, there are 10 possible ways one die shows a 2. Of these 10 ways, two of them total 6. The probability of a total of 6 given one die shows a 2 is 2/10.

35. (a) $75/135 = 5/9 = 0.666$ (b) 0.350
     (c) 0.400
     (d) Find P(Out of state and male or out of state and female) =
$P(O \mid M) \cdot P(M) + P(O \mid F) \cdot P(F) = (0.35)(60/135) + (0.40)(75/135) = 0.156 + 0.222 = 0.378$

     (e) $\dfrac{P(F \cap O)}{P(O)} = \dfrac{P(F) \cdot P(O \mid F)}{P(O)} = \dfrac{(5/9)(0.40)}{0.378} = \dfrac{2/9}{0.378} = 0.588$

37. (a) $C(9, 2)/C(20, 2) = 36/190 = 0.189$
     (b) $\dfrac{11}{20} \times \dfrac{9}{19} = 0.261$
     (c) $\dfrac{11}{20} \times \dfrac{9}{19} + \dfrac{9}{20} \times \dfrac{11}{19} = 0.521$

39. (a) $\dfrac{5}{22} \times \dfrac{4}{21} \times \dfrac{17}{20} = \dfrac{17}{462} = 0.037$ (b) $\dfrac{5}{22} \times \dfrac{4}{21} \times \dfrac{3}{20} = \dfrac{1}{154} = 0.006$
     (c) $\dfrac{17}{22} \times \dfrac{16}{21} \times \dfrac{15}{20} = \dfrac{34}{77} = 0.442$
     (d) Probability of at least one has done homework is 1 - probability no one has done homework $= 1 - 0.006 = 0.994$

41. $P(\text{forged}) = 1/10{,}000$ $P(\text{postdated}) = 0.05$
$P(\text{postdated} \mid \text{forged}) = 0.80$
Find $P(\text{forged} \mid \text{postdated}) = \dfrac{P(\text{forged and postdated})}{P(\text{postdated})}$
Since $P(\text{postdated} \mid \text{forged}) = \dfrac{P(\text{forged and postdated})}{P(\text{forged})}$,
then $0.80 = \dfrac{P(\text{forged and postdated})}{0.0001}$
$P(\text{forged and postdated}) = 0.00008$
$P(\text{forged} \mid \text{postdated}) = \dfrac{0.00008}{0.05} = 0.0016$

Section 7.4 Conditional Probability

43. P(scholarship and continues) = P(scholarship)P(continues | scholarship) = (0.30)(0.90) = 0.27

45. **(a)** $\dfrac{1}{9} \times \dfrac{1}{9} \times \dfrac{1}{9} = \dfrac{1}{729}$

(b) The probability is zero since any three numbers add to more than 5.

47. **(a)** 1/10 **(b)** $\left(\dfrac{9}{10}\right)\left(\dfrac{8}{9}\right) = \dfrac{8}{10} = \dfrac{4}{5}$

(c) Anita draws the unsolved problem and Al doesn't or Anita does not draw the unsolved problem and Al does not draw the unsolved problem, $p = \dfrac{1}{10} \times \dfrac{9}{9} + \dfrac{9}{10} \times \dfrac{8}{9} = \dfrac{9}{10}$

49. Let A represent those who passed the aptitude test and D represent those who passed the drug test.

(a) Find P(A ∩ D).
$$P(A \cap D) = \dfrac{42}{96} = 0.438$$

(b) Find P(A)
$$P(A) = \dfrac{54}{96} = 0.563$$

(c) Find P(D | A)
$$P(D \mid A) = \dfrac{P(A \cap D)}{P(A)} = \dfrac{42}{54} = 0.778$$

51. P(Eligible) = 0.75, P(Handicap | Eligible) = 0.05
(a) P(Eligible and Handicap) = 0.75(0.05) = 0.0375
(b) P(Not eligible, and eligible but not handicap) = 0.25(0.75)(0.95) = 0.178

53. The reduced sample space is F. P(E|F) is the fraction of the area of F that is occupied by E ∩ F and is represented by the shaded area.

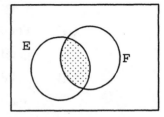

P(E'|F) is the fraction of the area of F that
is occupied by E' ∩ F and is represented by
the shaded area.

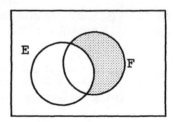

P(E'|F) is the fraction of the area of F that
is occupied by E' ∩ F and is represented by
the shaded area.

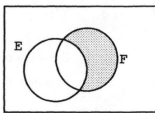

The area representing P(E|F) + P(E'|F) covers all of F so P(E|F) + P(E'|F) = 1

55. **(a)**   A wager is good if the probability of winning is greater than one-half.

      (b)   **(i)**   The color of the first card is irrelevant. The bet is won if the second card is a different color.

$$P(\text{2 different colors}) = 1 \times \frac{5}{9} = \frac{5}{9} = 0.556 \quad \text{This is a good}$$

wager.

           (ii) $P(\text{2 different numbers}) = 1 \times \frac{8}{9} = \frac{8}{9} = 0.889 \quad \text{This is a good}$

wager.

      (c)   **(i)** $P(\text{3 different colors}) = 1 \times \frac{10}{14} \times \frac{5}{13} = 0.275 \quad \text{This is not a good}$

wager.

           (ii) $P(\text{3 different numbers}) = 1 \times \frac{12}{14} \times \frac{9}{13} = 0.593 \quad \text{This is a good}$

wager.

      (d)   **(i)** $P(\text{4 different colors}) = 1 \times \frac{15}{19} \times \frac{10}{18} \times \frac{5}{17} = 0.129 \quad \text{This is not a}$

good wager.

           (ii) $P(\text{4 different numbers}) = 1 \times \frac{16}{19} \times \frac{12}{18} \times \frac{8}{17} = 0.264 \quad \text{This is not a}$

good wager.

57. The problem gives the following information.
The sample space is the set of all registered voters
n(Registered voters) = 84,349
n(people who voted) = 15,183 + 16,026 = 31,209
n(Democrats who voted) = 15,183
n(Republicans who voted) = 16,026
n(voted for Arnold) = 16,827
n(voted for Betros) = 14,382
P(voted for Arnold | Democrat) = 0.39
P(voted for Betros | Republican) = 0.32

(a) $\quad$ P(Registered and did not vote) = $\dfrac{84,349 - 31,209}{84,349} = \dfrac{53,140}{84,349} = 0.630$

(b) $\quad$ P(Democrat who voted) = $\dfrac{15,183}{84,349} = 0.180$

(c) $\quad$ P(voted for Betros) = $\dfrac{14,382}{84,349} = 0.171$

(d) $\quad$ P(voted for Arnold | voted) = $\dfrac{16,827}{31,209} = 0.539$

(e) $\quad$ We want to find P(Republican | voted for Betros).
First of all, we are restricted to those who voted, and we need
n(Republican and voted for Betros), and n(voted for Betros).
Those numbers are 5128 = 0.32 × 16,026 and 14,382, respectively.

$\qquad$ Thus, P(Republican | Voted for Betros) = $\dfrac{5128}{14382} = 0.357$

Section 7.5

1. $\quad$ **(a)** $\quad$ **(i)** $\quad$ Not mutually exclusive
$\qquad\qquad$ **(ii)** $\quad$ Independent, since P(E) = 40/160 = 1/4 and P(E|F) = 10/40 = 1/4

$\qquad$ **(b)** $\quad$ **(i)** $\quad$ Not mutually exclusive
$\qquad\qquad$ **(ii)** $\quad$ Dependent, since P(E) = 40/160 = 1/4 and P(E|F) = 10/50 = 1/5

$\qquad$ **(c)** $\quad$ **(i)** $\quad$ Not mutually exclusive
$\qquad\qquad$ **(ii)** $\quad$ Independent, since P(E) = 50/100 = 1/2 and P(E|F) = 30/60 = 1/2

$\qquad$ **(d)** $\quad$ **(i)** $\quad$ Mutually exclusive
$\qquad\qquad$ **(ii)** $\quad$ Dependent, since P(E « F) = 0 ≠ P(E)P(F) = $\dfrac{1}{16}$

$\qquad$ **(e)** $\quad$ **(i)** $\quad$ Mutually exclusive
$\qquad\qquad$ **(ii)** $\quad$ Dependent, since P(E) = 30/120 = 1/4 and P(E|F) = 0/40 = 0

$\qquad$ **(f)** $\quad$ **(i)** $\quad$ Mutually exclusive
$\qquad\qquad$ **(ii)** $\quad$ Independent, since P(E) = 0/80 = 0 and P(E|F) = 0/20 = 0

Section 7.5 Independent Events

3. Independent, since $(0.3)(0.5) = 0.15 = P(E \cap F)$

5. Dependent, since $(0.3)(0.7) \neq 0.20 = P(E \cap F)$

7. **(a)** $P(F \mid E) = \dfrac{P(E \cap F)}{P(E)} = \dfrac{0.21}{0.35} = 0.60$

$P(E \cup F) = P(E) + P(F) - P(E \cap F) = 0.35 + 0.60 - 0.21 = 0.74$

(b) Since $P(F) = P(F \mid E) = 0.60$, E and F are independent.

9. Since E and F are independent, $P(E \cup F) = P(E) + P(F) - P(E)P(F)$

$= 0.3 + 0.8 - 0.3 \times 0.8 = 0.86$

11. **(a)** Not mutually exclusive since one ring has both a diamond and a ruby.

(b) Independent since $P(E) = 2/4 = 1/2$ and $P(E \mid F) = 1/2$

13. Dependent, since $P(A \cap S) = 10/100 = 1/10$ and $P(A)P(S) = (15/100)(20/100)$

$= 3/100$

15. Since the items were taken from different bags, the events are independent.

P(Lauren gets granola and Anna gets raisins) $= \dfrac{6}{14} \times \dfrac{12}{21} = \dfrac{12}{49} = 0.245$

17. $(1/4)(1/5) = 1/20$

19. **(a)** $(0.2)(0.3) = 0.06$ **(b)** $(0.8)(0.7) = 0.56$

(c) $(0.2)(0.7) + (0.8)(0.3) = 0.14 + 0.24 = 0.38$

21. **(a)** $(0.5)(0.6)(0.8) = 0.24$ **(b)** $(0.5)(0.4)(0.2) = 0.04$

23. **(a)** $0.3 + 0.5 - (0.3)(0.5) = 0.8 - 0.15 = 0.65$

(b) $0.2 + 0.6 - (0.2)(0.6) = 0.8 - 0.12 = 0.68$

(c) $0.4 + 0.6 - (0.4)(0.6) = 1 - 0.24 = 0.76$

25. **(a)** Find $1 - P(\text{neither shows up}) = 1 - (0.4)(0.2) = 1 - 0.08 = 0.92$

(b) Find $1 - P(\text{both show up}) = 1 - (0.6)(0.8) = 1 - 0.48 = 0.52$

27. $1 - (0.7)(0.1) = 1 - 0.07 = 0.93$

Section 7.5 Independent Events

29. P(Male) = 18/30, P(mathematics major) = 10/30, P(Male and mathematics majors) = 6/30

P(Male)P(Mathematics major) = $\dfrac{18}{30} \times \dfrac{10}{30} = \dfrac{6}{30}$ = P(Male and mathematics major) so "Male" and "Mathematics major" are independent.

31. $0.8 \times 0.8 \times 0.8 \times 0.2 \times 0.2 = 0.020$

33. **(a)**

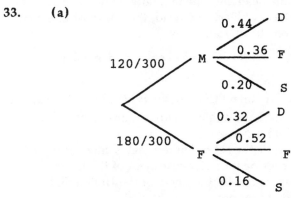

    (b) Find the probability of each branch that ends in F and add.
p = (120/300)(0.36) + (180/300)(0.52) = 0.456
    (c) (120/300)(0.44) = 0.176
    (d) P(F) = 0.456, P(F | W) = 0.52, so dependent

35. P(F)P(G) = (1/2)(3/13) =3/26, P(F ∩ G) = 6/52 = 3/26, so independent

37. **(a)** $(0.5)^{12} = 0.000244$ **(b)** $(0.90)^{12} = 0.282$

39. This succeeds when the nickel turns up heads and the quarter turns up tails twice or the nickel and quarter turn up tails.
(3/5)(1/3)(1/3) + (2/5)(1/3) = 1/15 + 2/15 = 3/15 = 1/5

41. P(met verbal) = $\dfrac{2850}{3200}$ = 0.891   P(met verbal | met math) = $\dfrac{2795}{2965}$ = 0.943
These data indicate that meeting the verbal and meeting the math requirements are dependent since P(met verbal) ≠ P(met verbal | met math)

43. P(Female) = $\dfrac{135}{297}$ = 0.4545   P(Female | Science) = $\dfrac{40}{88}$ = 0.4545
The events female student and science major are independent.

45. Denote at most one head by H and at least one head and one tail by B. Let's compare P(H) and P(H | B).

The four tosses can turn up in $2^4 = 16$ ways. At most one head occurs when all four are tails (1 way) or one head and 3 tails (4 ways). Thus, $P(H) = \dfrac{5}{16}$

Of the 16 possible outcomes at least one head and one tail fail only when the two cases of all heads or all tails occur. So B can occur in 14 ways. Of the 14 ways we can obtain at least one head and one tail, how many have at most one head? Since all of these have at least one head, then at most one head occurs when there is exactly one head, four cases.

Thus, $P(H | B) = \dfrac{4}{14}$ Since $P(H) \neq P(H \mid B)$ the events are dependent.

47. If E and F are mutually exclusive, then $P(E \cap F) = 0$. If $P(E)P(F) = 0$, then one of P(E), P(F) is zero. Thus, E and F (mutually exclusive) are independent when at least one of P(E) or P(F) is zero.

Example: Let the sample space be the set of United States citizens and let E = females who are currently president or vice-president of the United States. Let F = females who are currently members of the United States Senate. One person is selected at random.

Since E is empty, $p(E) = 0$. Since F is not empty, $P(F) \neq 0$. However E and F are mutually exclusive so $P(E \cap F) = 0$.

Thus, $P(E)P(F) = 0 \times P(F) = 0 = P(E \cap F)$ so E and F are independent.

49. A and C may or may not be independent. Here are two examples, the first A and C are not independent, and the second A and C are independent.

First, let A, B, and C be related as shown in the Venn diagram with the number of elements in each region shown.

From the diagram we obtain the following

$n(S) = 200$
$n(A) = 50$
$n(B) = 40$
$n(C) = 30$
$n(A \cap B) = 10$
$n(B \cap C) = 6$
$n(A \cap C) = 0$

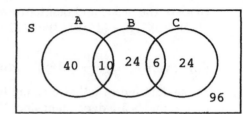

$P(A) = \dfrac{50}{200}$ $\quad P(B) = \dfrac{40}{200}$ $\qquad P(C) = \dfrac{30}{200}$

$P(A \cap B) = \dfrac{10}{200}$ $P(A \cap C) = 0$ $\qquad P(B \cap C) = \dfrac{6}{200}$

$P(A)P(B) = \dfrac{50}{200} \times \dfrac{40}{200} = \dfrac{20}{400} = \dfrac{10}{200} = P(A \cap B)$ so A and B are independent.

$P(B)P(C) = \dfrac{40}{200} \times \dfrac{30}{200} = \dfrac{12}{400} = \dfrac{6}{200} = P(B \cap C)$ so B and C are independent.

$P(A)P(C) = \dfrac{50}{200} \times \dfrac{30}{200} = \dfrac{15}{400} \neq P(A \cap C) = 0$ so A and C are not independent.

For the second example use the following Venn diagram.

From this diagram we obtain

$n(S) = 200$

$n(A) = 50$

$n(B) = 40$

$n(C) = 40$

$n(A \cap B) = 10$

$n(A \cap C) = 10$

$n(B \cap C) = 8$

$P(A) = \dfrac{50}{200} \quad P(B) = \dfrac{40}{200} \quad P(C) = \dfrac{40}{200}$

$P(A \cap B) = \dfrac{10}{200} \quad P(A \cap C) = \dfrac{10}{200} \qquad P(B \cap C) = \dfrac{8}{200}$

$P(A)P(B) = \dfrac{50}{200} \times \dfrac{40}{200} = \dfrac{20}{400} = \dfrac{10}{200} = P(A \cap B)$

so A and B are independent.

$P(B)P(C) = \dfrac{40}{200} \times \dfrac{40}{200} = \dfrac{16}{400} = \dfrac{8}{200} = P(B \cap C)$

so B and C are independent.

$P(A)P(C) = \dfrac{50}{200} \times \dfrac{40}{200} = \dfrac{20}{400} = \dfrac{10}{200} = P(A \cap C)$ so A and C are independent.

This example shows that it is possible for A and C to be independent when both A and B are independent and B and C are independent.

53. For Ron: P(no malfunction on any given day) = 1 - 0.01 = 0.99.

P(no malfunction for 30 days) = 0.99^{30} = 0.74.

For Angela: P(no malfunction on any given day) = 1 - 0.0005 = 0.9995.

Let x = number of days.

P(no malfunction for x days) = 0.9995^x which she wants to be greater than 0.99. Find the intersection of y = 0.9995^x and y = 0.99. It ocurs at x = 20.1. Angela can go 20 days and the probability of malfunction is less than 0.01.

55. There are more people than months so at least 2 must share the same month.

57. There are 2^6 = 64 ways to answer the questions. Since there are more students than ways to answer the questions, at least 2 students must give the same set of answers.

Section 7.6

1.

3. **(a)** $P(E_1 | F) = 0.7/0.9 = 7/9 = 0.778$

    (b) $P(E_1 | F) = \dfrac{(0.75)(0.40)}{(0.75)(0.40) + (0.25)(0.10)}$

                     $= \dfrac{0.3}{0.3 + 0.025} = 0.923$

5. **(a)** $P(E_1 \cap F) = (5/12)(2/5) = 1/6,$

            $P(E_2 \cap F) = (4/12)(1/4) = 1/12,$

            $P(E_3 \cap F) = (3/12)(1/3) = 1/12$

    (b) $P(F) = 1/6 + 1/12 + 1/12 = 1/3$

    (c) $P(E_1 | F) = \dfrac{\frac{1}{6}}{\frac{1}{3}} = \dfrac{1}{2}$ $P(E_3 | F) = \dfrac{\frac{1}{12}}{\frac{1}{3}} = \dfrac{1}{4}$

7. **(a)** $P(F) = 37/84$ **(b)** $P(E_1) = 32/84 = 8/21$

    (c) $P(E_2) = 52/84 = 13/21$ **(d)** $P(F | E_1) = 15/32$

    (e) $P(F | E_2) = 22/52 = 11/26$ **(f)** $P(E_1 | F) = 15/37$

    (g) $P(E_1 \cap F) = 15/84 = 5/28$ **(h)** $\dfrac{P(E_1 \cap F)}{P(F)} = \dfrac{15/84}{37/84} = \dfrac{15}{37}$

    (i) $\dfrac{P(E_1)P(F | E_1)}{P(E_1)P(F | E_1) + P(E_2)P(F | E_2)} = \dfrac{\left(\frac{32}{84}\right)\left(\frac{15}{32}\right)}{\left(\frac{32}{84}\right)\left(\frac{15}{32}\right) + \left(\frac{52}{84}\right)\left(\frac{22}{52}\right)} =$

          $\dfrac{\left(\frac{15}{84}\right)}{\left(\frac{15}{84}\right) + \left(\frac{22}{84}\right)} = \dfrac{15}{37}$

9. We find P(Prod. | Part.) =
$$\frac{P(Prod.)P(Part. | Prod.)}{P(Prod.)P(Part. | Prod.) + P(Sales, Adm.)P(Part. | Sales, Adm)}$$
$$= \frac{0.72(0.64)}{0.72(0.64) + 0.28(0.40)} = \frac{0.4608}{0.4608 + 0.112} = 0.804$$

11. $P(II | D) = \dfrac{P(II)P(D | II)}{P(I)P(D | I) + P(II)P(D | II)} = \dfrac{(0.45)(0.04)}{(0.55)(0.03) + (0.45)(0.04)} = 0.522$

13. $P(H | W) = \dfrac{P(H) \ P(W | H)}{P(H) \ P(W | H) + P(Away) \ P(W | Away)} =$
$$\frac{(0.60)(0.80)}{(0.60)(0.80) + (0.40)(0.55)} = 0.686$$

15. Use the notation So = sophomore, J = junior, Sr = senior, and A = received an A. The following information is given
$$P(So) = \frac{10}{50} = 0.20, P(J) = \frac{25}{50} = 0.50, P(Sr) = \frac{15}{50} = 0.30,$$
$$P(A | So) = \frac{3}{10} = 0.30, P(A | J) = \frac{5}{25} = 0.20, P(A | Sr) = \frac{6}{15} = 0.40$$
We are to find P(J | A).
$$P(J | A) = \frac{P(J)P(A | J)}{P(J)P(A | J) + P(So)P(A | So) + P(Sr)P(A | Sr)} =$$
$$\frac{0.50(0.20)}{0.50(0.20) + 0.20(0.30) + 0.30(0.40)}$$
$$= \frac{0.1}{0.1 + 0.06 + 0.12} = 0.357$$
The probability a student who received an A is a junior is about 0.36.

17. **(a)**

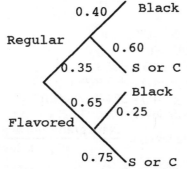

(b) P(Black coffee | regular) = 0.40
P(Black coffee and regular) = 0.35(0.40) = 0.14
P(Black coffee) = 0.35(0.40) + 0.65(0.25) = 0.14 + 0.1625 = 0.303

(c) P(Flavored coffee | cream or sugar)
= P(Flavored coffee and cream or sugar)/P(cream or sugar)

$$= \frac{0.65 \times 0.75}{0.35 \times 0.60 + 0.65 \times 0.75} = \frac{0.4875}{0.210 + 0.4875} = \frac{0.4875}{0.6975} = 0.699$$

19. Let A represent a driver involved in an accident, DE represent a driver with driver's ed, and no DE represent a driver with no DE.
We are given:
P(DE) = 0.80, P(no DE) = 0.20
P(A | DE) = 0.32, P(A | no DE) = 0.55
We are to find P(DE | A).

$$P(DE|A) = \frac{P(DE)P(A|DE)}{P(DE)P(A|DE) + P(no\ DE)P(A|no\ DE)} = \frac{0.80(0.32)}{0.80(0.32) + 0.20(0.55)} =$$

0.699

21. Let H = high school diploma, NH = no high school diploma, and U = unemployed. The given information is:

P(H) = 0.82, P(NH) = 0.18
P(U | H) = 0.04, P(U | NH) = 0.10
We are to find P(NH | U).

$$P(NH|U) = \frac{P(NH)P(U|NH)}{P(NH)P(U|NH) + P(H)P(U|H)}$$
$$= \frac{0.18(0.10)}{0.18(0.10) + 0.82(0.04)} = 0.354$$

The probability the person has no high school diploma is 0.35.

23.
$$P(hep|react) = \frac{P(Hep)P(react|Hep)}{P(Hep)P(react|Hep) + P(No Hep)P(react|No Hep)} =$$
$$= \frac{0.665}{0.671} = 0.991$$

25. **(a)** P(4th grade) = 678/1205 = 0.563
(b) P(5th grade) = 527/1205 = 0.437
(c) P(At grade level) = 666/1205 = 0.553
(d) P(At grade level | 4th grade) = 342/678 = 0.504
(e) P(At grade level | 5th grade) = 324/527 = 0.615
(f) P(4th grade and at grade level) = 342/1205 = 0.284
(g) P(4th grade | At grade level) = $\frac{P(4th\ grade\ and\ at\ grade\ level)}{P(grade\ level)}$

$$= \frac{342/1205}{666/1205} = \frac{342}{666} = 0.514$$

The computations can also be done as

$$\frac{P(\text{4th })P(\text{At grade } | \text{4th })}{P(\text{4th })P(\text{At grade } | \text{4th }) + P(\text{5th })P(\text{At grade } | \text{5th })}$$

$$= \frac{0.563(0.504)}{0.563(0.504) + 0.437(0.615)} = \frac{0.284}{0.553} = 0.514$$

27.   Let 1.5+ represent those who spent at least 1.5 hours on homework and 1.5-those who spent less.
We are given
$P(1.5+) = 0.43$, $P(1.5-) = 0.57$
$P(A \text{ or } B \mid 1.5+) = 0.78$, $P(A \text{ or } B \mid 1.5-) = 0.21$
Find $P(1.5+ \mid A \text{ or } B)$

$$P(1.5+ \mid A \text{ or } B) = \frac{P(1.5+)\, P(A \text{ or } B \mid 1.5+)}{P(1.5+)P(A \text{ or } B \mid 1.5+) + P(1.5-)P(A \text{ or } B \mid 1.5-)}$$

$$= \frac{0.43(0.78)}{0.43(0.78) + 0.57(0.21)} = 0.737$$

The probability an A or B student studied at least 1.5 hours is about 0.737.

29.   **(a)** $P(\text{fraud}) = (0.11)(0.20) + (0.89)(0.03) = 0.0487 = 4.87\%$

   (b) $P(\text{exceed} | \text{fraud}) = \dfrac{P(\text{exceed} \cap \text{fraud})}{P(\text{fraud})} = \dfrac{(0.11)(0.20)}{0.0487} = 0.452$

31.   We are given
$P(\text{So}) = 0.20, P(\text{Jr}) = 0.35, P(\text{Sr}) = 0.45$
$P(\text{Ac} | \text{So}) = 0.15, P(\text{Ac} | \text{Jr}) = 0.08, P(\text{Ac} | \text{Sr}) = 0.21$
Find $P(\text{Jr} | \text{Ac})$.

$$P(\text{Jr} | \text{Ac}) = \frac{P(\text{Jr})P(\text{Ac} | \text{Jr})}{P(\text{So})P(\text{Ac} | \text{So}) + P(\text{Jr})P(\text{Ac} | \text{Jr}) + P(\text{Sr})P(\text{Ac} | \text{Sr})}$$

$$= \frac{0.35(0.08)}{0.20(0.15) + 0.35(0.08) + 0.45(0.21)} = 0.184$$

About 18.4% of accounting majors were juniors.

33.   Let D = Democrat, R = Republican, and S = supports amendment. We are given
$P(D) = 0.45, P(R) = 0.55$
$P(S | D) = 0.80, P(S | R) = 0.20$
Find $P(R | S)$

$$P(R | S) = \frac{P(R)P(S | R)}{P(R)P(S | R) + P(D)P(S | D)} = \frac{0.55(0.20)}{0.55(0.20) + 0.45(0.80)} = \frac{0.11}{0.47} = 0.234$$

35. $P(Y | \text{def}) = \dfrac{(0.35)(5/350)}{(0.40)(1/40) + (0.35)(5/350) + (0.25)(2/250)} = \dfrac{0.005}{0.017} = 0.294$

37. We are given the following information
         p(perform satisfactorily | pass test) = 0.90
         p(not perform satisfactorily | pass test) = 0.10
         p(perform satisfactorily | not pass test) = 0.15
         p(not perform satisfactorily | not pass test) = 0.85
         p(passes test) = 0.65
     We wish to find p(pass test | perform satisfactorily)
     We draw a tree diagram that shows the information given.

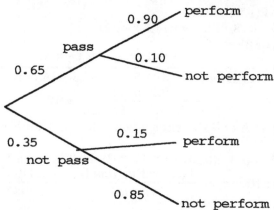

     Bayes' Rule states

$$p(\text{pass} \mid \text{performs}) = \frac{p(\text{pass and performs})}{p(\text{performs})}$$

     From the diagram there are two paths which terminate at performs
     satisfactorily so the probability of performing satisfactorily is the sum of
     those probabilities, that is,
         0.65(0.90) + 0.35(0.15) = 0.6375
     p(pass and perform satisfactorily) = 0.65(0.90) = 0.585

     Thus, P(pass | performs satisfactorily) = $\dfrac{0.585}{0.6375}$ = 0.918.

39. **(a)** P(A) = 0.75   **(b)**   P(D | A) = 0.45
     (c) P(N | B) = 0.10
     (d) P(N) = 0.75(0.15) + 0.25(0.10) = 0.1375
     (e) P(C) = 0.75(0.40) + 0.25(0.70) = 0.475
              P(D) = 0.75(0.45) + 0.25(0.20) = 0.3875

$$P(A \mid C) = \frac{0.75(0.40)}{0.475} = 0.632$$

$$P(B \mid C) = \frac{0.25(0.70)}{0.475} = 0.368$$

$$P(A \mid D) = \frac{0.75(0.45)}{0.3875} = 0.871$$

$$P(B \mid D) = \frac{0.25(0.20)}{0.3875} = 0.129$$

$$P(A \mid N) = \frac{0.75(0.15)}{0.1375} = 0.818$$

$$P(B \mid N) = \frac{0.25(0.10)}{0.1375} = 0.182$$

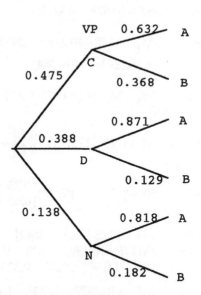

41. A and D are independent when $P(A) = P(A \mid D)$.

$P(A) = 0.25$

$$P(A \mid D) = \frac{P(A)P(D \mid A)}{P(A)P(D \mid A) + P(B)P(D \mid B)} = \frac{0.25(0.6)}{0.25(0.6) + 0.75x}$$

If $P(A) = P(A \mid D)$, then $0.25 = \dfrac{0.25(0.6)}{0.25(0.6) + 0.75x}$

$0.25(0.25(0.6) + 0.75x) = 0.25(0.6)$

$0.25(0.6) + 0.75x = 0.6$

$0.75x = 0.75(0.6)$

$x = 0.6$

A and D are independent when $P(D \mid B) = 0.6$.

Section 7.7

1.　**(a)** Is a transition matrix.
　　(b) Not a transition matrix because row 2 does not add to 1.
　　(c) Not a transition matrix because it is not square.
　　(d) A transition matrix.

3.　**(a)** 10%　　　　**(b)** 0.60　　　　**(c)** 0.90
　　(d) 40%

5.　**(a)** 0.4　　　　**(b)** 0.8　　　　**(c)** 0.3

7. $ST = [0.675 \quad 0.325]$

9. $M_1 = M_0T = [0.282 \quad 0.718]$ $\qquad$ $M_2 = M_1T = [0.282 \quad 0.718]T$
$= [0.32616 \quad 0.67384]$

11. $M_1 = M_0T = [0.175 \quad 0.375 \quad 0.45]$ $\qquad$ $M_2 = M_1T = [0.18 \quad 0.4075 \quad 0.4125]$

13. $ST = [2/3 \quad 1/3]\begin{bmatrix} 0.6 & 0.4 \\ 0.8 & 0.2 \end{bmatrix} = [2/3 \quad 1/3] = S$

15. $ST = [13/28 \quad 15/28]\begin{bmatrix} 0.25 & 0.75 \\ 0.65 & 0.35 \end{bmatrix} = [13/28 \quad 15/28] = S$

17. $MT = [0.25 \quad 0.33 \quad 0.42]$ $\qquad$ **19.** $\qquad$ Since $0.2 + x + 0.4 = 1, x = 0.4$
$(MT)T = [0.259 \quad 0.323 \quad 0.418]$
$((MT)T)T = [0.2577 \quad 0.3233 \quad 0.419]$
$MT^3 = [0.2577 \quad 0.3233 \quad 0.419]$

21. Next year: $[0.45 \quad 0.55]\begin{bmatrix} 0.75 & 0.25 \\ 0.15 & 0.85 \end{bmatrix} = [0.42 \quad 0.58]$ so 42% will contribute next year

In two years: $[0.42 \quad 0.58]\begin{bmatrix} 0.75 & 0.25 \\ 0.15 & 0.85 \end{bmatrix} = [0.402 \quad 0.598]$ so 40.2% will contribute in two years

23. $[0.6 \quad 0.4]\begin{bmatrix} 0.5 & 0.5 \\ 0.2 & 0.8 \end{bmatrix} = [0.38 \quad 0.62]$

$[0.38 \quad 0.62]\begin{bmatrix} 0.5 & 0.5 \\ 0.2 & 0.8 \end{bmatrix} = [0.314 \quad 0.686]$

$[0.314 \quad 0.686]\begin{bmatrix} 0.5 & 0.5 \\ 0.2 & 0.8 \end{bmatrix} = [0.2942 \quad 0.7058]$

$[0.2942 \quad 0.7058]\begin{bmatrix} 0.5 & 0.5 \\ 0.2 & 0.8 \end{bmatrix} = [0.28826 \quad 0.71174]$

$[0.28826 \quad 0.71174]\begin{bmatrix} 0.5 & 0.5 \\ 0.2 & 0.8 \end{bmatrix} = [0.286478 \quad 0.713522]$

$[0.286478 \quad 0.713522]\begin{bmatrix} 0.5 & 0.5 \\ 0.2 & 0.8 \end{bmatrix} = [0.2859434 \quad 0.7140566]$

It appears that $[0.286 \quad 0.714]$ is the steady-state matrix.
The steady-state matrix is actually $[2/7 \quad 5/7]$.

25. Find [x y z] such that $[x \quad y \quad z] \begin{bmatrix} 0.6 & 0.2 & 0.2 \\ 0.1 & 0.8 & 0.1 \\ 0.2 & 0.4 & 0.4 \end{bmatrix} = [x \quad y \quad z]$

This with $x + y + z = 1$ gives

$x + y + z = 1$
$-0.4x + 0.1y + 0.2z = 0$
$0.2x - 0.2y + 0.4z = 0$
$0.2x + 0.1y - 0.6z = 0$

$\begin{bmatrix} 1 & 1 & 1 & | & 1 \\ -0.4 & 0.1 & 0.2 & | & 0 \\ 0.2 & -0.2 & 0.4 & | & 0 \\ 0.2 & 0.1 & -0.6 & | & 0 \end{bmatrix} \quad \begin{bmatrix} 1 & 1 & 1 & | & 1 \\ 0 & 0.5 & 0.6 & | & 0.4 \\ 0 & -0.4 & 0.2 & | & -0.2 \\ 0 & -0.1 & -0.8 & | & -0.2 \end{bmatrix} \quad \begin{bmatrix} 1 & 0 & -7 & | & -1 \\ 0 & 0 & -3.4 & | & -0.6 \\ 0 & 0 & 3.4 & | & 0.6 \\ 0 & 1 & 8 & | & 2 \end{bmatrix}$

$\begin{bmatrix} 1 & 0 & 0 & | & 4/17 \\ 0 & 1 & 0 & | & 10/17 \\ 0 & 0 & 1 & | & 3/17 \\ 0 & 0 & 0 & | & 0 \end{bmatrix}$ so [4/17 10/17 3/17] is the steady-state matrix.

27. $[x\ y\ z] \begin{bmatrix} 0.3 & 0.2 & 0.5 \\ 0.4 & 0.6 & 0 \\ 0.1 & 0.8 & 0.1 \end{bmatrix} = [x\ y\ z]$

and $x + y + z = 1$ gives the augmented matrix

$\begin{bmatrix} 1 & 1 & 1 & | & 1 \\ -0.7 & 0.4 & 0.1 & | & 0 \\ 0.2 & -0.4 & 0.8 & | & 0 \\ 0.5 & 0 & -0.9 & | & 0 \end{bmatrix} \quad \begin{bmatrix} 1 & 1 & 1 & | & 1 \\ 0 & 1.1 & 0.8 & | & 0.7 \\ 0 & -0.6 & 0.6 & | & -0.2 \\ 0 & -0.5 & -1.4 & | & -0.5 \end{bmatrix}$

Divide row 3 by -0.6 and then interchange rows 2 and 3.

$\begin{bmatrix} 1 & 1 & 1 & | & 1 \\ 0 & 1 & -1 & | & 1/3 \\ 0 & 1.1 & 0.8 & | & 0.7 \\ 0 & -0.5 & -1.4 & | & -0.5 \end{bmatrix} \quad \begin{bmatrix} 1 & 0 & 2 & | & 2/3 \\ 0 & 1 & -1 & | & 1/3 \\ 0 & 0 & 1.9 & | & 1/3 \\ 0 & 0 & -1.9 & | & -1/3 \end{bmatrix} \quad \begin{bmatrix} 1 & 0 & 0 & | & 18/57 \\ 0 & 1 & 0 & | & 29/57 \\ 0 & 0 & 1 & | & 10/57 \\ 0 & 0 & 0 & | & 0 \end{bmatrix}$

so [18/57 29/57 10/57] is the steady-state matrix

29. $[x\ y\ z] \begin{bmatrix} 0.6 & 0.2 & 0.2 \\ 0.1 & 0.8 & 0.1 \\ 0.2 & 0.3 & 0.5 \end{bmatrix} = [x\ y\ z]$ and $x + y + z = 1$ gives

$\begin{bmatrix} 1 & 1 & 1 & | & 1 \\ -0.4 & 0.1 & 0.2 & | & 0 \\ 0.2 & -0.2 & 0.3 & | & 0 \\ 0.2 & 0.1 & -0.5 & | & 0 \end{bmatrix} \quad \begin{bmatrix} 1 & 1 & 1 & | & 1 \\ 0 & 0.5 & 0.6 & | & 0.4 \\ 0 & -0.4 & 0.1 & | & -0.2 \\ 0 & -0.1 & -0.7 & | & -0.2 \end{bmatrix} \quad \begin{bmatrix} 1 & 0 & -0.20 & | & 0.20 \\ 0 & 1 & 1.20 & | & 0.80 \\ 0 & 0 & 0.58 & | & 0.12 \\ 0 & 0 & -0.58 & | & -0.12 \end{bmatrix}$

$\begin{bmatrix} 1 & 0 & 0 & | & 0.2414 \\ 0 & 1 & 0 & | & 0.5517 \\ 0 & 0 & 1 & | & 0.2069 \\ 0 & 0 & 0 & | & 0 \end{bmatrix}$

The steady-state matrix = [0.2414 0.5517 0.2069] so at steady-state about 24.1% are in the Full-Meal plan, 55.2% are in the One-Meal plan, and 20.7% are in no plan. **31.** $[x \ y]\begin{bmatrix} 0.9 & 0.1 \\ 1 & 0 \end{bmatrix} = [x \ y]$ and $x + y = 1$ gives

$$\left[\begin{array}{cc|c} 1 & 1 & 1 \\ -0.1 & 0.1 & 0 \\ 0.1 & -0.1 & 0 \end{array}\right] \quad \left[\begin{array}{cc|c} 1 & 1 & 1 \\ 0 & 1.1 & 0.1 \\ 0 & -1.1 & -0.1 \end{array}\right] \quad \left[\begin{array}{cc|c} 1 & 0 & 10/11 \\ 0 & 1 & 1/11 \\ 0 & 0 & 0 \end{array}\right]$$

so $[10/11 \quad 1/11]$ is the steady-state matrix.

33. $[x \ y \ z]\begin{bmatrix} 1/3 & 1/3 & 1/3 \\ 1/2 & 1/2 & 0 \\ 0 & 1/4 & 3/4 \end{bmatrix} = [x \ y \ z]$ and $x + y + z = 1$ gives

$$\left[\begin{array}{ccc|c} 1 & 1 & 1 & 1 \\ -2/3 & 1/2 & 0 & 0 \\ 1/3 & -1/2 & 1/4 & 0 \\ 1/3 & 0 & -1/4 & 0 \end{array}\right] \quad \left[\begin{array}{ccc|c} 1 & 1 & 1 & 1 \\ 0 & 7/6 & 2/3 & 2/3 \\ 0 & -5/6 & -1/12 & -1/3 \\ 0 & -1/3 & -7/12 & -1/3 \end{array}\right] \quad \left[\begin{array}{ccc|c} 1 & 0 & 3/7 & 3/7 \\ 0 & 1 & 4/7 & 4/7 \\ 0 & 0 & 1/28 & 1/7 \\ 0 & 0 & -1/28- & 1/7 \end{array}\right]$$

$$\left[\begin{array}{ccc|c} 1 & 0 & 0 & 3/11 \\ 0 & 1 & 0 & 4/11 \\ 0 & 0 & 1 & 4/11 \\ 0 & 0 & 0 & 0 \end{array}\right] \qquad \text{The steady-state matrix is } [3/11 \quad 4/11 \quad 4/11]$$

35.

$$\begin{array}{c} \text{To} \\ \text{From} \end{array} \begin{array}{c} \\ 1 \\ 2 \\ 3 \\ 4 \\ 5 \end{array} \begin{array}{ccccc} 1 & 2 & 3 & 4 & 5 \\ \begin{bmatrix} 0 & 1/2 & 0 & 0 & 1/2 \\ 1/2 & 0 & 1/2 & 0 & 0 \\ 0 & 1/2 & 0 & 1/2 & 0 \\ 0 & 0 & 1/2 & 0 & 1/2 \\ 1/2 & 0 & 0 & 1/2 & 0 \end{bmatrix} \end{array}$$

37.

$$\left[\begin{array}{ccc|c} 1 & 1 & 1 & 1 \\ -1/2 & 1/4 & 0 & 0 \\ 1/2 & -1/2 & 1/2 & 0 \\ 0 & 1/4 & -1/2 & 0 \end{array}\right] \quad \left[\begin{array}{ccc|c} 1 & 1 & 1 & 1 \\ 0 & 3/4 & 1/2 & 1/2 \\ 0 & -1` & 0 & -1/2 \\ 0 & 1/4 & -1/2 & 0 \end{array}\right] \quad \left[\begin{array}{ccc|c} 1 & 1 & 1 & 1 \\ 0 & -1 & 0 & -1/2 \\ 0 & 3/4 & 1/2 & 1/2 \\ 0 & 1/4 & -1/2 & 01 \end{array}\right]$$

$$\left[\begin{array}{ccc|c} 1 & 0 & 1 & 1/2 \\ 0 & 1 & 0 & 1/2 \\ 0 & 0 & 1/2 & 1/8 \\ 0 & 0 & -1/2 & -1/8 \end{array}\right] \quad \left[\begin{array}{ccc|c} 1 & 0 & 0 & 1/4 \\ 0 & 1 & 0 & 1/2 \\ 0 & 0 & 1 & 1/4 \\ 0 & 0 & 0 & 0 \end{array}\right]$$

When the process reaches a steady state 25% of the flowers will be red, 50% of the flowers will be pink, and 25% of the flowers will be white.

39.   **(a)**   0.06   **(b)**   0.05   **(c)**   0.80

       (d)   0.04

Chapter 7 Review

1. No, because the probability assignments do not sum to 1.

3. P(medium-sized soft drink) = 146/360 = 0.406 since a total of 360 soft drinks were sold.

5. P(sports or band) = P(sports) + P(band) - P(both) = 1/3 + 2/5 - 1/6 = 17/30 = 0.567

7. Four students can be selected in C(10, 4) = 210 ways. Mark, Melanie, and two others can be selected in C(2, 2)C(8, 2) = 28 ways.

    The probability Mark and Melanie are among the four selected is $\dfrac{28}{210} = \dfrac{2}{15} = $ 0.133.

9. P(king or spade) = P(king) + P(spade) - P(king of spades)
    = 4/52 + 13/52 - 1/52 = 4/13

11. Let heads be considered a success. Then all heads up is $(1/2)^5 = 1/32$

13. (a) $\dfrac{10+8}{33} = \dfrac{18}{33} = \dfrac{6}{11} = 0.545$

    (b) $\dfrac{C(8,2)+C(15,2)}{C(33,2)} = \dfrac{28+105}{528} = \dfrac{133}{528} = 0.252$

15. (a) P(same kind) $= \dfrac{C(10,\ 2)}{C(40,\ 2)} + \dfrac{C(10,\ 2)}{C(40,\ 2)} + \dfrac{C(10,\ 2)}{C(40,\ 2)} + \dfrac{C(10,\ 2)}{C(40,\ 2)} = $
    $4(45/780) = 3/13 = 0.231$

    (b) P(different kinds) = 1 - P(same kinds) = 1 - 0.231 = 0.769

17. P(Even) = 20/52   P(Ten) = 4/52
    P(Even) P(Ten) = (20/52)(4/52) = 5/169   P(Even $\cap$ Ten) = 4/52
    Since P(Even) P(Ten) $\neq$ P(Even $\cap$ Ten), Even and Ten are not independent events .

19. (a) P(1, 2, 3, 4, in that order) = (1/6)(1/6)(1/6)(1/6) = 1/1296

    (b) P(1, 2, 3, 4, in any order) = Any one of the four on the first roll, any of the 3 remaining on second roll, etc. = (4/6)(3/6)(2/6)(1/6) = 24/1296 = 1/54

    (c) P(two even, then a 5, then a number less than 3) =
    (3/6)(3/6)(1/6)(2/6) = 18/1296 = 1/72

21. **(a)** P(passing math, history, and English, and failing chemistry)
= P(passing math) × P(passing history)× P(passing English) ×
P(failing chemistry) = (0.8)(0.3)(0.5)(0.3) = 0.036

(b) P(passing math and chemistry and failing history and English)
= P(passing math) × P(passing chemistry) × P(failing history) ×
P(failing English) = (0.8)(0.7)(0.7)(0.5) = 0.196

(c) P(passing all 4 subjects) = (0.8)(0.5)(0.3)(0.7) = 0.084

23. **(a)** P(male | abstainer)
$$= \frac{P(male)P(abstain | male)}{P(male)P(abstain | male) + P(female)P(abstain | female)}$$
$$= \frac{(0.45)(0.20)}{(0.45)(0.20)+(0.55)(0.40)} = \frac{0.09}{0.31} = 0.290$$

(b) P(female | heavy drinker)
$$= \frac{P(female)P(heavy \ drinker | female)}{P(female)P(heavy \ drink | female) + P(male)P(heavy \ drink | male)}$$
$$= \frac{(0.55)(0.05)}{(0.55)(0.05)+(0.45)(0.20)} = \frac{0.0275}{0.1175} = 0.234$$

25. **(a)** 0 since repetitions are not allowed   **(b)**   1/30

(c) P(64 or 46) = 2/30

(d)
First digit is	Number of 2-digit numbers
1	0
2	1
3	2
4	3
5	4
6	5
	15

There are 15 possible numbers so p = 15/30

(e) There are 5 numbers with 1 as the first digit. The only other numbers
less than 24 are 21 and 23 so p = 7/30

27. Let J, S, and G be the event of being a junior, senior, and
grad student, respectively. Let A be the event of receiving an A.
P(J) = 10/50 = 0.2   P(A | J) = 2/10 = 0.2
P(S) = 34/50 = 0.68   P(A | S) = 8/34 = 0.235
P(G) = 6/50 = 0.12   P(A | G) = 3/6 = 0.5

$$P(J \mid A) = \frac{P(J)\, P(A \mid J)}{P(J)\, P(A \mid J) + P(S)\, P(A \mid S) + P(G)\, P(A \mid G)} =$$

$$\frac{(0.2)(0.2)}{(0.2)(0.2) + (0.68)(0.235) + (0.12)(0.5)} = 0.154$$

29. **(a)** $2 \times 1 \times 3 \times 2 \times 1 = 12$

 (b) A semifinalist must be seated first, then a finalist, etc.
$3 \times 2 \times 2 \times 1 \times 1 = 12$

31. **(a)** $P(\text{First} < 3 \text{ and second} > 7) = \dfrac{2}{10} \times \dfrac{3}{10} = 0.06$

 (b) $P(\text{First} < 4 \text{ and second} < 4) = \dfrac{3}{10} \times \dfrac{3}{10} = 0.09$

33. **(a)** $P(\text{C or above}) = \dfrac{225}{420} = 0.536$

 (b) $P(\text{Low}) = \dfrac{68}{420} = 0.162$

 (c) $P(\text{Below C} \mid \text{Middle}) = \dfrac{118}{242} = 0.488$

 (d) $P(\text{High} \mid \text{C or above}) = \dfrac{98}{225} = 0.436$

35. $P(\text{Seat belt not used}) = \dfrac{94}{192} \qquad P(\text{Injuries}) = \dfrac{80}{192}$

 $P(\text{Seat belt not used and injuries}) = \dfrac{66}{192} = 0.344$

 $P(\text{Seat belt not used})P(\text{Injuries}) = \left(\dfrac{94}{192}\right)\left(\dfrac{80}{192}\right) = 0.204$

 Since $0.344 \ne 0.204$ injuries and seat belt not used are dependent.

37. $P(\text{Husband}) = 0.72, P(\text{Wife}) = 0.55, P(\text{Both}) = 0.35$

 (a) $P(\text{at least one}) = P(\text{Husband}) + P(\text{wife}) - P(\text{Both}) = 0.72 + 0.55 - 0.35$
$= 0.92$

 (b) $P(\text{neither}) = 1 - P(\text{at least one}) = 1 - 0.92 = 0.08$

Chapter 8
Statistics

Section 8.1

1.

	Number	Frequency
(a)	-1	2
(b)	2	3
(c)	4	4
(d)	6	2
(e)	8	1

3. Number of students who used swimming pool per day

Categories	Frequency
85-99	7
100-114	8
115-129	13
130-144	7
145-159	5

5.

Stem	Leaves
2	637
3	1839
4	41081
5	379

7.

Stem	Leaves
2	2031
2	796
3	134
3	57686
4	0131
4	856

9.
- **(a)** 67 students took the test.
- **(b)** 7 + 12 = 19 students scored below 70.
- **(c)** 14 + 8 = 22 students scored at least 80.
- **(d)** Cannot be determined.
- **(e)** 26 + 14 = 40 students scored between 69 and 90.
- **(f)** Cannot be determined.

11.

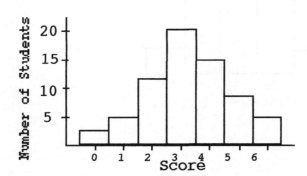

13.   Since the interval 0-10 includes all scores, use intervals of length 2 and count
       the number in each interval.

Interval	Frequency
0-2.0	3
2.01-4.0	15
4.01-6.0	7
6.01-8.0	3
8.01-10.0	2

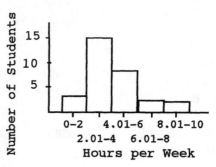

15.

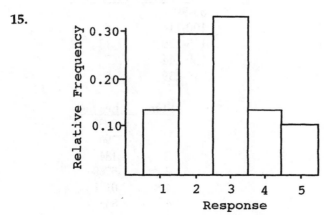

17.

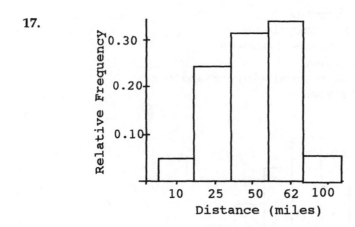

19.

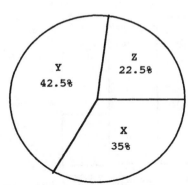

Brand of coffee preferred

21.

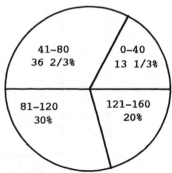

Concentration of ozone

23.

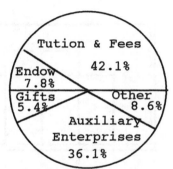

Income of Old Main University

25.

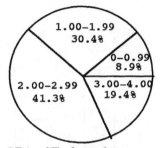

GPA of Tech students

29.

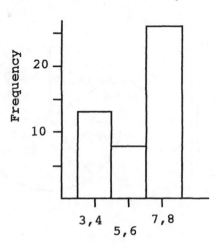

31.

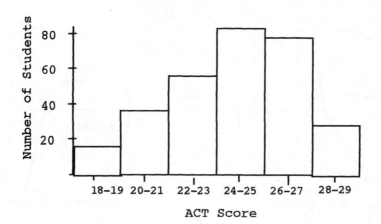

33.

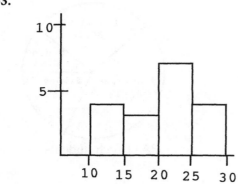

35.

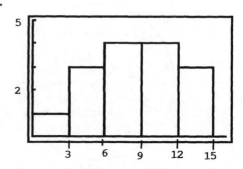

37.

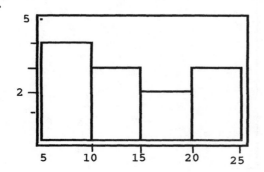

39.

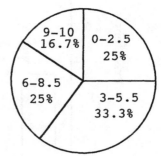

41.

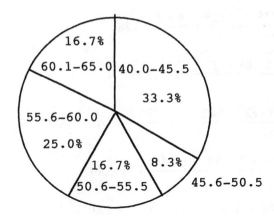

Section 8.2

1. $\mu = \dfrac{2 + 4 + 6 + 8 + 10}{5} = 6$ **3.** $\mu = \dfrac{2.1 + 3.7 + 5.9}{3} = 3.9$

5. $\mu = \dfrac{6 - 4 + 3 + 5 - 8 + 2}{6} = 0.667$

7. $\mu = \dfrac{5.9 + 2.1 + 6.6 + 4.7}{4} = 4.825$

9. $\mu = \dfrac{80 + 76 + 92 + 64 + 93 + 81 + 57 + 77}{8} = 77.5$

11. $\mu = \dfrac{90.5 + 89.2 + 78.4 + 91.0 + 84.2 + 73.5 + 88.7}{7} = 85.07 \text{ mi/hr}$

13. 9   **15.** $\dfrac{21 + 25}{2} = 23$ **17.** $(9 + 14)/2 = 11.5$

19. 65, 68, 72, 72, 72, 81, 86, 90, 98   Median is 72

21. 5   **23.** 1 and 5   **25.** No mode

27. $\dfrac{2 \times 96 + 3 \times 91 + 7 \times 85 + 13 \times 80 + 12 \times 75 + 10 \times 70 + 8 \times 60 + 5 \times 50}{60} =$
73.83

29. $\dfrac{0 \times 430 + 1 \times 395 + 2 \times 145 + 3 \times 25 + 4 \times 5}{1000} = 0.78$

31. $\dfrac{2.5 \times 28 + 8.0 \times 47 + 15.5 \times 68 + 25.5 \times 7}{150} = 11.19$

33. $\dfrac{95 \times 8 + 84.5 \times 15 + 74.5 \times 22 + 64.5v11 + 52 \times 5}{61} = 76$

35. $\dfrac{139.50 + 141.25 + 140.75 + 138.50 + 132.00}{5} = \$138.40$

37. $\dfrac{3.60 + 3.57 + 3.90 + 3.85 + 4.00 + 4.15 + 4.25 + 4.40}{8} = 3.965$

Mean = \$3.965 rounded to \$3.97

For median, arrange in order: 3.57, 3.60, 3.85, 3.90, 4.00, 4.15, 4.25, 4.40

$\dfrac{3.90 + 4.00}{2} = 3.95$, median

39.   **(a)**   5   **(b)**   2

   (c) $\dfrac{1 \times 5 + 2 \times 9 + 3 \times 3 + 4 \times 1}{18} = 2$ **(d)**   2

41.   Let x = the unknown grade

$78 = \dfrac{72 + 88 + 81 + 67 + x}{5}$

$390 = 72 + 88 + 81 + 67 + x$

$x = 82$

43. $\dfrac{21 \times 22000 + 9 \times 26000 + 14 \times 29000 + 26 \times 32000 + 13 \times 34000 + 8 \times 39000}{91} =$

\$29,538.46

45. $\dfrac{8(27450) + 10(31400)}{18} = \dfrac{533600}{18} = \$29,644.44$

47.   Put the scores in order: 65, 77, 82, 93
Since there are an odd number of test scores, the median, 82, must be the third number in the ordered group of test scores. So the fifth test score must be 82 or larger.

49. $\dfrac{3(1.27) + 2(1.34)}{5} = \dfrac{6.49}{5} = \$1.30$

51. $\mu = \dfrac{4(1) + 28(2) + 95(3) + 24(4) + 4(5)}{155} = 2.974$

55. **(a)** Mean per capita income = \$21,427
Mean number below poverty level = 34.5 million

57. 69.35   **59.** 22.3

61. Mean = 5, median = 5   **63.** Mean = 42.04, median = 42.3

Using Your TI-83

1. Mean = 6.43, median = 7   **3.** Mean = 8.675, median = 7.9

5. Mean = 2.25

Using Excel

1. Mean = 6.43, median = 7   **3.** Mean = 8.675, median = 7.9

Section 8.3

1. $\mu = \dfrac{19 + 10 + 15 + 20}{4} = 16$

$var = \dfrac{(19-16)^2 + (10-16)^2 + (15-16)^2 + (20-16)^2}{4} = 15.5$

$\sigma = \sqrt{15.5} = 3.94$

3. $\mu = \dfrac{4 + 8 + 9 + 10 + 14}{5} = 9$

$var = \dfrac{(4-9)^2 + (8-9)^2 + (9-9)^2 + (10-9)^2 + (14-9)^2}{5} = 10.4$

$\sigma = \sqrt{10.4} = 3.22$

5. $\mu = \dfrac{17 + 39 + 54 + 22 + 16 + 46 + 25 + 19 + 62 + 50}{10} = 35$

$(17 - 35)^2 + (39 - 35)^2 + (54 - 35)^2 + (22 - 35)^2 + (16 - 35)^2 + (46 - 35)^2 +$
$(25 - 35)^2 + (19 - 35)^2 + (62 - 35)^2 + (50 - 35)^2 = 2662$

$$\text{var} = \frac{2662}{10} = 266.2, \qquad \sigma = \sqrt{266.2} = 16.32$$

7. $\quad \mu = \dfrac{-8 - 4 - 3 + 0 + 1 + 2}{6} = -2$

variance =

$$\frac{(-8+2)^2 + (-4+2)^2 + (-3+2)^2 + (0+2)^2 + (1+2)^2 + (2+2)^2}{5} = 14,$$

$s = \sqrt{14} = 3.74$

9. $\quad \mu = \dfrac{-3 + 0 + 1 + 4 + 5 + 8 + 10 + 11}{8} = 36/8 = 4.5$

var =

$$\frac{(-3 - 4.5)^2 + (0 - 4.5)^2 + (1 - 4.5)^2 + (4 - 4.5)^2 + (5 - 4.5)^2 + (8 - 4.5)^2 + (10 - 4.5)^2 + (11 - 4.5)^2}{7}$$

$= 174/7 = 24.86, \ s = 4.99$

11. $\quad \mu = \dfrac{2 \times 1 + 5 \times 2 + 2 \times 3 + 3 \times 4 + 8 \times 5}{20} = 70/20 = 3.5$

$$\text{variance} = \frac{2(1 - 3.5)^2 + 5(2 - 3.5)^2 + 2(3 - 3.5)^2 + 3(4 - 3.5)^2 + 8(5 - 3.5)^2}{20} =$$

$43/20 = 2.15$

$\sigma = 1.47$

13. $\quad \mu = \dfrac{10(5.65) + 12(5.90) + 8(6.00) + 5(6.24)}{35} = 206.5/35 = 5.90$

$$\text{var} = \frac{10(-0.25)^2 + 12(0)^2 + 8(0.10)^2 + 5(0.34)^2}{34} = 1.283/34 = 0.0377$$

$s = 0.194$

15. $\quad \mu = \dfrac{5(5) + 12(15.5) + 8(25.5)}{25} = 415/25 = 16.6$

$$\text{var} = \frac{5(-11.6)^2 + 12(-1.1)^2 + 8(8.9)^2}{25} = 1321/25 = 52.84$$

$\sigma = 7.27$

17. $\quad \mu = \dfrac{14(12.5) + 18(18) + 18(23)}{50} = 913/50 = 18.26$

$$s^2 = \frac{14(-5.76)^2 + 18(-.26)^2 + 18(4.74)^2}{49} = 17.76$$

$s = \sqrt{17.76} = 4.21$

Section 8.3 Measures of Dispersion

19. $\mu = 217/7 = 31$ $\sigma = \sqrt{68/7} = \sqrt{9.714} = 3.12$

21. **(a)** $Z = (180 - 160)/16 = 20/16 = 1.25$ **(b)** $Z = -10/16 = -0.625$
 (c) $Z = 0/16 = 0$
 (d) $1 = (x - 160)/16$ so $x = 16 + 160 = 176$
 (e) $-0.875 = (x - 160)/16$ so $x = 16(-0.875) + 160 = 146$

23. Minimum = 3, Q1 = 6, median = 11.5, Q3 = 16, maximum = 20

25. The five-point summary is {3, 5.5, 11, 16, 20}

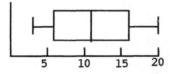

27. The five-point summary is {4.2, 7.5, 11.6, 17.2, 23.8}

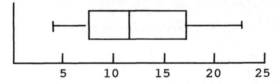

29. **(a)** <u>Lathe I</u>
 variance
$$= \frac{(0.501-0.5)^2 + (0.503-0.5)^2 + (0.495-0.5)^2 + (0.504-0.5)^2 + (0.497-0.5)^2}{5}$$
$$= 0.000012$$
$$\sigma = \sqrt{0.000012} = 0.00346, \text{ for Lathe I.}$$
<u>Lathe II</u>
 variance =
$$\frac{(0.502-0.5)^2 + (0.497-0.5)^2 + (0.498-0.5)^2 + (0.501-0.5)^2 + (0.502-0.5)^2}{5}$$
$$= 0.000044$$
$$\sigma = \sqrt{0.000044} = 0.00210, \text{ for Lathe II.}$$
 (b) Lathe II is more consistent because σ is smaller.

31.   Let x be the number of 2's in the set. Let y be the number of 3's in the set.
        Let 8 - x - y be the number of 4's in the set.

(a) $$\frac{2x + 3y + 4(8 - x - y)}{8} = 3$$

$$0.5 = \sqrt{\frac{x(2 - 3)^2 + y(3 - 3)^2 + (8 - x - y)(4 - 3)^2}{8}}$$

These can be rewritten as $2x + y = 8$

$0.25 = \dfrac{x + (8 - x - y)}{8}$ or $2 = 8 - y$

so $y = 6$, the number of 3's

$2x + y = 8$

$2x + 6 = 8$

$2x = 2$

$x = 1$, the number of 2's

$8 - x - y = 8 - 1 - 6 = 1$, the number of 4's

The set of numbers is 2, 3, 3, 3, 3, 3, 3, 4

(b)   The first equation is the same, $2x + y = 8$

$1 = \dfrac{x + (8 - x - y)}{8}$ or $8 = 8 - y$ so $y = 0$, the number of 3's

$2x + 0 = 8$

$x = 4$, the number of 2's

$8 - x - y = 8 - 4 = 4$, the number of 4's

The set of numbers is 2, 2, 2, 2, 4, 4, 4, 4

(c)   No, the largest possible value of the sum of the squared deviations
        is 8, for which $\sigma = 1$.

33.   30% scored above her so $110(0.30) = 33$ scored above. She ranked 34th out of
        110.

35.   2% scored higher so $0.02(1545) = 30.9$ or 31 scored higher

37.   On the first test his z-score was $z_1 = (86 - 72)/8 = 14/8 = 1.75$, on the second
        test his z-score was
        $z_2 = (82 - 62)/12 = 20/12 = 1.67$
        The 86 was the better score because he scored higher above the mean.

39. $z_R = (19 - 18)/1 = 1/1 = 1.0$

        $z_C = (64 - 59)/3 = 5/3 = 1.67$

        The runner's time was one z-score higher than the mean so the runner was
        slower than average. The cyclist's time was 1.67 z-scores above the mean so
        the cyclist was even slower than the average. The runner had the better
        performance.

43. $\mu = 19.26$ $\sigma = 3.79,$ $s = 4.23$

45. $\mu = 3.5$ $\sigma = 1.47,$ $s = 1.50$

47. $\mu = 5.9$ $\sigma = 0.191,$ $s = 1.94$

49. $\mu = 18.26$ $\sigma = 4.17,$ $s = 4.21$

51.   Mean = 4935, standard deviation = 1503.6

53.   Mean = 8.26 million, standard deviation = 2.11 million

55.   Mean = 8.44, population standard deviation = 5.96, sample standard deviation = 6.33.

57.   **(a)**   At least $1 - 1/9 = 8/9$, about 89%.
         (b)   At least $1 - 1/(1.5)^2 = 0.555$, about 56%.
         (c)   At least $1 - 1/(2.5)^2 = 0.84$, about 84%.

59.   **(a)**   23 of the 24, about 96%, fall in the interval. This is consistent since it is more than 75%.
         (b)   22 of the 24, about 92%, fall in the interval. This is consistent since it is more than 56%.

61.   The scores 1 standard deviation below and above the mean are 449 and 1029. 16 scores, 75%, fall in this interval. This is 7% higher than the Empirical Rule, but this suggests the data might be roughly bell-shaped.

Using the TI-83

1.   Mean = 4.17, population standard deviation = 2.41, sample standard deviation = 2.64

3.   Mean = 8.18, population standard deviation = 2.66, sample standard deviation = 2.79

1.   Five point summary = {2, 3.5, 6, 11.5, 14}

Using Excel

1. Range = 16, sample standard deviation = 5.4556, population standard deviation = 5.2017.

3. {17, 25, 33.5, 37, 48}

Section 8.4

1. | | |
    |---|---|
    | HHH | X = 3 |
    | HHT | X = 2 |
    | HTH | X = 2 |
    | THH | X = 2 |
    | TTH | X = 1 |
    | THT | X = 1 |
    | HTT | X = 1 |
    | TTT | X = 0 |

3. | | |
    |---|---|
    | Ann, Betty | X = 2 |
    | Ann, Jason | X = 1 |
    | Ann, Tom | X = 1 |
    | Betty, Jason, | X = 1 |
    | Betty, Tom | X = 1 |
    | Jason, Tom | X = 0 |

5. X = 0, 1, 2, or 3

7. (a) X = 0, 1, 2, 3, or 4
    (b) X = 0, 1, 2, or 3

9. (a) Discrete   (b) Continuous
    (c) Continuous   (d) Discrete

11. (a) Continuous   (b) Discrete
    (c) (i)   Discrete   (ii) Continuous
        (iii) Discrete

13. Yes, since sum is 1 and the fractions are nonnegative and not greater than 1.

15. | X | P(X) |
    |---|------|
    | 0 | 1/8 |
    | 1 | 3/8 |
    | 2 | 3/8 |
    | 3 | 1/8 |

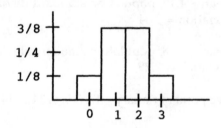

17.

X	P(X)
0	0.02
1	0.68
2	0.07
3	0.08
4	0.10
5	0.01
8	0.02
10	0.02

19.

X	P(X)
0	8/75
1	49/75
2	13/75
3	4/75
4	1/75

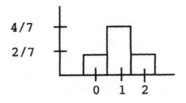

21.

	X	P(X)		
(Red and black)	0	$\dfrac{4(6)}{C(10,\ 2)}$	$=\dfrac{24}{45}$	$=0.533$
(2 black)	5	$\dfrac{C(6,\ 2)}{C(10,\ 2)}$	$=\dfrac{15}{45}$	$=0.333$
(2 red)	10	$\dfrac{C(4,\ 2)}{C(10,\ 2)}$	$=\dfrac{6}{45}$	$=0.133$

23. **(a)** $X \in \{1,\ 2,\ 3,...\}$ **(b)** Discrete

25. $X = 0$: $C(5,\ 0) \cdot C(6,\ 2) = (1)(15) = 15$
$X = 1$: $C(5,\ 1) \cdot C(6,\ 1) = (5)(6) = 30$
$X = 2$: $C(5,\ 2) \cdot C(6,\ 0) = (10)(1) = 10$

27.

X	P(X)		
0	$\dfrac{C(4,\ 0)\ C(4,\ 2)}{C(8,\ 2)}$	$=\dfrac{6}{28}$	$=\dfrac{3}{14}$
1	$\dfrac{C(4,\ 1)\ C(4,\ 1)}{C(8,\ 2)}$	$=\dfrac{4 \times 4}{28}$	$=\dfrac{4}{7}$
2	$\dfrac{C(4,\ 2) \times C(4,\ 0)}{C(8,\ 2)}$	$=\dfrac{6}{28}$	$=\dfrac{3}{14}$

29.

X	P(X)
0	$\dfrac{C(7,\,2)}{C(10,\,2)}=\dfrac{21}{45}=\dfrac{7}{15}$
1	$\dfrac{C(7,\,1)\,C(3,\,1)}{C(10,\,2)}=\dfrac{21}{45}=\dfrac{7}{15}$
2	$\dfrac{C(3,\,2)}{C(10,\,2)}=\dfrac{3}{45}=\dfrac{1}{15}$

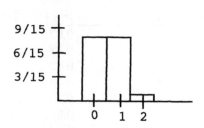

31. **(a)**

X	P(X)
0	$\dfrac{C(4,\,2)}{C(6,\,2)}=\dfrac{6}{15}=\dfrac{2}{5}$
1	$\dfrac{C(4,\,1)C(2,\,1)}{C(6,\,2)}=\dfrac{8}{15}$
2	$\dfrac{C(2,\,2)}{C(6,\,2)}=\dfrac{1}{15}$

(b)

X	P(X)
0	$\dfrac{C(5,\,2)}{C(6,\,2)}=\dfrac{10}{15}=\dfrac{2}{3}$
1	$\dfrac{C(5,\,1)\,C(1,\,1)}{C(6,\,2)}=\dfrac{5}{15}=\dfrac{1}{3}$

33.

X	P(X)
1	$24/50 = 12/25$
5	$12/50 = 6/25$
10	$8/50 = 4/25$
20	$2/50 = 1/25$
50	$1/50$
100	$1/50$

35. $X = 0$, 1 sequence (Gold, Gold);
$X = 1$, 2 sequences (Gold, Green, Gold; or Green, Gold, Gold)
$X = 2$, 3 sequences (Green, Green; or Green, Gold, Green; or Gold, Green, Green)

37. **(a)** A person's weight   **(b)** Shirt size, suit size
     (c) Shoe size   **(d)** Number of cavities

39. The first card can be either color, so the probability the second card is the same color = 3/7, the probability the second card is a different color = 4/7.

X	P(X)	
2	$3/7$	(2nd card same color)
3	$(4/7)(3/6) = 2/7$	(2nd card a different color)
4	$(4/7)(3/6)(3/5) = 6/35$	(2nd and 3rd cards a different color)
5	$(4/7)(3/6)(2/5)(3/4) = 3/35$	(2nd thru 4th a different color)

6      $(4/7)(3/6)(2/5)(1/4)(3/3) = 1/35$ (2nd thru 5th cards a different color)

41.

X	P(X)
1	1/2 (H)
2	$(1/2)(1/2) = 1/4$ (TH)
3	$(1/2)(1/2)(1/2) + (1/2)(1/2)(1/2) = 1/4$ (TTH or TTT)

43.

X	P(X)
0	$(0.4)^2 = 0.160$ (BB)
1	$(0.4)(0.6)(0.4) + (0.6)(0.4)(0.4) = 0.192$ (BAB or ABB)
2	$(0.6)^2 + (0.6)(0.4)(0.6) + (0.4)(0.6)(0.6) = 0.648$ (AA or ABA or BAA)

45. Let C = the color of the first card, and D = the other color.

X   P(X)
(# cards drawn)

1   0

2   5/11 $CC\,(1 \times \dfrac{5}{11} = \dfrac{5}{11})$

3   3/11 $CDD\,(1 \times \dfrac{6}{11} \times \dfrac{5}{10} = \dfrac{3}{11})$

4   4/33 $CDCC\,(1 \times \dfrac{6}{11} \times \dfrac{5}{10} \times \dfrac{4}{9} = \dfrac{4}{33})$

5   5/66 $CDCDD\,(1 \times \dfrac{6}{11} \times \dfrac{5}{10} \times \dfrac{5}{9} \times \dfrac{4}{8} = \dfrac{5}{66})$

6   5/154 $CDCDCC\,(1 \times \dfrac{6}{11} \times \dfrac{5}{10} \times \dfrac{5}{9} \times \dfrac{4}{8} \times \dfrac{3}{7} = \dfrac{5}{154})$

7   5/231 $CDCDCDD\,(1 \times \dfrac{6}{11} \times \dfrac{5}{10} \times \dfrac{5}{9} \times \dfrac{4}{8} \times \dfrac{4}{7} \times \dfrac{3}{6} = \dfrac{5}{231})$

8   2/231 $CDCDCDCC\,(1 \times \dfrac{6}{11} \times \dfrac{5}{10} \times \dfrac{5}{9} \times \dfrac{4}{8} \times \dfrac{4}{7} \times \dfrac{3}{6} \times \dfrac{2}{5} = \dfrac{2}{231})$

9   1/154 $CDCDCDCDD\,(1 \times \dfrac{6}{11} \times \dfrac{5}{10} \times \dfrac{5}{9} \times \dfrac{4}{8} \times \dfrac{4}{7} \times \dfrac{3}{6} \times \dfrac{3}{5} \times \dfrac{2}{4} = \dfrac{1}{154})$

10   1/462 $CDCDCDCDCC\,(1 \times \dfrac{6}{11} \times \dfrac{5}{10} \times \dfrac{5}{9} \times \dfrac{4}{8} \times \dfrac{4}{7} \times \dfrac{3}{6} \times \dfrac{3}{5} \times \dfrac{2}{4} \times \dfrac{1}{3} = \dfrac{1}{462})$

11   1/462 CDCDCDCDCDD

$$\left(1 \times \dfrac{6}{11} \times \dfrac{5}{10} \times \dfrac{5}{9} \times \dfrac{4}{8} \times \dfrac{4}{7} \times \dfrac{3}{6} \times \dfrac{3}{5} \times \dfrac{2}{4} \times \dfrac{2}{3} \times \dfrac{1}{2} = \dfrac{1}{462}\right)$$

12   1/462 CDCDCDCDCDCD

$$\left(1 \times \dfrac{6}{11} \times \dfrac{5}{10} \times \dfrac{5}{9} \times \dfrac{4}{8} \times \dfrac{4}{7} \times \dfrac{3}{6} \times \dfrac{3}{5} \times \dfrac{2}{4} \times \dfrac{2}{3} \times \dfrac{1}{2} \times \dfrac{1}{1} = \dfrac{1}{462}\right)$$

(The probability the cards alternate in color)

Section 8.5

1. $E(X) = 0.3(3) + 0.2(8) + 0.1(15) + 0.4(22) = 12.8$

3. $E(X) = 0.2(150) + 0.2(235) + 0.2(350) + 0.2(410) + 0.2(480) = 325$

5. $E(X) = (8)(1/2) + (2)(1/2) = \$5$

7.   The probability of two heads is $(1/2)(1/2) = 1/4$ as is the probability of two tails so the probability of the same faces is $1/2$. The probability of different faces is $1/2$.
       (a) $E(X) = (1)(1/2) + 0(1/2) = \$0.50$ so a charge of $\$0.50$ will break even.
       (b) $\$1 + \$0.50 = \$1.50$

9.   **(a)** $(0.15)(20) + (0.65)(130) + (0.20)(700) = \$227.50$
       (b) $(125)(\$227.50) = \$28,437.50$

11. $\mu = 0.4(100) + 0.5(140) + 0.1(210) = 131$
      $\sigma^2 = 0.4(-31)^2 + 0.5(9)^2 + 0.1(79)^2 = 1049$
      $\sigma = 32.39$

13. $\mu = 0.4(10) + 0.2(30) + 0.3(50) + 0.1(90) = 34$
      $\sigma^2 = 0.4(-24)^2 + 0.2(-4)^2 + 0.3(16)^2 + 0.1(56)^2 = 624$
      $\sigma = 24.98$

15. $X = 0, 1,$ or 2 for the number of defective toys that can be selected.
      $E(X) = 0 + (1)\dfrac{4 \times 11}{C(15,2)} + (2)\dfrac{C(4,2)}{C(15,2)} = \dfrac{44}{105} + \dfrac{12}{105} = \dfrac{56}{105} = 0.533$

17.   The probability of a person's card being drawn is $1/8$ and the value of the pot is $\$8$ so the expected value is $(1/8)(\$8) = \$1$.

19.   The probability Herbert selects the same five numbers is $1/C(25, 5) = 1/53,130$. If he wins, his net winnings are $\$999$. His expected value is
      $\dfrac{1}{53,130}(999) = \dfrac{999}{53,130} = 0.0188$, almost 2 cents.

21. $E(X_A) = (0.60)(50,000) + (0.30)(0) + (0.10)(-70,000) = \$23,000$
      $E(X_B) = (0.55)(100,000) + (0.45)(-60,000) = \$28,000$
      B appears to be more profitable.

Section 8.5 Expected Value

23. $E(X) = 0.75(0) + 0.10(1) + 0.06(2) + 0.04(3) + 0.04(4) + 0.01(5) = 0.55$

25. Two numbers can be selected from the five numbers in $C(5, 2) = 10$ ways. The scores 1, 2, 3, 4, and 5 can be obtained in the following ways.

Score	Pairs
5	(1, 4), (2, 3)
1	(1, 2), (1, 3), (1, 5)
2	(2, 4), (2, 5)
3	(3, 4), (3, 5)
4	(4, 5)

(a) $E(X) = 1(3/10) + 2(2/10) + 3(2/10) + 4(1/10) + 5(2/10) = 2.7$

(b) $\mu = 2.7 \quad \sigma^2 = 2.21 \quad \sigma = 1.487$

27. The probability he gets no hits is $(0.6)(0.6) = 0.36$.
The probability he gets one hit is $(0.4)(0.6) + (0.6)(0.4) = 0.48$.
The probability he gets two hits is $(0.4)(0.4) = 0.16$.
$$E(X) = 0(0.36) + 1(0.48) + 2(0.16) = 0.8.$$
His expected earnings are $0.80.

29. **(a)** $E(X) = 0.05(-23) + 0.95(18) = \$15.95$

(b) $(150,000)(15.95) = \$2,392,500$

31. Let x = amount received by a player.
$E(X) = 0.3x + 0.7(-5) = 0.3x - 3.5$ which is -\$0.20 since the casino plans to average \$0.20 per game.
$$\text{For } 0.3x - 3.5 = -0.20$$
$$x = 11$$
The payoff for winning is \$11.

33. **(a)** The probability a player wins \$1.00 is $\dfrac{18}{38}$ and the probability of losing \$1.00 is $\dfrac{20}{38}$ so the expected value is $E = \dfrac{18}{38}(1) + \dfrac{20}{38}(-1) = -0.0526$

The expected value is about negative 5 cents per play for the player, so a player can expect to lose an average of about 5 cents per play.

(b) Since the expected value is negative to the player, the game favors the house.

(c) For the game to be fair a payoff of x dollars would give
$$\frac{18}{38}x + \frac{20}{38}(-1) = 0$$
$$18x = 20$$
$$x = 1.1111$$
A payoff of \$1.11 would make the game fair.

Section 8.5 Expected Value

(d)   For a bet on green the probability of winning x dollars is $\frac{2}{38}$ and the probability of losing one dollar is $\frac{36}{38}$. To be a fair game

$$\frac{2}{38}x + \frac{36}{38}(-1) = 0$$
$$\text{so } 2x = 36$$
$$x = 18$$

A payoff of \$18 for winning a green bet makes the game fair.

35. $[4 \ 8 \ 10 \ 6 \ 3]\begin{bmatrix} 0.15 \\ 0.10 \\ 0.25 \\ 0.30 \\ 0.20 \end{bmatrix} = [6.3]$ The expected value is 6.3.

37.

$[140 \ \ 150 \ \ -75 \ \ -50 \ \ 200 \ \ -250]\begin{bmatrix} 0.12 \\ 0.08 \\ 0.24 \\ 0.16 \\ 0.22 \\ 0.18 \end{bmatrix} = [1.8]$ The expected value is 1.8.

39.   **(a)**   The probability of two games is $2(0.5)(0.5) = 0.5$.
The probability of three games is $4(0.5)^3 = 0.5$.
The expected number of games is $0.5(2) + 0.5(3) = 2.5$.

(b)   The probability of two games is $0.6(0.6) + 0.4(0.4) = 0.52$.
The probability of three games is $0.4(0.6)(0.6) + 0.6(0.4)(0.6) + 2(0.4)^2(0.6) = 0.48$.
The expected number of games is $0.52(2) + 0.48(3) = 2.48$

41.   **(a)**   4 games in $C(3, 3) = 1$ way
5 games in $C(4, 3) = 4$ ways
6 games in $C(5, 3) = 10$ ways
7 games in $C(6, 3) = 20$ ways.

(b)   Expected number of games $= 2(0.5)^4(4) + 8(0.5)^5(5) + 20(0.5)^6(6) + 40(0.5)^7(7)$
$= 0.5 + 1.25 + 1.875 + 2.1875 = 5.8125$ rounded to 5.8 games

43. $E(X) = 19.8$

Section 8.6

1. $10(0.0429)(0.4225) = 0.181$

3. $210(0.0256)(0.0467) = 0.251$

5. $792(0.1160)(0.00064) = 0.059$

7. $C(8, 3)(0.25)^3(0.75)^5$
$= 56(0.0156)(0.2373) = 0.208$

9. $C(5, 4)(0.1)^4(0.9)^1 = 0.00045$

11. $P(X = 3) = C(5, 3)(0.25)^3(0.75)^2 = 0.0879$

13.

X	P(X successes)
0	0.2401
1	0.4116
2	0.2646
3	0.0756
4	0.0081

15.

X	P(X)
0	$C(5,0)(0.3)^0(0.7)^5 = 0.1681$
1	$C(5,1)(0.3)^1(0.7)^4 = 0.3602$
2	$C(5,2)(0.3)^2(0.7)^3 = 0.3087$
3	$C(5,3)(0.3)^3(0.7)^2 = 0.1323$
4	$C(5,4)(0.3)^4(0.7)^1 = 0.0284$
5	$C(5,5)(0.3)^5(0.7)^0 = 0.0024$

17.

X	P(X)
0	$C(5,0)(0.4)^0(0.6)^5 = 0.0778$
1	$C(5,1)(0.4)^1(0.6)^4 = 0.2592$
2	$C(5,2)(0.4)^2(0.6)^3 = 0.3456$
3	$C(5,3)(0.4)^3(0.6)^2 = 0.2304$
4	$C(5,4)(0.4)^4(0.6)^1 = 0.0768$
5	$C(5,5)(0.4)^5(0.6)^0 = 0.0102$

19.

X	P(X)
0	$C(4,0)(1/6)^0(5/6)^4 = 0.4823$
1	$C(4,1)(1/6)^1(5/6)^3 = 0.3858$
2	$C(4,2)(1/6)^2(5/6)^2 = 0.1157$
3	$C(4,3)(1/6)^3(5/6)^1 = 0.0154$
4	$C(4,4)(1/6)^4(5/6)^0 = 0.0008$

21.

X	P(X)
0	$C(4,0)(1/2)^4 = 0.0625$
1	$C(4,1)(1/2)^4 = 0.2500$
2	$C(4,2)(1/2)^4 = 0.3750$
3	$C(4,3)(1/2)^4 = 0.2500$
4	$C(4,4)(1/2)^4 = 0.0625$

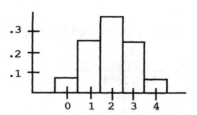

23.

X	P(X)
0	$C(4,0)(0.6)^0(0.4)^4 = 0.0256$
1	$C(4,1)(0.6)^1(0.4)^3 = 0.1536$
2	$C(4,2)(0.6)^2(0.4)^2 = 0.3456$
3	$C(4,3)(0.6)^3(0.4)^1 = 0.3456$
4	$C(4,4)(0.6)^4(0.4)^0 = 0.1296$

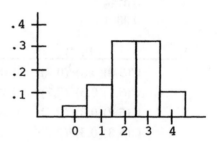

25.

success $P=0.7(0.7)(0.3)$

success $P=0.7(0.3)(0.7)$

success $P =0.3(0.7)(0.7)$

P(two hits in 3 attempts) = 3(0.7)(0.7)(0.3) = 0.441

27. $n=5, p=0.8, x=3$ $P(x=3) = C(5, 3)(0.8)^3(0.2)^2 = 0.2048$

29. $n = 3, P = 1/4, x = 2$ $\qquad$ $P(x = 2) = C(3, 2)(1/4)^2(3/4)^1 = 0.1406$

31. $n = 6, P = 1/6$ $\;$ $P(x = 4) = C(6, 4)(1/6)^4(5/6)^2 = 0.0080$

33. **(a)** $\quad n = 4, p = 1/6,$ $\quad P(x = 0) = C(4, 0)(1/6)^0(5/6)^4 = 0.4823$

$\quad$ **(b)** $\quad n = 4, p = 1/6,$ $\quad P(x = 1) = C(4, 1)(1/6)^1(5/6)^3 = 0.3858$

$\quad$ **(c)** $\quad n = 4, p = 1/6,$ $\quad P(x = 2) = C(4, 2)(1/6)^2(5/6)^2 = 0.1157$

$\quad$ **(d)** $\quad n = 4, p = 1/6,$ $\quad P(x = 3) = C(4, 3)(1/6)^3(5/6)^1 = 0.0154$

$\quad$ **(e)** $\quad n = 4, p = 1/6,$ $\quad P(x = 4) = C(4, 4)(1/6)^4(5/6)^0 = 0.00077$

35. $n = 6, p = 1/2, P(x = 4) = C(6, 4)(1/2)^4(1/2)^2 = 0.2344$

37. $P(X = 40) = C(90, 40)(0.4)^{40}(0.6)^{50}$

39. $P(X = 35) = C(75, 35)(0.25)^{35}(0.75)^{40}$

41. This is a repeated trials with
$n = 6, X = 4, p = 0.8$
$P(X = 4) = C(6, 4)(0.8)^4(0.2)^2$
$\qquad = 15(0.4096)(0.04) = 0.2456$
The probability 4 of the next 6 customers
will tip 15% or more is 0.2456

43. $n = 5$ and $p = 0.5$

X	P(X heads)
0	0.0313
1	0.1563
2	0.3125
3	0.3125
4	0.1563
5	0.0313

45. This is a repeated trials with $n = 5, x = 3, p = 0.4$
$P(X = 3) = C(5,3)(0.4)^3(0.6)^2 = 10(0.064)(0.36) = 0.2304$
The probability Meka will receive more than 40% for three books is 0.2304.

47. This is repeated trials with $n = 5, p = 0.6$
We are to find $P(X = 3) + P(X = 4) + P(X = 5)$ which is
$C(5, 3)(0.6)^3(0.4)^2 + C(5, 4)(0.6)^4(0.4) + C(5, 5)(0.6)^5$
$= 10(0.216)(0.16) + 5(0.1296)(0.4) + 1(0.07776) = 0.6826$
The probability at least three will raise their HDL by 20% is about 0.68.

49. $n = 5, p = 1/6$
$P(X \geq 3) = P(X = 3) + P(X = 4) + P(X = 5) =$
$C(5, 3)(1/6)^3(5/6)^2 + C(5, 4)(1/6)^4(5/6)^1 + C(5, 5)(1/6)^5(5/6)^0 = 0.0355$

Section 8.6 Binomial Experiments

51. $n = 8, p = 1/2,$
$P(X \geq 5) = P(X = 5) + P(X = 6) + P(X = 7) + P(X = 8)$
$= C(8, 5)(1/2)^5(1/2)^3 + C(8, 6)(1/2)^6(1/2)^2 + C(8, 7)(1/2)^7(1/2)$
$+ C(8, 8)(1/2)^8 = 0.3633$

53. **(a)** $n = 10, p = 0.5, \quad P(X = 8) = C(10, 8)(0.5)^8(0.5)^2 = 0.0439$
  (b) $n = 10, p = 0.5, \quad P(X \geq 8) = P(X = 8) + P(X = 9) + P(X = 10)$
  $= C(10, 8)(0.5)^8(0.5)^2 + C(10, 9)(0.5)^9(0.5)^1 + C(10, 10)(0.5)^{10} = 0.0547$
  (c) $n = 10, p = 0.5,$
  $P(X \leq 2) = P(X = 0) + P(X = 1) + P(X = 2)$
  $= C(10, 0)(0.5)^0(0.5)^{10} + C(10, 1)(0.5)^1(0.5)^9 + C(10, 2)(0.5)^2(0.5)^8$
  $= 0.0547$

55. $n = 6, p = 1/6 \; P(X = 4) = C(6, 4)(1/6)^4(5/6)^2 = 0.0080$

57. $n = 5, p = 0.36$
$P(X \geq 2) = 1 - P(X < 2) = 1 - [P(X = 0) + P(X = 1)]$
$= 1 - [C(5, 0)(0.36)^0(0.64)^5 + C(5, 1)(0.36)^1(0.64)^4] = 0.5906$

59. **(a)** $n = 4, p = 0.357$
The probability of at least one hit is
$P(X = 1) + P(X = 2) + P(X = 3) + P(X = 4)$
$= C(4,1)(0.357)^1(0.643)^3 + C(4,2)(0.357)^2(0.643) +$
$C(4,3)(0.357)^3(0.643) + C(4,4)(0.357)^4$
$= 4(0.357)(0.2658) + 6(0.1274)(0.4134) + 4(0.0455)(0.643) + (0.01624)$
$= 0.3796 + 0.3160 + 0.1170 + 0.0162 = 0.8288$
The probability of at least one hit is about 0.83.

  (b) The probability of at least one hit in 56 consecutive games is $(0.83)^{56}$
  $= 0.000029$
  This feat, as unlikely as it is, is considered by some to be the greatest
  accomplishment in baseball.

61. $p = \dfrac{1}{2}, n = 10, X = 5 \qquad P(X = 5) = C(10, 5)(0.5)^5(0.5)^5 = 0.2461$

63. Let n = number of games.
$1 - P(X = 0) = 1 - C(n, 0)(0.12)^0(0.88)^n = 1 - 0.88^n$
Graph $y = 1 - 0.88^X$ and $y = 0.75$ to determine the smallest integer such that
$1 - 0.88^X \geq 0.75$. The curves cross between $X = 10$ and $X = 11$ so Dan should plan
to limit himself to 11 games per day.

65. **(a)** $n = 4, p = 0.6$

X	P(X successes)
0	0.0256
1	0.1536
2	0.3456
3	0.3456
4	0.1296

(b) $n = 4, p = 0.4$

X	P(X successes)
0	0.1296
1	0.3456
2	0.3456
3	0.1536
4	0.0256

(c) $n = 5, p = 0.5$

X	P(X successes)
0	0.0313
1	0.1563
2	0.3125
3	0.3125
4	0.1563
5	0.0313

(d) $n = 8, p = 0.3$

X	P(X successes)
0	0.0576
1	0.1977
2	0.2965
3	0.2541
4	0.1361
5	0.0467
6	0.0100
7	0.0012
8	0.0001

67. $C(220, 75)(0.65)^{75}(0.35)^{145} = 8.3 \times 10^{-21}$, practically zero.

69. $n = 30, p = 0.2$

(a) $P(X \geq 1) = 1 - P(X = 0) = 1 - C(30, 0)(0.2)^0(0.8)^{30} = 1 - (0.8)^{30} = 0.9988$

(b) $P(X \geq 5) = 1 - P(X = 0) - P(X = 1) - P(X = 2) - P(X = 3) - P(X = 4)$

$= 1 - (0.8)^{30} - C(30, 1)(0.2)(0.8)^{29} - C(30, 2)(0.2)^2(0.8)^{28}$

$- C(30, 3)(0.2)^3(0.8)^{27} - C(30, 4)(0.2)^4(0.8)^{26} = 0.7447$

71.

X	P(X)
0	0.0010
1	0.0098
2	0.0439
3	0.1172
4	0.2051
5	0.2461
6	0.2051
7	0.1172
8	0.0439
9	0.0098
10	0.0010

73.

X	P(X)
0	0.1335
1	0.3115
2	0.3115
3	0.1730
4	0.0577
5	0.0115
6	0.0013
7	0.0001

Section 8.7 Normal Distribution

75. p = 0.5, n = 4

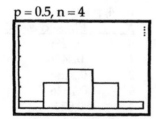

77. p = 0.4, n = 5

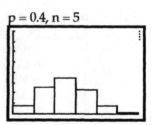

79. p = 0.7, n = 8

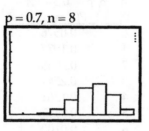

81. 0.1128 **83.** 0.0865 **85.** 0.0183

Using Your TI-83

1.

X	P(X)
0	0.1785
1	0.3845
2	0.3105
3	0.1115
4	0.0150

3.

X	P(X)
0	0.0024
1	0.0284
2	0.1323
3	0.3087
4	0.3602
5	0.1681

Section 8.7

1. $z = (3.1 - 4.0)/0.3 = -3$

3. $z = (10.1 - 10.0)/2.0 = 0.05$

5. $z = (2.65 - 0)/1.0 = 2.65$

7. $A = 0.1915$

9. $A = 0.0987$

11. $A = 0.364$

13. $A = 0.2734$

15. $z = 0.46$ so $A = 0.1772 = 17.72\%$

17. $z = 0.38$ so $A = 0.1480 = 14.8\%$

19. $z = -1.24$ so $A = 0.3925 = 39.25\%$

21. $z = -2.9$ so $A = 0.4981 = 49.81\%$

23. For $z = 1.25$, $A = 0.3944$ so the total area is $A = 2(0.3944) = 0.7888$

25. For each z, $A = 0.4861$ so the total area is
$A = 2(0.4861) = 0.9722$

27. $z = 0.65$
$A = 2(0.2422) = 0.4844 = 48.44\%$

29. $z = 0.38$
$A = 2(0.1480) = 0.2960 = 29.6\%$

31. $A = A_1 + A_2 = 0.2258 + 0.3997 = 0.6255$
33. $A = A_1 + A_2 = 0.2881 + 0.4974 = 0.7855$

35. $A = 0.5 - 0.4032 = 0.0968$

37. $A = 0.5 - 0.4918 = 0.0082$

39. $1 - 2(0.3051) = 0.3898 = 38.98\%$

41. $1 - 2(0.4332) = 0.1336 = 13.36\%$

43. $z = (98 - 85)/5 = 2.60$, $A = 0.4953$

45. $z = (80 - 85)/5 = -1.00$, $A = 0.3413$

47. $z_1 = (220 - 226)/12 = -0.5$, $z_2 = (235 - 226)/12 = 0.75$ so $A_1 + A_2 = 0.1915 + 0.2734 = 0.4649$

49. $z_1 = (211 - 226)/12 = -1.25$, $z_2 = (241 - 226)/12 = 1.25$ so $2(A) = 2(0.3944) = 0.7888$

51. $z_1 = (144 - 140)/8 = 0.5$, $z_2 = (152 - 140)/8 = 1.5$ so $A_2 - A_1 = 0.4332 - 0.1915 = 0.2417$

53. $z_1 = (146 - 140)/8 = 0.75$, $z_2 = (156 - 140)/8 = 2$ so $A_2 - A_1 = 0.4773 - 0.2734 = 0.2039$

55. $z_1 = (80 - 75)/5 = 1$, $z_2 = (85 - 75)/5 = 2$ so $A_2 - A_1 = 0.4773 - 0.3413 = 0.1360$

57. $z = (76 - 75)/5 = 0.2$ so $0.5 - 0.0793 = 0.4207$

59. $z_1 = (70 - 75)/5 = -1$, $z_2 = (80 - 75)/5 = 1$ so $1 - 2(0.3413) = 0.3174$

61. $z_1 = (155 - 168)/10 = -1.3$, $z_2 = (169 - 168)/10 = 0.1$ so $A_1 + A_2 = 0.4032 + 0.0398$
$= 0.4430 = 44.3\%$

63. $z = (172 - 168)/10 = 0.4$ so $0.5 + 0.1554 = 0.6554 = 65.54\%$

65. $z = (173 - 168)/10 = 0.5$ so $A = 0.5 - 0.1915 = 0.3085 = 30.85\%$

67. $z = (184 - 168)/10 = 1.6$ so $A = 0.5 + 0.4452 = 0.9452 = 94.52\%$

69. Since 8% of the scores are to the right of z, $A = 0.5 - 0.08 = 0.42$ is the area between the mean and z. The z that corresponds to $A = 0.42$ is $z = 1.41$

71. Since 86% of the scores lie to the left of z, z is above the mean and 36% of the scores lie between the mean and z. The value of z that corresponds to $A = 0.36$ is $z = 1.08$

73. Since 91% of the scores lie between z and -z, one-half of the scores, 45.5%, lie between the mean and z. The value of z that corresponds to $A = 0.455$ is $z = 1.70$.

75. $\mu = (50)(0.4) = 20$,
$\text{var} = (50)(0.4)(0.6) = 12$, $\sigma = \sqrt{12} = 3.46$

77. $\mu = (600)(0.52) = 312$,
$\text{var} = (600)(0.52)(0.48) = 149.76$, $\sigma = \sqrt{149.76} = 12.24$

79. $\mu = (470)(0.08) = 37.6$,
$\text{var} = (37.6)(0.92) = 34.592$, $\sigma = \sqrt{34.592} = 5.88$

81. $np = 50(0.7) = 35 \geq 5$, $nq = (50)(0.3) = 15 \geq 5$ so the normal distribution is a good estimate.

Section 8.7 Normal Distribution

83. $np = (40)(0.9) = 36$, $nq = (40)(0.1) = 4 < 5$ so the normal distribution is not a good estimate.

85. $np = (25)(0.5) = 12.5 \geq 5$, $nq = (25)(0.5) = 12.5 \geq 5$ so the normal distribution is a good estimate.

87. For the normal distribution to be a reasonable approximation, both $np \geq 5$ and $nq \geq 5$

For $np \geq 5$, $n(0.35) \geq 5$ and $n \geq \dfrac{5}{0.35} = 14.29$

For $nq \geq 5$, $n(0.65) \geq 5$ and $n \geq \dfrac{5}{0.65} = 7.69$

Thus, $n \geq 14.29$ so the smallest integer value of n is 15.

89. $\mu = (50)(0.7) = 35$, $\sigma = \sqrt{(35)(0.3)} = \sqrt{10.5} = 3.24$

   (a)   For $x_1 = 39.5$, $z_1 = (39.5 - 35)/3.24 = 1.39$, $A_1 = 0.4177$

             For $x_2 = 40.5$, $z_2 = (40.5 - 35)/3.24 = 1.70$, $A_2 = 0.4554$

             $P(X = 40) = A_2 - A_1 = 0.4554 - 0.4177 = 0.0377$

   (b)   For $x_1 = 27.5$, $z_1 = (27.5 - 35)/3.24 = -2.31$, $A_1 = 0.4896$

             For $x_2 = 28.5$, $z_2 = (28.5 - 35)/3.24 = -2.01$, $A_2 = 0.4778$

             $P(X = 28) = A_2 - A_1 = 0.4896 - 0.4778 = 0.0118$

   (c)   For $x_1 = 31.5$, $z_1 = (31.5 - 35)/3.24 = -1.08$, $A_1 = 0.3599$

             For $x_2 = 32.5$, $z_2 = (32.5 - 35)/3.24 = -0.77$, $A_2 = 0.2794$

             $P(X = 32) = A_2 - A_1 = 0.3599 - 0.2794 = 0.0805$

91. $\mu = 15(0.4) = 6$, $\sigma = \sqrt{15(0.4)(0.6)} = \sqrt{3.6} = 1.90$

   (a)   For $x_1 = 4.5$, $z_1 = (4.5 - 6)/1.90 = -0.79$, so $A_1 = 0.2852$

             For $x_2 = 7.5$, z_2, $(7.5 - 6)/1.90 = 0.79$, so $A_2 = 0.2852$

             $P(4 < X < 8) = 0.2852 + 0.2852 = 0.5704$

   (b)   For $x_1 = 3.5$, $z_1 = (3.5 - 6)/1.90 = -1.32$, so $A_1 = 0.4066$

             For $x_2 = 8.5$, $z_2 = (8.5 - 6)/1.90 = 1.32$, so $A_2 = 0.4066$

             $P(4 \leq X \leq 8) = 0.4066 + 0.4066 = 0.8132$

   (c)   For $x_1 = 6.5$, $z_1 = (6.5 - 6)/1.90 = 0.26$, so $A_1 = 0.1026$

             For $x_2 = 8.5$, $z_1 = 1.32$, so $A_2 = 0.4066$

             $P(7 \leq X \leq 8) = 0.4066 - 0.1026 = 0.3040$

93. $\mu = 8, \sigma = 2$

   (a)   For $x = 5.5$, $z = (5.5 - 8)/2 = -1.25$, so $A = 0.3944$
              $P(X > 5) = 0.5000 + 0.3944 = 0.8944$

   (b)   For $x = 4.5$, $z = (4.5 - 8)/2 = -1.75$, so $A = 0.4599$
              $P(X \geq 5) = 0.5000 + 0.4599 = 0.9599$

   (c)   For $x = 9.5$, $z = (9.5 - 8)/2 = 0.75$, so $A = 0.2734$
              $P(X > 9) = 0.5000 - 0.2734 = 0.2266$

95. $\mu = 7.2, \sigma = 2.24$

   (a)   For $x = 9.5$, $z = (9.5 - 7.2)/2.24 = 1.03$, so $A = 0.3485$
              $P(X < 10) = 0.5000 + 0.3485 = 0.8485$

   (b)   For $x = 10.5$, $z = (10.5 - 7.2)/2.24 = 1.47$, so $A = 0.4292$
              $P(X \leq 10) = 0.5000 + 0.4292 = 0.9292$

   (c)   For $x = 5.5$, $z = (5.5 - 7.2)/2.24 = -0.76$, so $A = 0.2764$
              $P(X < 6) = 0.5000 - 0.2764 = 0.2236$

97.   **(a)** $z = (3.75 - 3.15)/0.75 = 0.8$ so $0.5 - A = 0.5 - 0.2881 = 0.2119 = 21.19\%$
     (b) 0.2119

99.   **(a)** $z_1 = (120 - 110)/12 = 0.83$, $A_1 = 0.2967$
              $z_2 = (125 - 110)/12 = 1.25$, $A_2 = 0.3944$ so $A_2 - A_1$
              $= 0.3944 - 0.2967 = 0.0977$

   (b) $z = (100 - 110)/12 = -0.83$, $A = 0.2967$ so $0.5 - A = 0.5 - 0.2967 = 0.2033$
   (c) $z_1 = (105 - 110)/12 = -0.42$, $A_1 = 0.1628$
              $z_2 = (115 - 110)/12 = 0.42$ so $2(A) = 2(0.1628) = 0.3256$

101. $\mu = 60, \sigma = 4.90$

   (a)   For $x_1 = 49.5$, $z_1 = (49.5 - 60)/4.90 = -2.14$, so $A_1 = 0.4838$
              For $x_2 = 75.5$, $z_2 = (75.5 - 60)/4.90 = 3.16$, so $A_2 = 0.4992$
              $P(50 \leq X \leq 75) = 0.4838 + 0.4992 = 0.9830$

   (b)   For $x = 75.5$, $z = 3.16$, so $A = 0.4992$
              $P(X > 75) = 0.5000 - 0.4992 = 0.0008$

   (c)   For $x = 49.5$, $z = -2.14$, so $A = 0.4838$
              $P(X \leq 50) = 0.5000 - 0.4838 = 0.0162$

103. $p = 1/3$, find $P(6 \leq X \leq 8)$. $\mu = (20)(1/3) = 6.67$,
      $\sigma = \sqrt{(20)(1/3)(2/3)} = \sqrt{4.44} = 2.11$
      For $x_1 = 5.5$, $z_1 = (5.5 - 6.67)/2.11 = -0.55$, so $A_1 = 0.2088$
      For $x_2 = 8.5$, $z_2 = (8.5 - 6.67)/2.11 = 0.87$, so $A_2 = 0.3079$
      $P(6 \leq X \leq 8) = 0.2088 + 0.3079 = 0.5167$

105. $n = 100, p = 0.5, \mu = 50, \sigma = \sqrt{25} = 5$

    (a)   Find $P(49.5 \le X \le 50.5)$

$$x_1 = 49.5, z_1 = \frac{49.5 - 50}{5} = -0.10, A_1 = 0.0398$$

$$x_2 = 50.5, z_2 = \frac{50.5 - 50}{5} = 0.10, A_2 = 0.0398$$

$P(X = 50) = 0.0398 + 0.0398 = 0.0796.$

    (b)   Find $P(44.5 \le X \le 55.5)$

$$x_1 = 44.5, z_1 = \frac{-5.5}{5} = -1.10, A_1 = 0.3643$$

$$x_2 = 55.5, z_2 = \frac{5.5}{5} = 1.10, A_2 = 0.3643$$

$P(44.5 \le X \le 55.5) = 0.3643 + 0.3643 = 0.7286.$

    (c)   Find $P(48.5 \le X \le 51.5)$

$$x_1 = 48.5, z_1 = \frac{-1.5}{5} = -0.30, A_1 = 0.1179$$

$$x_2 = 51.5, z_2 = \frac{1.5}{5} = 0.30, A_2 = 0.1179$$

$P(48.5 \le X \le 51.5) = 0.1179 + 0.1179 = 0.2358.$

107. $p = 1/4$, find $P(20 \le X \le 24)$.

$\mu = (64)(1/4) = 16, \sigma = \sqrt{(16)(3/4)} = \sqrt{12} = 3.46$

For $x_1 = 19.5$, $z_1 = (19.5 - 16)/3.46 = 1.01$, so $A = 0.3438$

For $x_2 = 24.5$, $z_2 = (24.5 - 16)/3.46 = 2.46$, so $A = 0.4931$

$P(20 \le X \le 24) = 0.4931 - 0.3438 = 0.1493$

109. **(a)**   The area between the mean and the A cutoff is $0.5 - 0.12 = 0.38$ so $z = 1.18$. Let $x =$ the cutoff score. Then $1.18 = \frac{x - 66}{17}$ and $x = 86.06$ rounded to 86.

    (b)   The area between the mean and the A cutoff is $0.5 - 0.06 = 0.44$ so $z = 1.56$. Let $x =$ the cutoff score. Then $1.56 = \frac{x - 66}{17}$ gives $x = 92.52$ which we round to 93.

111. **(a)** $z_1 = -0.75, z_2 = 0.75$ so $A = 2(0.2734) = 0.5468$

    (b) $z_1 = -90/40 = -2.25, z_2 = 30/40 = 0.75$ so $A = 0.4878 + 0.2734 = 0.7612$

    (c)   0.50 score less than 300

$z = 26/40 = 0.65$ so $A = 0.5 + 0.2422 = 0.7422.$

0.7422 of the students score less than 326

    (d)   Let x be the cutoff score. Since 10% of the area under the normal curve is above x, then $A = 0.5 - 0.1 = 0.4$ is the area between the mean

Section 8.7 Normal Distribution

and x. For $A = 0.4$, $z = 1.28$ so $1.28 = \dfrac{x - 300}{40}$ which gives $x = 351.2$. A student must score 352 or higher to be in the upper 10%.

(e) $z_1 = 12/40 = 0.3$, $z_2 = 24/40 = 0.6$ so the area between the two scores is $0.2258 - 0.1179 = 0.1079$ which also is the probability of scoring between 312 and 324.

113.   For $x = 128$, $z = 0.8$, $A = 0.2881$ so the probability a student scores below 128 is $0.5 + 0.2881 = 0.7881$ and the probability both score less is $(0.7881)^2 = 0.6211$

115.   For $z = (1140 - 1050)/50 = 1.8$, $A = 0.4641$ which gives 0.0359 as the area above 1140. The probability the device lasts at least 1140 hours is 0.0359.

117.   Estimate the probability that $x \leq 250$.
The probability a passenger will show is 0.90.
$\mu = (270)(0.9) = 243$, $\sigma = \sqrt{(243)(0.1)} = \sqrt{24.3} = 4.93$
For $x = 250.5$, $z = (250.5 - 243)/4.93 = 1.52$
$P(X \leq 250) = 0.5 + 0.4357 = 0.9357$

119.   Find $P(X \geq 65)$ $\mu = (250)(0.3) = 75$, $\sigma = \sqrt{(75)(0.7)} = \sqrt{52.5} = 7.25$
For $x = 64.5$, $z = (64.5 - 75)/7.25 = -1.45$, so $A = 0.4265$
$P(X \geq 65) = 0.5 + 0.4265 = 0.9265$

121.   Let x = number correct, then $90 - x$ = number incorrect.
(a)   Points earned $= 3x - 1(90 - x) \geq 98$, $4x \geq 188$, $x \geq 47$
(b) $\mu = (90)(1/2) = 45$, $\sigma = \sqrt{(45)(1/2)} = \sqrt{22.5} = 4.74$
      For $x = 46.5$, $z = (46.5 - 45)/4.74 = 0.32$
      $P(X \geq 47) = 0.5 - 0.1255 = 0.3745$

123.   (a) $P(X = 5) + P(X = 6) + P(X = 7) = 0.1859 + 0.2066 + 0.1771 = 0.5696$
       (b)   In Exercise 91 (a) $P(4 < x < 8) = 0.5704$.
              The normal distribution gives a slightly higher estimate.

125.   (a) $P(X = 11) + P(X = 12) + P(X = 13) = 0.1876 + 0.2501 + 0.2309 = 0.6686$
       (b) $\mu = 12$, $\sigma = 1.549$
              The area under the normal curve between $x_1 = 10.5$ and $x_2 = 13.5$ is found as:
$$z_1 = \frac{10.5 - 12}{1.549} = -0.97 \text{ and } A_1 = 0.3340 \qquad z_2 = \frac{13.5 - 12}{1.549} = 0.97 \text{ and } A_2 = 0.3340$$
              $P(10 < X < 14) = 0.3340 + 0.3340 = 0.6680$
       (c)   The normal distribution estimate is quite close.

Section 8.7 Normal Distribution

127. **(a)** $\mu = 30, \sigma = 3.873$

(b) We want the integers in the interval 30 ± 3.873 which are in the interval $27 \le x \le 33$

(c) $P(27 \le X \le 33) = P(X = 27) + 0... + P(X = 33)$
$= 0.0763 + 0.0900 + 0.0993 + 0.1026 + 0.0993 + 0.0900 + 0.0763 = 0.6338$

129. **(a)** $0.4332 - 0.1915 = 0.2417$ **(b)** $0.4192 - 0.2258 = 0.1934$

(c) $0.4032 - 0.2580 = 0.1452$ **(d)** $0.3849 - 0.2881 = 0.0968$

(e) $0.3643 - 0.3159 = 0.0484$ **(f)** $0.3531 - 0.3289 = 0.0242$

(g) $0.3438 - 0.3389 = 0.0049$

131. No, the mean is the midpoint of the data in a normal distribution.

133. Yes, the mean and median are approximately equal and near the midpoint.

135. As a whole, the curve does not have the symmetry needed for a normal distribution. The portion obtained by deleting the 7 rightmost bars is somewhat normal in shape so data in that region might be considered approximately normal.

137. **(a)** Mean = 69.91, median = 73, 25th percentile = 62, 75th percentile = 82.

(b) $\sigma = 17.37$

(c) The interval $69.91 \pm \sigma$ is (52.54, 87.28) This interval contains 72% of the scores which is fairly consistent with 68% in a normal distribution.
The interval $69.91 \pm 2\sigma$ is (35.17, 104.65) which is really (35.17, 100) since 100 is the maximum score. This interval contains 94% of the scores which is consistent with 95.4% in a normal distribution.
The interval $69.91 \pm 3\sigma$ is (17.8, 122.02) which is really (17.8, 100) since 100 is the maximum score. This interval contains 98.9% of the scores which is consistent with 99.7% in a normal distribution.

(d)

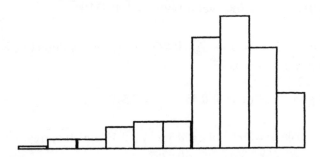

Using Your TI-83

1.   0.6997   3.   0.7625

Using Excel

1.   0.9088   3.   0.5074   5.   0.1499

7.   0.8783

Section 8.8

1.   **(a)**   S.E. $= \sqrt{\dfrac{0.45(0.55)}{50}} = 0.070$ **(b)**   S.E. $= \sqrt{\dfrac{0.32(0.68)}{100}} = 0.047$

   (c)   S.E. $= \sqrt{\dfrac{0.20(0.80)}{400}} = 0.020$

3. $\overline{p} = 0.30$, $n = 50$, For $c = 0.90$ A $= 0.45$ which corresponds to $z = 1.65$
   SE $= \sqrt{(0.30)(0.70)/50} = 0.0648$
   E $= (1.65)(0.0648) = 0.1069$
   0.30 ± 0.1069 gives the interval $0.1931 < x < 0.4069$

5. $\overline{p} = 0.30$, $n = 50$, $c = 0.98$, A $= 0.49$ which corresponds to $z = 2.33$
   SE $= \sqrt{(0.30)(0.70)/50} = 0.0648$
   E $= (2.33)(0.0648) = 0.1510$
   0.30 ± 0.1510 gives the interval $0.1490 < x < 0.4510$

7. $\overline{p} = 0.5$, $n = 400$, For $c = 0.95$, A $= 0.475$ which corresponds to $z = 1.96$
   SE $= \sqrt{(0.5)(0.5)/400} = 0.025$
   E $= (1.96)(0.025) = 0.049$
   0.5 ± 0.049 gives the interval $0.451 < x < 0.549$

9.   **(a)** $\overline{p} = 0.45$, $n = 300$, For $c = 0.90$, A $= 0.45$ which corresponds to $z = 1.65$
         SE $= \sqrt{(0.45)(0.55)/300} = 0.0287$
         E $= (1.65)(0.0287) = 0.0473$
         0.45 ± 0.0474 gives the interval $0.4026 < x < 0.4974$

Section 8.8 Estimating Bounds on a Proportion

(b) $\overline{p} = 0.45$, $n = 300$, For $c = 0.95$, $= 0.475$ which corresponds to $z = 1.96$
$SE = 0.0287$ $\qquad\qquad$ $E = (1.96)(0.0287) = 0.0563$
0.45 ± 0.0563 gives the interval $0.3937 < x < 0.5063$

11. $\overline{p} = 243/300 = 0.81$, $n = 300$, For $c = 0.98$, $A = 0.490$ which corresponds to $z = 2.33$
$SE = \sqrt{(0.81)(0.19) / 300} = 0.0226$ $\qquad$ $E = (2.33)(0.0226) = 0.0527$
0.81 ± 0.0527 gives the interval $0.7573 < x < 0.8627$

13. **(a)** The estimate for $\overline{p}$ is $\dfrac{144}{195} = 0.738 = 73.8\%$.

(b) $SE = \sqrt{\dfrac{0.738(0.262)}{195}} = 0.0315$
For a 95% confidence interval $z = 1.96$.
The error bounds are
$\qquad$ $0.738 + 1.96(0.0315) = 0.7997$
and $\qquad$ $0.738 - 1.96(0.0315) = 0.6763$
The 95% confidence interval is $0.6763 < x < 0.7997$.

(c) At the 99% confidence level $z = 2.58$.
The error bounds are
$0.738 + 2.58(0.0315) = 0.8193$
$0.738 - 2.58(0.0315) = 0.6567$
The 99% confidence interval is $0.6567 < x < 0.8193$.

15. **(a)** $E = 1.96 \sqrt{\dfrac{0.27(0.73)}{300}} = 0.0502$
Error bounds: $0.27 - 0.0502 = 0.2198$
$\qquad\qquad\qquad$ $0.27 + 0.0502 = 0.3202$

(b) $E = 1.96 \sqrt{\dfrac{0.27(0.73)}{1000}} = 0.0275$
Error bounds: $0.27 - 0.0275 = 0.2425$
$\qquad\qquad\qquad$ $0.27 + 0.0275 = 0.2975$

(c) The larger sample yields a smaller confidence interval.

17. **(a)** $\overline{p} = 0.28$, $n = 200$, For $c = 0.95$ $A = 0.475$ which corresponds to $z = 1.96$
$SE = \sqrt{(0.28)(0.72) / 200} = 0.0317$
$E = (1.96)(0.0317) = 0.0621$
0.28 ± 0.0621 gives the interval $0.2179 < x < 0.3421$

(b) $SE = \sqrt{(0.28)(0.72) / 400} = 0.0224$
$E = (1.96)(0.0224) = 0.0440$
0.28 ± 0.0440 gives the interval $0.2360 < x < 0.3240$

Section 8.8 Estimating Bounds on a Proportion

19. $\overline{p} = 144/320 = 0.45$, $n = 320$, For $c = 0.95$, $A = 0.475$ which corresponds to $z = 1.96$

$SE = \sqrt{(0.45)(0.55)/320} = 0.0278$

$E = (1.96)(0.0278) = 0.0545$

0.45 ± 0.0545 gives the interval $0.3955 < x < 0.5045$

21. $\overline{p} = 0.54$, $n = 60$, For $c = 0.90$, $= 0.45$ which corresponds to $z = 1.65$

$SE = \sqrt{(0.54)(0.46)/60} = 0.0643$

$E = (1.65)(0.0643) = 0.1061$

0.54 ± 0.1061 gives the interval $0.4339 < x < 0.6461$

23. $E = 0.06$, $c = 0.95$ corresponds to $z = 1.96$ so

$0.06 = 1.96\sqrt{0.25/n}$

$0.000937 = 0.25/n$,

$n = 266.8$ rounded to 267

25. $E = 0.02$, $c = 0.95$ which corresponds to $z = 1.96$ so

$0.02 = 1.96\sqrt{0.25/n}$

$0.000104123 = 0.25/n$

$n = 2401$

27. In each case $\overline{p} = 0.65$ and $z = 1.96$.

(a) $SE = \sqrt{\dfrac{0.65(0.35)}{300}} = 0.0275$

The error bounds are
$0.65 + 1.96(0.0275) = 0.7040$
$0.65 - 1.96(0.0275) = 0.5960$

(b) $SE = \sqrt{\dfrac{0.65(0.35)}{360}} = 0.0251$

The error bounds are
$0.65 + 1.96(0.0251) = 0.6993$
$0.65 - 1.96(0.0251) = 0.6007$

(c) $SE = \sqrt{\dfrac{0.65(0.35)}{400}} = 0.0238$

The error bounds are
$0.65 + 1.96(0.0238) = 0.6967$
$0.65 - 1.96(0.238) = 0.6033$

(d) $SE = \sqrt{\dfrac{0.65(0.35)}{460}} = 0.0222$

The error bounds are
$0.65 + 1.96(0.0222) = 0.6936$
$0.65 - 1.96(0.0222) = 0.6064$

(e) $SE = \sqrt{\dfrac{0.65(0.35)}{500}} = 0.0213$

The error bounds are
$0.65 + 1.96(0.0213) = 0.6918$
$0.65 - 1.96(0.0213) = 0.6082$

(f) $SE = \sqrt{\dfrac{0.65(0.35)}{1000}} = 0.0151$

The error bounds are
$0.65 + 1.96(0.0151) = 0.6796$
$0.65 - 1.96(0.0151) = 0.6204$

Section 8.8 Estimating Bounds on a Proportion

29. $E = 0.0785$, $c = 0.95$ which corresponds to $z = 1.96$ so

$0.0785 = 1.96 \sqrt{0.25 / n}$

$0.001604 = 0.25/n$

$n = 155.8$ rounded to 156

31. $\overline{p}$ is the midpoint of 0.5007 and 0.5393, $\overline{p} = 0.52$.
The upper bound is

$$0.52 + 1.65 \sqrt{\frac{0.52(0.48)}{n}} \quad \text{where } z = 1.65 \text{ corresponds to } A = 0.45$$

Thus, $0.5393 = 0.52 + 1.65 \sqrt{\dfrac{0.2496}{n}}$

$$\frac{0.0193}{1.65} = \sqrt{\frac{0.2496}{n}}$$

$$n = 0.2496 \left(\frac{1.65}{0.0193} \right)^2 = 1824.3 \text{ rounded up to } 1825.$$

The sample contained 1825 students.

33. The set of sample means forms a normal distribution with

$p = 0.60$ as the mean, $n = 300$, $SE = \sqrt{\dfrac{0.60(0.40)}{300}} = 0.02828$ as the standard error

and we want to find the area under the normal curve below 0.54.

For $\overline{p} = 54\%$ $z = \dfrac{0.54 - 0.60}{0.02828} = -2.12$

The area under the normal curve between the mean and $z = -2.12$ is 0.4830.
The area less than $z = -2.12$ is $0.5000 - 0.4830 = 0.0170$.
The probability that less than 54% of the sample favor the fees is 0.017.

35. For all random samples of size 500, the proportions of out of state students in the samples form a normal distribution with the population proportion $p = 0.45$ as the mean. The probability a sample proportion is ≥ 0.50 is the fraction of the area above 0.50 for the normal curve with mean 0.45 and

$s = \sqrt{\dfrac{0.45(0.55)}{500}}$

$p = 0.45$, $n = 500$, $SE = \sqrt{\dfrac{0.45(0.55)}{500}} = 0.02225$

For $\overline{p} = 0.50$ $z = \dfrac{0.50 - 0.45}{0.02225} = 2.25$

$A = 0.4878$ corresponds to $z = 2.25$

Thus the probability $\overline{p} > 0.50$ is $0.5000 - 0.4878 = 0.0122$.
The probability that at least 50% are from out-of-state is 0.0122.

37. $SE = \sqrt{\dfrac{0.46(0.54)}{1498}} = 0.012877$

The margin of error ± 03 comes from $\pm z \times SE$ so $0.03 = z(0.012877)$

$z = 2.33$

$z = 2.33$ corresponds to $A = 0.4901$ since the confidence extends on both sides of

$\overline{p}$, $2 \times 0.4901 = 0.9802 = 98\%$ gives the 98% confidence level.

39. The sample proportion of defective labels is

$$\overline{p} = \frac{6}{1600} = 0.00375 = 0.375\%$$

$$SE = \sqrt{\frac{0.00375(0.99625)}{1600}} = 0.00153$$

At the 95% confidence level the error bounds are
$0.00375 + 1.96(0.00153) = 0.00675$ and $0.00375 - 1.96(0.00153) = 0.000755$
So the true proportion of defective labels is in the interval $0.000755 < p < 0.00675$ with probability 0.95, or equivalently, between 0.0755% and 0.675%. The manager is 95% confident that less than 0.7% of the labels are defective so the machine should not be shut down.

41. We are looking for $\overline{p}$ such that $\overline{p} = 0.68 + z\sqrt{\dfrac{0.68(0.32)}{4500}}$ where z corresponds

to an area $0.5000 - .0808 = 0.4192$ (0.0808 of the area is above z under the standard normal curve.)
For $A = 0.4192$ $z = 1.40$ so

$\overline{p} = 0.68 + 1.40\sqrt{\dfrac{0.68(0.32)}{4500}} = 0.68 + 1.40(0.006954) = 0.6897$

The sample proportion would exceed 0.6897 only about 8.08% of the time.

Chapter 8 Review

1.

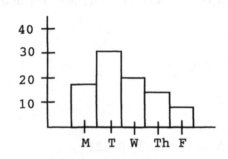

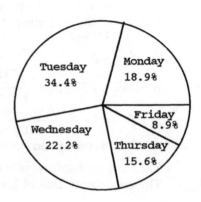

3. (a) 6

(b) $(9+12)/2 = 21/2 = 10.5$

(c) $(1+3)/2 = 2$

5. $\mu = \dfrac{54(1.5) + 32(3.5) + 12(5.5)}{98} = 259/98 = 2.64$

7. $90.22/n = 3.47$

9. $X \in \{0, 1, 2, 3\}$

$n = 90.22/3.47 = 26$

11. $E(X) = 0.05(0) + 0.20(1) + 0.15(2) + 0.20(3) + 0.25(4) + 0.15(5) = 2.85$

13. $E(X) = 0.20(1) + 0.32(2) + 0.21(3) + 0.15(4) + 0.12(5) = 2.67$

15. For $x = 400$, $z = (400 - 350)/25 = 2$

$0.5 - A = 0.5 - 0.4773 = 0.0227$

17.

X	P(X)
0	$C(4, 0)(0.25)^0(0.75)^4 = 0.3164$
1	$C(4, 1)(0.25)^1(0.75)^3 = 0.4219$
2	$C(4, 2)(0.25)^2(0.75)^2 = 0.2109$
3	$C(4, 3)(0.25)^3(0.75)^1 = 0.0469$
4	$C(4, 4)(0.25)^4(0.75)^0 = 0.0039$

19. Use the normal curve to estimate the probabilities.

$\mu = (80)(0.65) = 52$, $\sigma = \sqrt{(52)(0.35)} = \sqrt{18.2} = 4.266$

(a) For $x = 50.5$, $z = (50.5 - 52)/4.266 = -0.35$ and $A = 0.1368$

$P(X > 50) = 0.5 + 0.1368 = 0.6368$

(b) For $x_1 = 64.5$, $z_1 = (64.5 - 52)/4.266 = 2.93$ and $A = 0.4983$

For $x_2 = 65.5$, $z_2 = (65.5 - 52)/4.266 = 3.16$ and $A = 0.4992$

$P(X = 65) = A_2 - A_1 = 0.4992 - 0.4983 = 0.0009$

(c) For $x_1 = 55.5$, $z_1 = (55.5 - 52)/4.266 = 0.82$ and $A = 0.2939$

For $x_2 = 59.5$, $z_2 = (59.5 - 52)/4.266 = 1.76$ and $A = 0.4608$

$P(55 < X < 60) = A_2 - A_1 = 0.4608 - 0.2939 = 0.1669$

21. $216 - 43 + 1 = 174$ was the number who placed the same or lower so $174/216 = 0.805$ shows she was at the 81st percentile.

23.

X	P(X)
1	$20/800 = 0.025$
2	$160/800 = 0.200$
3	$370/800 = 0.4625$
4	$215/800 = 0.2688$
5	$35/800 = 0.0438$

25.

X	Number of outcomes	
0	10	C(5, 8)
1	30	C(3, 1)C(5, 2)
2	15	C(3, 2)C(5, 1)
3	1	C(3, 3)

27.

x	Number of ways
1	1
2	1
3	4

29.

$$\mu = \frac{8(3) + 13(8.5) + 6(12.5) + 15(17.5)}{42} = 11.238$$

31. **(a)** $\sqrt{\dfrac{(0.35)(0.65)}{60}} = 0.0616$

(b) $\sqrt{\dfrac{(0.64)(0.36)}{700}} = 0.0181$

(c) $\sqrt{\dfrac{(0.40)(0.60)}{950}} = 0.0159$

33. $\mu = 22/30 = 0.733$

$$0.733 \pm 1.96\sqrt{\frac{0.733(0.267)}{30}}$$

$$0.575 < p < 0.891$$

Chapter 9
Game Theory

Section 9.1

1.

$$\begin{array}{cc} & C \\ & \begin{array}{cc} H & T \end{array} \\ R \begin{array}{c} H \\ T \end{array} & \left[\begin{array}{cc} -1 & -0.5 \\ -0.5 & 2 \end{array}\right] \end{array}$$

3.

$$\begin{array}{cc} & C \\ & \begin{array}{cc} 1 & 2 \end{array} \\ R \begin{array}{c} 1 \\ 2 \end{array} & \left[\begin{array}{cc} 2 & -3 \\ -3 & 4 \end{array}\right] \end{array}$$

5. (a) R receives 10 from C   (b) 10
        (c) 10   (d) 8

7.

Row
Min
$$\left[\begin{array}{cc} -3 & -5 \\ 2 & -1 \end{array}\right]\begin{array}{c} -5 \\ -1 \end{array}$$
Col. Max   2   -1

Strictly determined, the saddle
point is (2, 2), value = -1, and
the solution is row 2 and column 2.

9.

Row
Min
$$\left[\begin{array}{ccc} 1 & 2 & 3 \\ 4 & -5 & 1 \\ 2 & 6 & 3 \end{array}\right]\begin{array}{c} 1 \\ -5 \\ 2 \end{array}$$
Col. Max   4   6   3

Not strictly determined

11.

Row
Min
$$\left[\begin{array}{ccc} 2 & 0 & 1 \\ 0 & -3 & 4 \\ 3 & -2 & 0 \end{array}\right]\begin{array}{c} 0 \\ -3 \\ -2 \end{array}$$
Col. Max   3   0   4

Strictly determined, the saddle point is (1, 2),
value = 0, and the solution is row 1 and column 2.

13. (a)

Row
Min
$$\left[\begin{array}{ccc} -3 & 2 & 1 \\ 0 & 2 & 3 \\ -1 & -4 & 2 \end{array}\right]\begin{array}{c} -3 \\ 0 \\ -4 \end{array}$$
Col. Max   0   2   3

Strictly determined, saddle point (2, 1), value = 0, solution row 2,
column 1.

261

(b)

$$\begin{array}{c} & & \text{Row} \\ & & \text{Min} \\ \begin{bmatrix} -1 & 2 & 3 \\ 4 & -1 & 0 \\ 0 & 1 & -1 \end{bmatrix} & \begin{matrix} -1 \\ -1 \\ -1 \end{matrix} \\ \text{Col. Max} \quad 4 \quad 2 \quad 3 \end{array}$$

Not strictly determined because the largest row minimum does not equal the smallest column maximum.

(c)

$$\begin{array}{c} & & \text{Row} \\ & & \text{Min} \\ \begin{bmatrix} 1 & -2 & 1 \\ 5 & 7 & 3 \\ -1 & 3 & -4 \end{bmatrix} & \begin{matrix} -2 \\ 3 \\ -4 \end{matrix} \\ \text{Col. Max} \quad 5 \quad 7 \quad 3 \end{array}$$

Strictly determined, saddle point (2, 3), value = 3, solution row 2 and column 3.

(d) Strictly determined, saddle point (2, 2), value = 0, solution row 2, column 2.

15.

$$\begin{array}{c} & & \text{Row} \\ & & \text{Min} \\ \begin{bmatrix} 10 & -5 & 25 \\ -20 & -15 & 10 \\ -10 & -10 & 5 \end{bmatrix} & \begin{matrix} -5 \\ -20 \\ -10 \end{matrix} \\ \text{Col. Max} \quad 10 \quad -5 \quad 25 \end{array}$$

ReMark should choose option 1, the sports highlights, and Century should choose option 2, the game show. The value is -5, ReMark should expect to lose 5 points in the ratings.

17. (a)

$$\begin{array}{c} & & \text{Row} \\ & & \text{Min} \\ \begin{bmatrix} 85 & 120 & 150 \\ 60 & 165 & 235 \\ 70 & 150 & 175 \end{bmatrix} & \begin{matrix} 85 \\ 60 \\ 70 \end{matrix} \\ \text{Col. Max} \quad 85 \quad 165 \quad 235 \end{array}$$

The saddle point is (1, 1) so the farmer should plant milo.

(c) Total income over five years
Milo: 85 + 3(120) + 150 = 595
Corn: 60 + 3(165) + 235 = 790
Wheat: 70 + 3(150) + 175 = 695
In the long term corn appears to be the best and wheat second.

Section 9.2

1. $\quad E = PAQ = \begin{bmatrix} 2/3 & 1/3 \end{bmatrix} \begin{bmatrix} 20 & 12 \\ 8 & 30 \end{bmatrix} \begin{bmatrix} 1/4 \\ 3/4 \end{bmatrix} = \begin{bmatrix} 16 & 18 \end{bmatrix} \begin{bmatrix} 1/4 \\ 3/4 \end{bmatrix} = 17.5$

3. $\quad E = \begin{bmatrix} 2/5 & 3/5 \end{bmatrix} \begin{bmatrix} 3 & -6 \\ -2 & 4 \end{bmatrix} \begin{bmatrix} 2/3 \\ 1/3 \end{bmatrix} = \begin{bmatrix} 0 & 0 \end{bmatrix} \begin{bmatrix} 2/3 \\ 1/3 \end{bmatrix} = 0$

5. **(a)** $\begin{bmatrix} 1 & 0 & 0 \end{bmatrix} \begin{bmatrix} -15 & 40 & 25 \\ 25 & -10 & -5 \\ 45 & 20 & -15 \end{bmatrix} \begin{bmatrix} 0 \\ 1 \\ 0 \end{bmatrix} = 40$

  (b) $\begin{bmatrix} 1/2 & 1/2 & 0 \end{bmatrix} \begin{bmatrix} -15 & 40 & 25 \\ 25 & -10 & -5 \\ 45 & 20 & -15 \end{bmatrix} \begin{bmatrix} 1/2 \\ 0 \\ 1/2 \end{bmatrix} = \begin{bmatrix} 5 & 15 & 10 \end{bmatrix} \begin{bmatrix} 1/2 \\ 0 \\ 1/2 \end{bmatrix} = 7.5$

  (c) $\begin{bmatrix} 1/5 & 2/5 & 2/5 \end{bmatrix} \begin{bmatrix} -15 & 40 & 25 \\ 25 & -10 & -5 \\ 45 & 20 & -15 \end{bmatrix} \begin{bmatrix} 1/3 \\ 1/3 \\ 1/3 \end{bmatrix} = \begin{bmatrix} 25 & 12 & -3 \end{bmatrix} \begin{bmatrix} 1/3 \\ 1/3 \\ 1/3 \end{bmatrix} = \dfrac{34}{3}$

  (d) $\begin{bmatrix} 0.3 & 0.1 & 0.6 \end{bmatrix} \begin{bmatrix} -15 & 40 & 25 \\ 25 & -10 & -5 \\ 45 & 20 & -15 \end{bmatrix} \begin{bmatrix} 0.2 \\ 0.2 \\ 0.6 \end{bmatrix} = \begin{bmatrix} 25 & 23 & -2 \end{bmatrix} \begin{bmatrix} 0.2 \\ 0.2 \\ 0.6 \end{bmatrix} = 8.4$

7. $\quad P1 = \dfrac{20 - (-5)}{15 + 20 - 10 - (-5)} = \dfrac{25}{30} = \dfrac{5}{6}$ and $p2 = \dfrac{1}{6}$

$\quad q1 = \dfrac{20 - 10}{30} = \dfrac{1}{3}$ and $q2 = \dfrac{2}{3}$

Row strategy $= \begin{bmatrix} \dfrac{5}{6} & \dfrac{1}{6} \end{bmatrix}$ and column strategy $= \begin{bmatrix} \dfrac{1}{3} & \dfrac{2}{3} \end{bmatrix}$

$E = \dfrac{15(20) - 10(-5)}{30} = \dfrac{350}{30} = 11.67$

9. **(a)** Row 1 and row 2 dominate row 3, and column 2 dominates columns 1, 3, and 4, so row 3 and columns 1 and 3 can be deleted leaving $\begin{bmatrix} -5 \\ 0 \end{bmatrix}$

  (b) Column 1 dominates column 4, so the matrix reduces to $\begin{bmatrix} 5 & 2 & -2 \\ 3 & 1 & 13 \\ 1 & 3 & 6 \end{bmatrix}$

11. Row 1 dominates Row 2 and Column 1 dominates Column 2 so Row 2 and Column 2 can be deleted giving $\begin{bmatrix} 6 & 3 \\ 5 & 8 \end{bmatrix}$

$P_1 = \dfrac{8 - 5}{6 + 8 - 5 - 3} = \dfrac{3}{6} = \dfrac{1}{2}$ and $p_3 = \dfrac{1}{2}$ Row strategy $= \begin{bmatrix} \dfrac{1}{2} & 0 & \dfrac{1}{2} \end{bmatrix}$

$q_1 = \dfrac{8 - 3}{6} = \dfrac{5}{6}$ and $q_2 = \dfrac{1}{6}$ Column strategy $= \begin{bmatrix} \dfrac{5}{6} & 0 & \dfrac{1}{6} \end{bmatrix}$

$E = \dfrac{6(8) - 5(3)}{6} = \dfrac{33}{6} = 5.5$

13 **(a)** $E = \dfrac{6(4) - 3(8)}{6 + 4 + 3 + 8} = \dfrac{0}{21} = 0$. This is a fair game.

(b) $E = \dfrac{5(3) - 1(2)}{5 + 3 - 2 - 1} = \dfrac{13}{5}$. This is not a fair game.

(c) $E = \dfrac{3(6) - 2(-9)}{3 + 6 + 9 - 2} = \dfrac{36}{16}$. This is not a fair game.

15. $\begin{bmatrix} 10 & -5 & 25 \\ -20 & 15 & 10 \\ -10 & -10 & 5 \end{bmatrix}$

Since each entry in Row 1 is greater than the corresponding entry in Row 3, we can remove Row 3 from consideration. Since Column 1 dominates Column 3, we can remove Column 3 from consideration. We then have the reduced payoff matrix

$$\begin{array}{cc} & C \\ R & \begin{bmatrix} 10 & -5 \\ -20 & 15 \end{bmatrix} \end{array}$$

For ReMark $p_1 = \dfrac{15 - (-20)}{10 + 15 - (-5) - (-20)} = \dfrac{35}{50} = \dfrac{7}{10}$, $p_2 = \dfrac{3}{10}$ and

$P = \begin{bmatrix} \dfrac{7}{10} & \dfrac{3}{10} & 0 \end{bmatrix}$.

For Century $q_1 = \dfrac{15 - (-5)}{50} = \dfrac{20}{50} = \dfrac{2}{5}$, $q_2 = \dfrac{3}{5}$, and $Q = \begin{bmatrix} \dfrac{2}{5} & \dfrac{3}{5} & 0 \end{bmatrix}$

$E = \dfrac{10(15) - (-5)(-20)}{50} = \dfrac{50}{50} = 1$

ReMark should show the sports highlights 70% of the time, the mystery drama 30% of the time, and drop the variety show. Century should air the talk show 40% of the time, the game show 60% of the time and drop the educational documentary.

Using these strategies ReMark can expect an average gain of 1 rating point.

17.   **(a)**   Expected survival time $= \begin{bmatrix} 1 & 0 \end{bmatrix} \begin{bmatrix} 25 & 30 \\ 5 & 35 \end{bmatrix} \begin{bmatrix} 0.60 \\ 0.40 \end{bmatrix} = \begin{bmatrix} 25 & 30 \end{bmatrix} \begin{bmatrix} 0.60 \\ 0.40 \end{bmatrix}$

$= 27$ years

(b)   Expected survival time $= \begin{bmatrix} 0 & 1 \end{bmatrix} \begin{bmatrix} 25 & 30 \\ 5 & 35 \end{bmatrix} \begin{bmatrix} 0.60 \\ 0.40 \end{bmatrix} = \begin{bmatrix} 5 & 35 \end{bmatrix} \begin{bmatrix} 0.60 \\ 0.40 \end{bmatrix}$

$= 17$ years

(c)   No surgery is the better option when expected survival time with no surgery is greater than expected survival time with surgery.

$$\begin{bmatrix} 0 & 1 \end{bmatrix} \begin{bmatrix} 25 & 30 \\ 5 & 35 \end{bmatrix} \begin{bmatrix} q_1 \\ 1-q_1 \end{bmatrix} > \begin{bmatrix} 1 & 0 \end{bmatrix} \begin{bmatrix} 25 & 30 \\ 5 & 35 \end{bmatrix} \begin{bmatrix} q_1 \\ 1-q_1 \end{bmatrix}$$

$$35 - 30q_1 > 30 - 5q_1$$

$$5 > 25q_1 \qquad \frac{5}{25} > q_1$$

No surgery is the better option when $q_1 < 0.20$.

19.   **(a)**   Expected survival $= \begin{bmatrix} 1 & 0 \end{bmatrix} \begin{bmatrix} 20 & 22 \\ 3 & 25 \end{bmatrix} \begin{bmatrix} 0.70 \\ 0.30 \end{bmatrix} = 20.6$ years

(b)   Expected survival $= \begin{bmatrix} 0 & 1 \end{bmatrix} \begin{bmatrix} 20 & 22 \\ 3 & 25 \end{bmatrix} \begin{bmatrix} 0.70 \\ 0.30 \end{bmatrix} = 9.6$ years

(c)   Surgery is the better options when

$$\begin{bmatrix} 1 & 0 \end{bmatrix} \begin{bmatrix} 20 & 22 \\ 3 & 25 \end{bmatrix} \begin{bmatrix} q_1 \\ 1-q_1 \end{bmatrix} > \begin{bmatrix} 0 & 1 \end{bmatrix} \begin{bmatrix} 20 & 22 \\ 3 & 25 \end{bmatrix} \begin{bmatrix} q_1 \\ 1-q_1 \end{bmatrix}$$

$$20q_1 + 22(1 - q_1) > 3q_1 + 25(1 - q_1)$$

$$22 - 2q_1 > 25 - 22q_1$$

$$20q_1 > 3 \qquad q_1 > 3/20 = 0.15$$

Surgery is the better option when the probability of malignancy is greater than 0.15.

Chapter 9 Review

1. **(a)**
$$\begin{bmatrix} 5 & -1 \\ 2 & 4 \\ 2 & 4 \end{bmatrix}\begin{matrix} -1 \\ 2 \end{matrix}$$
This game is not strictly determined.

(b)
$$\begin{bmatrix} 1 & 3 & 9 \\ 7 & 4 & 8 \\ -5 & 3 & 4 \\ 7 & 4 & 9 \end{bmatrix}\begin{matrix} 1 \\ 4 \\ -5 \end{matrix}$$
This game is strictly determined. The (2, 2) location is the saddle point. The value of the game is 4.

(c)
$$\begin{bmatrix} 140 & 210 \\ 300 & 275 \\ 300 & 275 \end{bmatrix}\begin{matrix} 140 \\ 275 \end{matrix}$$
This game is strictly determined. The (2, 2) location is the saddle point. The value of the game is 275.

(d)
$$\begin{bmatrix} -6 & 2 & 9 & 1 \\ 5 & -4 & 0 & 2 \\ 4 & 2 & 8 & 3 \\ 5 & 2 & 9 & 3 \end{bmatrix}\begin{matrix} -6 \\ -4 \\ 2 \end{matrix}$$
This game is strictly determined. The (3, 2) location is the saddle point. The value of the game is 2.

3. **(a)** This game is not strictly determined, so there is no solution.
   (b) This game is strictly determined with value 4. The saddle point is at location (2,1). So the solution consists of the offense adopting strategy 2 and the defense adopting strategy 1.

5. **(a)** $E = \begin{bmatrix} 0.3 & 0.7 \end{bmatrix}\begin{bmatrix} 5 & 9 \\ 11 & 2 \end{bmatrix}\begin{bmatrix} 0.6 \\ 0.4 \end{bmatrix} = \begin{bmatrix} 9.2 & 4.1 \end{bmatrix}\begin{bmatrix} 0.6 \\ 0.4 \end{bmatrix} = 7.16$

   (b) $E = \begin{bmatrix} 0.5 & 0.5 \end{bmatrix}\begin{bmatrix} -2 & 6 \\ 3 & 9 \end{bmatrix}\begin{bmatrix} 0.1 \\ 0.9 \end{bmatrix} = \begin{bmatrix} 0.5 & 7.5 \end{bmatrix}\begin{bmatrix} 0.1 \\ 0.9 \end{bmatrix} = 6.8$

   (c) $E = \begin{bmatrix} 0.1 & 0.4 & 0.5 \end{bmatrix}\begin{bmatrix} -3 & 2 & 1 \\ 4 & -2 & 5 \\ 3 & 1 & 2 \end{bmatrix}\begin{bmatrix} 0.2 \\ 0.2 \\ 0.6 \end{bmatrix} = \begin{bmatrix} 2.8 & -0.1 & 3.1 \end{bmatrix}\begin{bmatrix} 0.2 \\ 0.2 \\ 0.6 \end{bmatrix} = 2.4$

7. $P1 = \dfrac{210 - 175}{250 + 210 - 140 - 175} = \dfrac{35}{145} = \dfrac{35}{145} = 0.24 \quad P2 = 0.76$

   The farmer should plant 24% of the crop in the field and 76% in the greenhouse.

Chapter 10
Logic

Section 10.1

1. **(a)** Statement. It is a true declarative sentence.
    (b) Statement. It is a false declarative sentence.
    (c) Not a statement. It is a question.
    (d) Not a statement. It is an opinion.

3. **(a)** Statement. It is a true declarative sentence.
    (b) Statement. It is a false declarative sentence.
    (c) Statement. It is a true declarative sentence.
    (d) Not a statement. It is a command.
    (e) Not a statement. It is an opinion.

5. **(a)** Neither   **(b)**   Disjunction
    (c) Conjunction   **(d)**   Disjunction

7. **(a)** Betty has blonde hair or Angela has dark hair.
    (b) Angela does not have dark hair.
    (c) Betty has blonde hair and Angela has dark hair.
    (d) Betty does not have blonde hair and Angela does not have dark hair.

9. **(a)** $p \wedge q$ **(b)** $p \vee q$ **(c)** $(\sim p) \wedge q$

11. **(a)** False because $4^2 = 15$ is false.
    (b) False because the first part is false.
    (c) True because both statements making up the conjunction are true.

13. **(a)** False because both parts are false.
    (b) True because both parts are true.
    (c) True because the first part is true.

15. **(a)** I do not have six one-dollar bills in my wallet.
   (b) Roy cannot name all 50 states.
   (c) A quorum was present for the meeting.

17. **(a)** I drink coffee at breakfast and I eat salad for lunch and I like a dessert after dinner.
   (b) I drink coffee at breakfast, or I eat salad for lunch and I like a dessert after dinner.
   (c) I drink coffee at breakfast and I eat salad for lunch, and I do not like a dessert after dinner.
   (d) I drink coffee at breakfast and I eat salad for lunch, or I drink coffee at breakfast and I like dessert after dinner.

19. **(a)** T because all parts are T   **(b)** F because ~p is F
   (c) F because ~p is F   **(d)** F because ~r is F
   (e) T because p ∨ q is T

21.

p	q	~p	~q	~p ∧ ~q
T	T	F	F	F
T	F	F	T	F
F	T	T	F	F
F	F	T	T	T

23. **(a)**

p: Jane brings chips.   q: Tony brings drinks
r: Hob brings cookies.   s: Ingred brings chips.
t: Alex brings drinks.   u: Hester brings drinks.
v: Alice brings cookies.   w: Jenn brings cookies.

The statement is (p ∧ q ∧ r) ∨ (s ∧ (t ∨ u) ∧ (v ∨ w))

   (b) The statements have the truth values: p = T, q = F, r = F, s = T, t = T, u = T, v = F, w = F
   Substituting these values in the symbolic statement we have
   (T ∧ F ∧ F) ∨ (T ∧ (T ∨ T) ∧ (F ∨ F)) which reduces to F ∨ (T ∧ T ∧ F)
   which is F.

Section 10.2

1. **(a)** If I have $5.00, then I can rent a video.
   (b) If I can rent a video, then I have $5.00.

3. **(a)** True because the hypothesis and the conclusion are both true.
   (b) True because the hypothesis is false.
   (c) False because the hypothesis is true and the conclusion is false.

5.   Converse: If I live in Colorado, then I live in Denver.
        Inverse: If I do not live in Denver, then I do not live in Colorado.
        Contrapositive: If I do not live in Colorado, then I do not live in Denver.

7.   **(a)**   False because the components have different truth values, true and false, respectively.
       (b)   True because both components are true.
       (c)   True because both components are true.

9.

p	q	p∧q	~(p∧q)
T	T	T	F
T	F	F	T
F	T	F	T
F	F	F	T

11.

p	q	~q	p∧~q
T	T	F	F
T	F	T	T
F	T	F	F
F	F	T	F

13.

p	q	~p	~q	~p ∨ ~q
T	T	F	F	F
T	F	F	T	T
F	T	T	F	T
F	F	T	T	T

15.

p	q	q∨p	~p→(q∨p)
T	T	T	T
T	F	T	T
F	T	T	T
F	F	F	F

17.

p	q	r	p∨q	(p∨q)∧r
T	T	T	T	T
T	F	T	T	T
F	T	T	T	T
F	F	T	F	F
T	T	F	T	F
T	F	F	T	F
F	T	F	T	F
F	F	F	F	F

19.

p	q	r	p→q	q→r	(p→q)∧(q→r)
T	T	T	T	T	T
T	F	T	F	T	F
F	T	T	T	T	T
F	F	T	T	T	T
T	T	F	T	F	F
T	F	F	F	T	F
F	T	F	T	F	F
F	F	F	T	T	T

21.

p	q	r	p∨q	(p∨q)↔r
T	T	T	T	T
T	F	T	T	T
F	T	T	T	T
F	F	T	F	F
T	T	F	T	F
T	F	F	T	F
F	T	F	T	F
F	F	F	F	T

23.

p	~p	~(~p)	~(~p)↔p
T	F	T	T
F	T	F	T

25. **(a)**   p: your insurance company paid the provider directly for part of your expenses.
q: you paid only the amount that remained.
r: include on line 1 only the amount you paid.
The statement is (p ∧ q) → r.

(b)   p: you leave line 65 blank.
q: the IRS will figure the penalty.
r: the IRS will send you the bill.
The statement is p → (q ∧ r).

(c)   p: you changed your name because of marriage, divorce, etc.
q: you made estimated tax payments using your former name.
r: attach a statement to the front of Form 1040 explaining all the payments you and your spouse made in 1997.
s: attach the service center where you made the payments.
t: attach the name and SSN under which you made the payments.
The statement is (p ∧ q) → (r ∧ s ∧ t).

Section 10.3

1.

p	q	p→q	~q→~p	
T	T	T	T	
T	F	F	F	
F	T	T	T	
F	F	T	T	Equivalent

3.

p	q	~p	p∧q	~p∨(p∧q)	~p∨q
T	T	F	T	T	T
T	F	F	F	F	F
F	T	T	F	T	T
F	F	T	F	T	T

Equivalent

5.

p	q	r	~p	q∧r	~p∨(q∧r)	p→(q∧r)
T	T	T	F	T	T	T
T	F	T	F	F	F	F
F	T	T	T	T	T	T
F	F	T	T	F	T	T
T	T	F	F	F	F	F
T	F	F	F	F	F	F
F	T	F	T	F	T	T
F	F	F	T	F	T	T

Equivalent

7.

p	q	p→q	~(p→q)	p	
T	T	T	F	T	
T	F	F	T	T	
F	T	T	F	F	
F	F	T	F	F	Not equivalent

9.

p	q	r	q∨r	p∧q	p∧(q∨r)	(p∧q)∨r
T	T	T	T	T	T	T
T	F	T	T	F	T	T
F	T	T	T	F	F	T
F	F	T	T	F	F	T
T	T	F	T	T	T	T
T	F	F	F	F	F	F
F	T	F	T	F	F	F
F	F	F	F	F	F	F

Not equivalent

11.   Let p represent the statement "The exceptions above apply," and let q
        represent the statement "Use Form 2210." Then statement (a) can be
        represented by ~p → q and statement (b) can be represented by p ∨ q. Make a
        truth table and compare the truth values of ~p → q and p ∨ q.

p	q	~p	~p→q	p∨q
T	T	F	T	T
T	F	F	T	T
F	T	T	T	T
F	F	T	F	F

Since the truth values of ~p → q and p ∨ q are identical, the statements are
equivalent.

13.   p: I will buy a jacket.
        q: I will buy a shirt.
        r: I will buy a tie.
        Then statement (a) is p ∧ (q ∨ r)
        (b) is (p ∧ q) ∨ (p ∧ r)
        We form a truth table and compare the truth values of the two statements.

p	q	r	q∨r	p∧(q∨r)	p∧q	p∧r	(p∧q)∨(p∧r)
T	T	T	T	T	T	T	T
T	F	T	T	T	F	T	T
F	T	T	T	F	F	F	F
F	F	T	T	F	F	F	F
T	T	F	T	T	T	F	T
T	F	F	F	F	F	F	F
F	T	F	T	F	F	F	F
F	F	F	F	F	F	F	F

Since the truth values of p ∧ (q ∨ r) and (p ∧ q) ∨ (p ∧ r) are identical, they
are equivalent.

Section 10.4

1.   p: Eat your beans.
        q: You may have dessert.

        p → q

        p_____

        q   Valid, Law of Detachment

3. p: You do not study.
    q: You cannot do the homework.
    r: You cannot pass the course.

    $p \rightarrow q$

    $q \rightarrow r$

    $p \rightarrow r$ Valid, syllogism

5. p: You eat your beans.
    q: You may have dessert.

    $p \rightarrow q$

    $\sim p$

    $\sim q$ Not valid. See Example 5.

7. p: The ice is six inches thick.
    q: Shelley will go skating.

    $p \rightarrow q$

    $\sim q$

    $\sim p$

    Valid, indirect reasoning

9. p: Inflation increases.
    q: The price of new cars will increase.
    r: More people will buy used cars.

    $p \rightarrow q$

    $q \rightarrow r$

    $p \rightarrow r$ Valid, Syllogism

11. Check $[(p \rightarrow q) \wedge (q \wedge r)] \rightarrow (p \vee r)$

p	q	r	$p \rightarrow q$	$q \wedge r$	$p \vee r$	$[(p \rightarrow q) \wedge (q \wedge r)] \rightarrow (p \vee r)$
T	T	T	T	T	T	T
T	F	T	F	F	T	T
F	T	T	T	T	T	T
F	F	T	T	F	T	T
T	T	F	T	F	T	T
T	F	F	F	F	T	T
F	T	F	T	F	F	T
F	F	F	T	F	F	T

Valid

13.   Check $[(p \wedge q) \wedge (p \rightarrow \sim q)] \rightarrow (p \wedge \sim q)$

p	q	p∧q	p→~q	p∧~q	[(p∧q)∧(p→~q)]→(p∧~q)
T	T	T	F	F	T
T	F	F	T	T	T
F	T	F	T	F	T
F	F	F	T	F	T

Valid

15.   Check $[(q \rightarrow r) \wedge (\sim p \vee q) \wedge p] \rightarrow r$

p	q	r	q→r	~p∨q	[(q→r)∧(~p∨q)∧p]→r
T	T	T	T	T	T
T	F	T	T	F	T
F	T	T	T	T	T
F	F	T	T	T	T
T	T	F	F	T	T
T	F	F	T	F	T
F	T	F	F	T	T
F	F	F	T	T	T

Valid

17.   Check $[(p \rightarrow q) \wedge (p \rightarrow r)] \rightarrow (q \wedge r)$

p	q	r	p→q	p→r	q∧r	[(p→q)∧(q→r)]→(q∧r)
T	T	T	T	T	T	T
T	F	T	F	T	F	T
F	T	T	T	T	T	T
F	F	T	T	T	F	F
T	T	F	T	F	F	T
T	F	F	F	F	F	T
F	T	F	T	T	F	F
F	F	F	T	T	F	F

Not valid

19.   Check $[(p \rightarrow q) \wedge (q \rightarrow r) \wedge \sim q] \rightarrow \sim r$

p	q	r	p→q	q→r	~q	r	[(p→q)∧(q→r)∧~q]→~r
T	T	T	T	T	T	F	T
F	T	T	T	T	F	F	T
F	F	T	T	T	T	F	F
T	T	F	T	F	F	T	T
T	F	F	F	T	F	F	T
T	F	T	F	T	T	T	T
F	T	F	T	F	F	T	T
F	F	F	T	T	T	T	T

Not valid

21. This argument is of the form
Premise: $p \rightarrow q$

$$\frac{\sim q}{}$$

Conclusion: $\sim p$
so it is Indirect Reasoning.

23. This argument is of the form
Premise: $p \rightarrow q$

$$\frac{q \rightarrow r}{}$$

Conclusion: $p \rightarrow r$
so it is a Syllogism.

25. p: The KOT's have a party
     Friday night.
   q: I will go.
   This argument is of the form
   Premise: $p \rightarrow q$

$$\frac{p}{}$$

   Conclusion: q

27. p: I trim the hedge.
   q: I may go to the movie.
   The argument can be represented
   as
   Premise: $p \rightarrow q$

$$\frac{p}{}$$

   Conclusion: q
   Valid, Law of Detachment.

29. This is of the form

$$p \vee q$$
$$\frac{\sim p}{q}$$

   This form is the disjunctive
   syllogism so it is valid.

31. This argument is of the form

$$p \rightarrow q$$
$$\frac{\sim q}{\sim p}$$

   This is indirect reasoning,
   so it is valid.

33. p: The class votes for an oral exam.
   q: The class votes for a take home exam.
   This argument is of the form

$$p \vee q$$
$$\frac{\sim p}{q}$$

   This is a disjunctive syllogism, so it is valid.

Chapter 10 Review

1. (a) Statement   (b) Not a statement
    (c) Not a Statement   (d) Statement

3. (a) Rhonda is not sick today.
    (b) Rhonda is sick today and she has a temperature.
    (c) Rhonda is not sick today and Rhonda does not have a temperature.
    (d) Rhonda is sick today or she has a temperature.

5.   **(a)**   True   **(b)**   True   **(c)**   Fals
         (d)   True

7.   **(a)**   T   **(b)**   T   **(c)**   T
         (d)   F

9.   **(a)**   True   **(b)**   False

11.   Inverse: "If I do not turn my paper in late, then I will not be penalized."
        Converse: "If I will be penalized, then I will turn my paper in late."
        Contrapositive: "If I will not be penalized, then I will not turn my paper in late."

13.

p	q	~p	p∧q	~p → (p∧q)
T	T	F	T	T
T	F	F	F	T
F	T	T	F	F
F	F	T	F	F

15.

p	q	p∨q	p∧(p∨q)
T	T	T	T
T	F	T	T
F	T	T	F
F	F	F	F

Since p and p∧(p ∨ q) have the same truth values they are logically equivalent.

17.   This argument is of the form

$$p \rightarrow q$$
$$\underline{\sim q\qquad}$$
$$\sim p$$

This is indirect reasoning, so it is valid.

19.   This is an argument of the form

$$p \rightarrow q$$
$$\underline{q \rightarrow r}$$
$$p \rightarrow r$$

so it is valid by syllogism.

21.   This argument is of the form

$$p \vee q$$
$$\underline{\sim p\qquad}$$
$$q$$

This form is a disjunctive syllogism so it is valid.

Appendix A
Algebra Review

Section A.1

1. $(-1)13 = -13$

3. $-(-23) = 23$

5. $(-5)(6) = -30$

7. $5(-7) = -35$

9. $-(7 - 2) = -5$

11. $21/(-3) = -7$

13. $(-4) + (-6) = -10$

15. $(-4)(2) = -8$

17. $5/3 + 4/3 = 9/3 = 3$

19. $12/5 - 3/5 = 9/5$

21. $2/3 + 3/4 = 8/12 + 9/12 = 17/12$

23. $5/6 - 7/4 = 10/12 - 21/12 = -11/12$

25. $2/5 + 1/4 = 8/20 + 5/20 = 13/20$

27. $4/7 - 3/5 = 20/35 - 21/35 = -1/35$

29. $(3/4)/(9/8) = (3/4) \times (8/9) = 2/3$

31. $(2/7)/(4/5) = (2/7) \times (5/4) = 5/14$

33. $(1/3)(1/5) = 1/15$

35. $2/5 \times 4/3 = 8/15$

37. $(-3/5)(-4/7) = 12/35$

39. $3/11 + 1/3 = 9/33 + 11/33 = 20/33$

41. $5/7 \div 15/28 = (5/7) \times (28/15) = 4/3$

43. $(5/8)/(1/3) = (5/8)(3) = 15/8$

45. $(3/4 + 1/5) \div (2/9) = (15/20 + 4/20)(9/2) = (19/20)(9/2) = 171/40$

47. $-2(3a + 11b) = -6a - 22b$ **49.** $-5(2a + 10b) = -10a - 50b$

Section A.2

1. $2x - 4 = -10$
$x = -3$ is a solution,
since $2(-3) - 4 = -6 - 4 = -10$

3. $2x - 3 = 5$
$2x = 8$
$x = 4$

5. $4x - 3 = 5$
$4x = 8$
$x = 2$

7. $7x + 2 = 3x + 4$
$4x = 2$
$x = 1/2$

9. $12x + 21 = 0$
$12x = -21$
$x = -21/12 = -7/4$

11. $3(x - 5) + 4(2x + 1) = 9$
$3x - 15 + 8x + 4 = 9$
$11x = 20$
$x = 20/11$

13. $\dfrac{2x + 3}{3} + \dfrac{5x - 1}{4} = 2$
$8x + 12 + 15x - 3 = 24$
$23x = 15$
$x = \dfrac{15}{23}$

15. $\dfrac{12x + 4}{2x + 7} = 4$
$12x + 4 = 8x + 28$
$4x = 24$
$x = 6$

17. **(a)** $y = 0.20(650) + 112 = \$242$
(b) $y = 0.20(1500) + 112 = \$412$
(c) Solve $0.20x + 112 = 302$
$0.20x = 190$
$x = 950$ miles

19. **(a)** $y = 0.42(42 - 8) = 0.42(34) = \$14.28$
(b) $y = 0.42(113 - 8) = 0.42(105) = \$44.10$
(c) Solve $0.42(x - 8) = 22.26$
$x - 8 = 53$
$x = 61$ pounds

Section A.3

1.

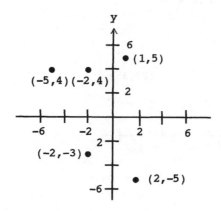

3.

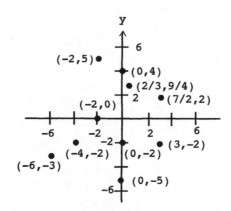

5.   **(a)**   Second quadrant: negative x-coordinates, positive y-coordinates
    (b)   Third quadrant: negative x-coordinates, negative y-coordinates
    (c)   Fourth quadrant: positive x-coordinates, negative y-coordinates

7.

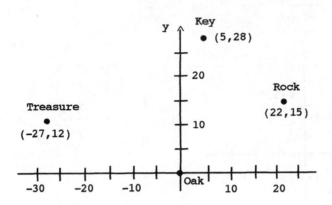

Section A.4

1. **(a)** $9 > 3$ is True, since $9 - 3 = 6$. **(b)** $4 > 0$ is True, since $4 - 0 = 4$.
(c) $-5 > 0$ is False, since $-5 - 0 = -5$.
(d) $-3 > -15$ is True, since $-3 - (-15) = 12$.
(e) $5/6 > 2/3$ is True, since $5/6 - 2/3 = 5/6 - 4/6 = 1/6$.

3. $3x - 5 < x + 4$
$2x < 9$
$x < 9/2$

5. $5x - 22 \le 7x + 10$
$-2x \le 32$
$x \ge -16$

6. $13x - 5 \le 7 - 4x$
$17x \le 12$
$x \le 12/17$

7. $3(2x + 1) < 9x + 12$
$6x + 3 < 9x + 12$
$-3x < 9$
$x > -3$

9. $3x + 2 \le 4x - 3$
$-x \le -5$
$x \ge 5$

11. $6x + 5 < 5x - 4$
$x < -9$

13. $3(x + 4) < 2(x - 3) + 14$
$3x + 12 < 2x - 6 + 14$
$x < -4$

15. $3(2x + 1) < -1(3x - 10)$
$6x + 3 < -3x + 10$
$9x < 7$
$x < 7/9$

Section A.4 Linear Inequalities

17. $-16 < 3x + 5 < 22$
$-21 < 3x < 17$
$-7 < x < 17/3$

19. $14 < 3x + 8 < 32$
$6 < 3x < 24$
$2 < x < 8$

21. $3x + 4 \leq 1$
$3x \leq -3$
$x \leq -1$
$(-\infty, -1]$

23. $-7x + 4 \geq 2x + 3$
$-9x \geq -1$
$x \leq 1/9$
$(-\infty, 1/9]$

25. $-45 < 4x + 7 \leq -10$
$-52 < 4x \leq -17$
$-13 < x \leq -17/4$
$(-13, -17/4$

27. $\dfrac{6x + 5}{-2} \geq \dfrac{4x - 3}{5}$
$30x + 25 \leq -8x + 6$
$38x \leq -19$
$x \leq -1/2$

29. $\dfrac{2}{3} < \dfrac{x + 5}{-4} \leq \dfrac{3}{2}$
$-8 > 3x + 15 \geq -18$
$-23 > 3x \geq -33$
$-23/3 > x \geq -11$

31. $75 \leq 35 + 5x < 90$
$40 \leq 5x < 55$
$8 \leq x < 11$
8, 9, or 10 correct answers

33. $85 \leq 3x + 25 \leq 100$
$60 \leq 3x \leq 75$
$20 \leq x \leq 25$
20, 21, 22, 23, 24, or 25